THEORETICAL CHEMISTRY
Periodicities in Chemistry and Biology

Volume 4

Contributors to This Volume

Britton Chance
Louis J. DeFelice
Robert L. DeHaan
Richard J. Field
Benno Hess
A. Ipaktchi
P. Ortoleva
G. Oster
William C. Troy
A. T. Winfree

THEORETICAL CHEMISTRY
Periodicities in Chemistry and Biology

VOLUME 4

EDITED BY

HENRY EYRING

Department of Chemistry
University of Utah
Salt Lake City, Utah

DOUGLAS HENDERSON

IBM Research Laboratory
Monterey and Cottle Roads
San Jose, California

ACADEMIC PRESS   New York   San Francisco   London   1978
A Subsidiary of Harcourt Brace Jovanovich, Publishers

ACADEMIC PRESS, INC.
111 Fifth Avenue, New York, New York 10003

United Kingdom Edition published by
ACADEMIC PRESS, INC. (LONDON) LTD.
24/28 Oval Road, London NW1 7DX

LIBRARY OF CONGRESS CATALOG CARD NUMBER: 75–21963

ISBN 0–12–681904–1

PRINTED IN THE UNITED STATES OF AMERICA

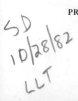

Contents

Stably Rotating Patterns of Reaction and Diffusion

A. T. Winfree

Chemistry of Inorganic Systems Exhibiting Nonmonotonic Behavior

Richard J. Field

Population Cycles

G. Oster and A. Ipaktchi

Mathematical Modeling of Excitable Media in Neurobiology and Chemistry

William C. Troy

Oscillating Enzyme Reactions

Benno Hess and Britton Chance

Oscillatory Properties and Excitability of the Heart Cell Membrane

Robert L. DeHaan and Louis J. DeFelice

Contents

Selected Topics from the Theory of Physico-Chemical Instabilities

P. Ortoleva

List of Contributors

Numbers in parentheses indicate the pages on which the authors' contributions begin.

Britton Chance (159), *Johnson Research Foundation, University of Pennsylvania Medical School, Philadelphia, Pennsylvania*

Louis J. DeFelice (181), *Department of Anatomy, Emory University, Atlanta, Georgia*

Robert L. DeHaan (181), *Department of Anatomy, Emory University, Atlanta, Georgia*

Richard J. Field (53), *Department of Chemistry, University of Montana, Missoula, Montana*

Benno Hess (159), *Max Planck Institut für Ernahrungsphysiologie, Dortmund, Federal Republic of Germany*

A. Ipaktchi (111), *Department of Mechanical Engineering, University of California, Berkeley, California*

P. Ortoleva (235), *Department of Chemistry, Indiana University, Bloomington, Indiana*

G. Oster (111), *Division of Entomology and Parasitology, College of Natural Resources, University of California, Berkeley, California*

William C. Troy (133), *Department of Mathematics, University of Pittsburgh, Pittsburgh, Pennsylvania*

A. T. Winfree (1), *Department of Biological Science, Purdue University, West Lafayette, Indiana*

Preface

In a sense, theoretical chemistry has existed as long as scientists have tried to understand chemical phenomena. However, it has been only recently that theoretical chemistry has grown into a mature field. Three developments, all relatively recent, have spurred this growth. First, at the end of the nineteenth century the foundations of statistical mechanics were laid. As a result, the bulk properties of chemical systems could be calculated from their microscopic dynamics. Second, during the first third of the twentieth century, quantum mechanics was developed, giving a satisfactory theory for the microscopic dynamics of chemical systems. Finally, in the past two decades, fast electronic computers have made accessible the full richness of quantum and statistical mechanics for the theoretical description of complex chemical systems.

Despite the maturity of theoretical chemistry, there are very few journals or review series devoted to all aspects of this field. It is hoped that this serial publication will fill, in part at least, this gap. Articles concerning all aspects of theoretical chemistry will be published in these volumes. Articles concerning experimental chemistry which pose or answer questions of theoretical interest may also be published from time to time.

In this volume a multisided phase of periodicity is presented. Winfree is concerned with the principles that relate the geometry of the visible activity patterns to the less obvious geometry of reaction kinetics. His procedure is analogous to Gibbs geometrizing of thermodynamics. He is chiefly concerned with biological periodicities. Field is concerned with chemical periodicities and how they arise and gets explicitly involved with a variety of chemical mechanisms. Troy is concerned with the mathematical approach to nerve impulse transmission and to an interpretation in depth of several other chemical reactions.

Hess and Chance give a historical introduction to a variety of biological periodicities controlled by enzymes and then develop models of glycolytic oscillations in more detail. They classify the enzyme functions in the latter case into promotion of four processes.

1. An input supplying the substrate
2. The primary oscillator enzyme
3. A feedback function
4. A sink

From DeHaan and DeFelice we get an in-depth picture of heart behavior together with an interpretation of this behavior at the molecular level. Ortoleva provides a very general overview of periodicities and shows they are wide ranging throughout nature. Finally, Oster and Ipaktchi treat the ecological problems associated with population growth and stability and illustrate the complexity of the matters requiring attention.

HENRY EYRING
DOUGLAS HENDERSON

Contents of Previous Volumes

Volume 2

Volume 3

THEORETICAL CHEMISTRY
Periodicities in Chemistry and Biology

Volume 4

Stably Rotating Patterns of Reaction and Diffusion

A. T. Winfree

Department of Biological Science
Purdue University, West Lafayette, Indiana

I. Introduction

During my first summer of graduate school, 12 years ago, I had the good fortune to work in Robert DeHaan's laboratory in the Carnegie Institute of Embryology in Baltimore. More than whatever I learned about heart cells, I remember Bob's recurrent enquiry: "What question are you asking?" That question never failed to stop frantic agitation, whether of tongue or of hands, dead in its tracks. What question *am* I asking? It is particularly important to answer clearly at the outset of this particular essay, which is less closely tied than I would like to either to quantitative experiment or to rigorous mathematics.

The question concerns a new class of phenomenon: a persistent rotating mode of activity in media that can alternatively lie stably inactive, except for giving a pulselike response to occasional stimuli. Such media include nerve membrane, heart muscle, the smooth muscle of stomach and gut, certain primitive embryos, and, since 1970, a feverishly studied inorganic chemical reaction.

And what am I asking about these self-organizing patches of rotary activity? At this stage, nothing rigorously quantitative, nothing steadfastly reductionistic, but something fuzzier, something like "How can we begin to think about them? What principles relate the geometry of the visible activity patterns to the less obvious geometry of reaction kinetics? How can such geometric principles be made at least compatible with physical chemistry? How can they be maneuvered to show us how rotors come about and how they cease to exist? What alternative modes of spatially self-organizing chemical activity can be anticipated?"

With mathematicians everywhere busily geometrizing everything from relativity theory through thermodynamics to classical mechanics, it would seem at least sporting to have a go at chemical dynamics (e.g., see Oster and Perelson, 1974). Perhaps the game will enrich its flavor by assimilating the rotating waves of chemical activity first discovered by Zhabotinsky (1970).

The question posed here, then, is "What aspects of the observed chemical oscillations and rotating structures can be understood, at least qualitatively, in geometric terms adaptable to a variety of underlying molecular mechanics?" In attempting an answer, we come upon a principle of potentially wider application in Section III,B,2. I use it only heuristically, but I suspect more rewarding sequelae await its proper mathematical development. The principle stems from Poincare's introduction of "phase portraits" (Section II,A), which catalyzed deeper understanding of ordinary differential equations, and therefore of dynamics in well-stirred reactors. No comparably intuitive and general method has emerged to save us from *partial* differential equations, such as inevitably arise in spatially distributed reactions. And there are good reasons why not. Convection and turbulence, surface absorption, electrostatic effects of ion concentrations, and so on, really do make the subject of reaction *morphology* a fairyland of mathematically complicated special phenomena [e.g., see Aris' book on reaction and diffusion interactions (1975)].

Ten years ago the special case of unstirred isothermal reaction in homogeneous phase, without interfacial exchanges, without hydrodynamic transport, attracted the attention of few specialists, and then mainly for instabilities of the spatially uniform solution. Even this interest is lost if the reactants have nearly equal diffusion coefficients. But all that has changed since the oscillating reaction of Belousov (1959) was parlayed into a spatially self-organizing reaction by Zaikin and Zhabotinsky (1970). The experimental convenience and dramatic quality of these phenomena have unleashed an explosive proliferation of work, both experimental and mathematical, on what seemed previously an unpromising special case: isothermal, isodiffusive, hydrodynamically inert, without surface phenomena.

And, to get to the point, it is in precisely that case that Poincare's phase portrait methods may once again provide a considerable aid to intuition, if introduced through the geometrical trick of Section III,B,2. I intend that to be the main point of this article, apart from its supportive context of allusions to specific models and experimental systems.

Let us begin with homogeneous reactions, involving only one point of physical space. We will quickly move to reactions distributed along a one-dimensional filament with molecular diffusion coupling adjacent volume elements. Then we consider new features that emerge if the filament is made into a ring, and finally fill in the ring to consider a two-dimensional disk of reacting medium. Geometry comes into its own with a vengeance in three-dimensional media, but this lovely subject goes beyond my ambitions for this volume.

II. Zero-Dimensional Situations

A. Composition Space

The chemical composition of any homogeneous solution can be described mathematically as a point in composition space. Composition space is the N-dimensional positive orthant of ordinary Euclidean space, with its N coordinate axes representing the positive concentrations of N chemical species involved in the pertinent reactions. To each composition there corresponds a net rate of synthesis, and degradation of each of the N substances. (These rates are obtained from the pertinent kinetic equations, including input and output conditions. In this discussion I regard all chemical systems as thermodynamically "open," at least in this sense: they consume energy-rich species whose depletion need not concern us on the time scale of interest.) Thus, a little arrow at each point in composition space points the direction and rate of change of composition. Those arrows collectively describe a flow (Fig. 1) that propels the composition of a reacting mixture

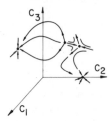

Fig. 1. Three concentration axes figuratively depict N-dimensional composition space. The origin represents concentration zero of each substance. The arrows sketchily evoke a flow in composition space, determined by the rates of change of all concentrations due to reaction, as a function of instantaneous composition. Several stagnation points of the flow (steady states of reaction and degradation of all substances) are depicted.

from almost any initial condition ultimately to some attractor. We say "almost" because every reaction must have at least one and may have more than one stagnant composition at which all the net rates of synthesis or degradation are zero (e.g., see Schmitz, 1975); and none of these need be an attractor (Wei, 1962). A stagnant composition *may* attract, in the sense that the flow arrows lead toward it from all nearby compositions. Such a point is stable against small perturbations of composition. Or it may *repel*, if there is at least one line along which a small perturbation will grow, i.e., if nearby arrows point outward in those directions. Mathematically, a point attracts if the real parts of all N eigenvalues of the linearized rate equations are negative. Otherwise, it repels (excepting the improbable case of purely imaginary eigenvalues).

Suppose the flow arrows not very far away on one side of an attracting stagnation point lead the composition initially farther away before turning around to funnel into the stagnation point (Fig. 2). Borrowing a term from

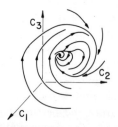

Fig. 2. As in Fig. 1, but the flow depicts an "excitable" reaction: an increase of C_3 or a decrease of C_2 from the attractor sets the system into a large excursion from the attractor.

physiologists, who have for a long time dealt with such models of biological excitability, such reactions are called "excitable." Nerve cells, for example, have an attracting steady state of ion balance across the cell membrane. It is stable to small perturbations, but a not-so-small perturbation results in an immediate and dramatic excursion further away from steady state before things settle down again (see chapter by Troy, this volume). At least one real chemical reaction in homogeneous solution has been found to exhibit excitable behavior, both in the laboratory and on paper (see Appendix, Rossler *et al.*, 1972, and chapter by Field, this volume). We will deal with the simplest geometrical caricature of this reaction shortly.

A slight local change in the flow field near the stagnation point turns the funnel trajectory into a closed loop (Fig. 3) (Franck, 1973, 1974). We then

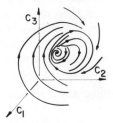

Fig. 3. As in Figs. 1 and 2, but the flow depicts a reaction that approaches a limit cycle along which its composition fluctuates regularly.

have a limit cycle: a cyclic change in composition that leads round and round periodically. In idealized reaction systems involving only two chemical species or only two reaction rates, this is quite a common situation. For example, if there is but one stagnation point and it repels, and if concentrations all decrease from arbitrarily high values as required in any reasonable chemical model, then there is not much else trajectories can do but to run in a ring circulating around the repelling stagnation point. Papers exhibiting such attracting limit cycles in two-dimensional composition spaces have become quite numerous (for reviews, see Higgins, 1967; Hess and Boiteux, 1971; Nicolis and Portnow, 1973; Goldbeter and Caplan, 1976).

Unfortunately, most chemically realizable reactions involve at least three independent variables. Limit cycles in such systems are much more difficult to exhibit analytically and have been exhibited in very few chemical models (e.g., Glass and Perez, 1974; Goldbeter, 1975; Hastings and Murray, 1975; Tyson and Kauffman, 1975; Cummings, 1976; Stanshine and Howard, 1976). A number of real systems are currently under intensive study for the likelihood that they would exhibit true limit cycles in a chemostat with suitable input of raw materials and overflow of product (see *Faraday Symp. Chem. Soc.* **9**, 1974).

Trajectories of more subtle geometry can arise once we abandon the early theorists' blue laws against having more than two variables. For example, in recent years an exciting literature has begun to blossom around the term "strange attractor" (Lorenz, 1963; Moore and Spiegel, 1966; May, 1976; May and Oster, 1976; Roessler, 1976a, 1976b, 1976c, 1977; Gilpin, 1977; Williams, 1977; Guckenheimer et al., 1977). A strange attractor is a region of two or more dimensions in composition space that attracts trajectories and from which no trajectories emerge. But, within a strange attractor the kinetics almost defies description. It does not settle down to a stable periodicity, even of a complex sort, nor to stationarity. Several theoretical examples under current study (May, 1976; May and Oster, 1976; Rossler, 1976a, 1976b, 1976c, 1977; Guckenheimer et al., 1977; Williams, 1977) seem to be perfectly feasible consequences of realistic kinetic equations: not at all mathematically pathological delicacies. The "origami" constructions of Rossler (1976a,b,c,d) make it all much more intuitively understandable, and even assist in deliberate construction of chaotic reaction schemes. Olsen and Degn (1977) give experimental evidence of a strange attractor in the reaction of H_2O_2 with peroxidase. Rossler and Wegmann (1977), and Schmitz, Graziani, and Hudson (1977) show what appears to be chaotic dynamics in the Belousov–Zhabotinsky reaction; Schmitz and Garrigan (1977) show the same in Pt-catalyzed H oxidation.

B. SKETCHING THE FLOW

One way to describe the complex kinetics of a reaction is to draw its trajectories as a flow in composition space. Such artwork is no substitute for understanding the mechanisms of molecular encounter and rearrangement. But neither does the molecular cause and effect description by itself tell us much about the dynamic behavior of a process involving many coupled reactions. This is what the flow diagram or "phase portrait" provides in one timeless glance. This geometric construction is a routine exercise for digital computer and plotter if the kinetic equations are known. But if they are not, as is more commonly the case in research, then the sketch must be made from fragments of data and anecdotal phenomenology.

In preparing such a sketch, the main questions to answer are as follows.

(i) How many independent variables (concentrations or reaction rates) are important in the sense that they all interact in the time scale of interest? This is the dimension N of composition space. (Quantities that change much more slowly may be treated as parameters affecting the rate laws, rather than as state variables.)

(ii) Does the reaction's composition stay on some lower dimensional surface embedded in the Euclidean N-space, except for negligibly brief transients? If so, what is the topology and shape of that surface? (See Feinn and Ortoleva, 1977.)

(iii) How many stagnation points (steady states) are there in addition to the one topologically required? A topological theorem assigns to each stagnation point an index $= \pm 1$, and asserts that the sum of the indices must be 1 (Glass, 1975). This fact is sometimes useful as a guarantee that a search for further stagnation points will be rewarded.

(iv) Which stagnation points are attractors?

(v) In the systems under study, are there any of the two known types of higher dimensional attractors, viz., limit cycles or strange attractors?

(vi) Each attractor has an attractor basin, an N-dimensional volume of composition space from which all trajectories lead to that attractor. How do their boundaries fit together geometrically? All the repelling stagnation points are in these boundaries.

Such questions can be answered numerically in any system whose kinetics is already completely solved. I would even argue that it is *necessary* to do so in order to understand the integrated behavioral consequences of what is already understood in the piecemeal terms of individual reactions kinetics. In cases with kinetics *not* yet completely solved (more commonly encountered in research work) a preliminary sketch often permits strong inferences about behavior not yet observed. The geometric relationships implicit in whatever behavior is known about a particular system may, for example, imply repelling steady states between the known attracting steady states, or a limit cycle may be implicit in the directions of local flows, or a crossing of trajectories may suggest an overlooked independent variable.

To help formulate such sketches, one might experimentally explore a reaction's composition space in search of attractor basins and their geometric relationships. While monitoring any convenient observable (e.g., O.D., pH, temperature, etc.), one can perturb the reaction from its steady state by injecting one reactant. This offsets the initial conditions parallel to the injectate concentration axis. If this line of obtainable initial conditions crosses an attractor basin boundary, then a second attractor will be discovered as the monitored observable takes up a new persistent mode of behavior (Fig. 4). A

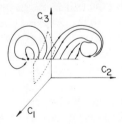

Fig. 4. Reaction flow as in Figs. 1–3, but containing an attracting stagnation point and an attracting limit cycle. Displacement of the initial state from the attracting stagnation point, along one coordinate direction, provides initial conditions for trajectories. These may go to either attractor. An attractor basin boundary surface divides the line. A portion of such a surface is outlined by dashes.

more comprehensive search requires a second, third, fourth ... syringe for a simultaneous fast jump of concentration in a perpendicular direction corresponding to a second, third, fourth ... substance. Working syringes in combination gives access to a correspondingly greater variety of initial conditions.

In the particular case of oscillating reactions, a simplified version of this technique naturally recommends itself. Suppose we have two separate volumes, V and $1 - V$, of oscillating reagent at phases $\phi 1$ in V and $\phi 2$ in $(1 - V)$. By suddenly combining them we instantly achieve initial conditions intermediate between the compositions of the separate volumes: each reactant's concentration becomes $VC(\phi_1) + (1 - V)C(\phi_2)$, where $C(\phi)$ is that reactant's concentration at phase ϕ (Fig. 5). By varying $\phi 1$, $\phi 2$, and V we

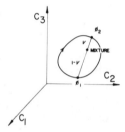

Fig. 5. Two volume elements on a limit cycle in composition space, combined, create a volume element with concentrations intermediate in proportions to volume ratio.

explore a convex three-dimensional volume of initial conditions and the trajectories issuing from it range through a four-dimensional volume of composition space (if composition space has that many dimensions). Topological considerations guarantee that some unique phenomena must be found in this volume (Winfree, 1974a). For example, by varying ϕ_1 and ϕ_2 at fixed $V = 1 - V = \frac{1}{2}$ one is topologically guaranteed access to a trajectory that never returns to the limit cycle and might therefore approach an otherwise unobservable stagnation point. Experiments in this format have been tried using *biological* limit cycle oscillators. Ghosh, Chance, and Pye (1971) combined two volumes of yeast cells, at two different phases in the 30-sec cycle of glycolytic oscillation. Variability of biological material limited the precision of the resetting observations and, in any event, the presence of semipermeable membranes undoubtedly restricted the free averaging of concentration. But this experiment does seem well-adapted for the oscillating chemical reactions currently under precise control and observation in many laboratories. By finding the coordinates of $\phi 1$ and $\phi 2$ that lead to the stagnation point, one might determine whether it is an attractor or a repellor. This distinction may determine the choice between kinetic models in the theoretical literature of glycolytic oscillations (Chance *et al.*, 1973; Aldridge, 1976), the Belousov–Zhabotinsky oscillator (Othmer, 1975; Clarke, 1976; Tyson, 1976, 1977; Stanshine and Howard, 1976; Graziani *et al.*, 1976), and of circadian clocks (Kalmus and Wigglesworth, 1960; Aldridge, 1976; Aldridge and Pavlidis, 1976).

C. SIMPLE MODELS

This discussion of the spatially homogeneous case concludes with a description of a simple model of an excitable and/or spontaneously oscillating medium. Numerical computations on this model in situations of 0, 1, and 2 dimensions of physical space constitute much of the foundation for my questions and speculations to follow.

The simplest model of an excitable reaction would have an attracting state and a nearby repelling state from which one trajectory goes directly to the attractor while another makes a large excursion enroute to the attractor. This one-dimensional ring-shaped state space has received substantial attention from mathematicians and biophysicists who are curious about heart and nerve viewed as excitable media (Wiener and Rosenblueth, 1946; Krinskii, 1968; Kuramoto and Yamada, 1976). For example, in 1946 Norbert Wiener attached the labels shown in Fig. 6 and proved a variety of theorems

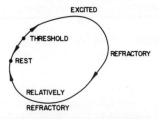

Fig. 6. Simplest continuous-state model of an excitable medium distinguishes states only along a one-dimensional path, as labeled. An attracting and a repelling stagnation point are necessary features on this excitation loop.

about heartbeat and its pathologies in terms of such a model. However, its lack of insides renders it impotent to support the rotating waves emphasized below. Since such waves do exist in a number of different excitable but not oscillatory systems (Stibitz and Rytand, 1968; Zhabotinsky and Zaikin, 1973; Durston, 1973; Allessie *et al.*, 1973, 1976, 1977), we are forced to take account of at least *two* state variables.

The one-dimensional ring-shaped composition space is implicitly embedded in a two-dimensional Euclidean composition space whose coordinates we might refer to substances A and B. For qualitative purposes, such a flow can be described simply as (see references mentioned below)

$$dA/dt = -A - B \text{ (plus } S \text{ if } A > 1)$$

$$dB/dt = kA. \tag{2.1}$$

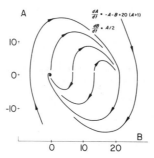

Fig. 7. To allow for states intermediate between states on the excitation loop of Fig. 6, we make use of a two-dimensional composition space. The idealized excitable reaction has a single globally attracting stagnation point. Beyond a nearby threshold, trajectories initially move farther away from the stagnation point. A simple numerical implementation is given in terms of a piecewise-linear kinetic equation in two variables. A and B are deviations from the stagnant composition, not absolute (positive) concentrations. This diagram is identical to Fig. 1 of Winfree 1974b and page 87 of Winfree 1974d, except that the units of A and of B are made 20 times smaller to place the threshold at unit concentration.

This notion of an excitable medium embodies just a few simple ideas (Fig. 7: c.f. Fig. 2).

(i) The kinetics are linear, but above unit concentration of A there is an additional constant rate of A synthesis. Geometrically, this is equivalent to cutting the linear flow diagram [(2.1) with $S = 0$] along the horizontal line $A = 1$ and offsetting the top part by S units to the right. Chemically, $A > 1$ is intended to represent the oxidizing phase of Z reagent (see Appendix) according to the FKN kinetics (Field, Koros, and Noyes, 1972), during which phase ferroin is converted from the orange form to blue.

(ii) The system has a single attracting stagnation point at the origin with eigenvalues $(-1 \pm i)/2$. If the extra synthesis rate S is not too large, *all* trajectories lead to the origin. (However, if S exceeds 33, with $k = \frac{1}{2}$, then the stagnation point's attractor basin shrinks to a finite disk surrounded by an attracting limit cycle as in Fig. 3.) The imaginary part of the eigenvalue is of little consequence except close to the origin. If $k < \frac{1}{4}$, so that both eigenvalues become real, it matters little to the overall shape of the flow because the top half ($A > 1$) of the linear phase portrait is laterally offset by S; the overall flow pattern resembles a clockwise rotation.

(iii) Trajectories diverge at the threshold $A = 1$. Thus initial conditions as near the origin as possible with $A > 1$ are on trajectories initially pointing upward, thus departing further from origin and threshold. Quantitatively, the divergence of flow is -1 everywhere, plus Kronecker's delta function along the threshold line $A = 1$.

This particular model has been studied by Andronov in connection with mechanical clocks (Andronov, Vitt, and Khaikin, 1966) and McKean (1970) as a model of excitable nerve. In its qualitative essentials it resembles a recent simplification of the Belousov–Zhabotinsky reaction kinetics (Tyson, 1976, 1977; Troy and Field, 1977). It also resembles Karfunkel's excitable enzyme-kinetic scheme (Karfunkel and Seelig, 1975) that also exhibits the rotating wave discussed below (Karfunkel, 1975). The remainder of this article describes a project carried out at Purdue University in 1973, before the "Oregonator" model of Belousov–Zhabotinsky kinetics (see articles by Field and Troy, this volume) achieved its current popularity. It might be worthwhile to repeat this study using the Oregonator instead of (2.1).

III. One-Dimensional Situations

A. MATHEMATICAL APPROACHES

Up to now we have ignored the consequences of molecular diffusion in physical space by dealing only with a homogeneous reaction in a very small or well-stirred volume. Let us now imagine a one-dimensional continuum of reacting medium, e.g., a burning fuse, a nerve fiber, or a capillary tube of Z reagent. Mathematically, this amounts to adding Laplacian operators to our model (2.1) to represent diffusion. Such piecewise-linear parabolic partial differential equations have been studied mathematically by Offner *et al.* (1940), McKean (1970), Rinzel and Keller (1973), Rinzel (1975a,b), Zaikin (1975), and others in context of propagating electrochemical disturbances in nerve fibers and in Z reagent. My approach here is exclusively graphical and numerical. This will be taken up in Section III,B.

1. Pseudowaves

We begin with a slight digression to imagine a temporally oscillating reaction arranged along a line, e.g., in a capillary tube. Let there be a monotone gradient of cycle phase along the tube, shallow enough so transport along gradients is negligible. If the reaction changes color during its cycle, then one end of the tube changes before the other and we see a wave of color sweeping along the tube with local velocity inversely proportional to the local steepness of the phase gradient. If the phase gradient spans several cycles, then there are several bands of color "moving" along the tube. They ostensibly "move," but not by actual propagation. Their "velocity" can be arbitrarily large in a region where the oscillation is spatially almost synchronous. Such patterns, first distinguished by Beck and Varadi (1971), have been called "pseudowaves" (Winfree, 1972).

2. Kinematic Waves

Now let there be a monotone gradient of cycle period along the tube, perhaps due to a temperature gradient. Because one end of the tube cycles faster than the other, the phase difference between the ends increases. The spatial gradient of phase along the tube grows ever steeper. If the reaction changes color during its cycle, we see waves of color sweeping along the tube with velocity inversely proportional to the local steepness of the phase gradient (Fig. 8). As the phase gradient steepens, the waves slow down. Models

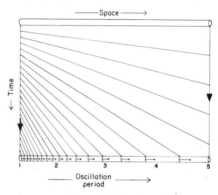

Fig. 8. At each point along a capillary tube (horizontally: space) a chemical event recurs periodically (downward: time) at intervals that increase linearly to the right. This gives the appearance of waves (diagonal lines) propagating at constant speed to the right, each wave slower than its predecessor.

of this sort have been studied analytically by Beck and Varadi (1971), Smoes and Dreitlein (1973), Kopell and Howard (1973a), and Thoenes (1973). Kopell and Howard (1973a,b) and Thoenes (1973) additionally exhibited such waves in the oscillating reaction of Belousov and Zhabotinsky. These waves are "kinematic" in the sense that their properties derive entirely from local reaction kinetics and the gradient of local period. Kinematic waves in no way involve coupling of adjacent volume elements by molecular diffusion. Kopell and Howard demonstrated this feature of their chemical "kinematic waves" by deploying an impermeable barrier: it had little effect on the apparent propagation.

3. Phase Waves

But as the phase gradient steepens, the waves slow down and pack closer together (Fig. 8). Molecular diffusion must eventually play a dominant role.

Its first effect, in the "weak diffusion approximation" (Rosen, 1976) is "phase diffusion," explored theoretically by Ortoleva and Ross (1973), Kuramoto and Tsuzuki (1976), and Ortoleva (1976). It turns out that so long as diffusion from nearby volume elements does not force the reaction far off its limit cycle, we can treat diffusion of reacting concentrations as diffusion of "phase" along the limit cycle. This phase diffusion approximation is good for shallow phase gradients in kinetics varying smoothly in time and staying close to a limit cycle. The waves I have studied in Z reagent unfortunately exhibit none of these features.

4. *Limit-Cycle Oscillator Waves*

Another well-developed mathematical approach to the study of reaction–diffusion waves was pioneered by Kopell and Howard (1973b). They postulate a smooth limit-cycle oscillation, with stagnation point either attracting or repelling, modeled by a particularly symmetric class of second-order ordinary differential equations that they call the "λ-ω" system. Adding a diffusion term, they solve that class of partial differential equation for propagating plane waves, and show their dispersion curves and stability criteria. The same class of equations has been studied by Ortoleva and Ross (1974), Dreitlein and Smoes (1974), Kuramoto and Tsuzuki (1976), Yamada and Kuramoto (1976), and Greenberg (1977). In the particular case of wave phenomena in the Belousov–Zhabotinsky reaction, we are once again debarred from employing these powerful tools, in this case by the fact that the wave phenomena in that reagent persist when its chemical parameters are altered to eliminate the homogeneous limit-cycle oscillation (Winfree, 1972).

5. *Dissipative Structures*

A fourth well-developed mathematical approach to the study of reaction–diffusion waves stems from the work of Turing (1952). This is the study of "dissipative structures" (Nicolis and Prigogine, 1977). In this approach to reaction–diffusion phenomena spatial structures, including waves, develop out of an instability of the spatially homogeneous solution. These "diffusive instabilities" arise from the inequalities among the several reactants' diffusion coefficients (Gmitro and Scriven, 1966; Othmer and Scriven, 1969, 1974; Segal and Jackson, 1972; DeSimone *et al.*, 1973; Nicolis, 1974; Balslev and Degn, 1975; Herschkowitz-Kaufman, 1975; Erneux and Herschkowitz-Kaufman, 1975; Erneux and Herschkowitz-Kaufman, 1977). Busse (1968), Herschkowitz-Kaufman (1970), Ross (1976), Thompson and Hunt (1977), and Defay *et al.* (1977) have invoked such models in explanation of periodic structures in the Belousov–Zhabotinsky reaction. Beck and Varadi (1971) were unable to reproduce Busse's experiment. Moreover, the experimental evidence presented by Busse and by Herschkowitz-Kaufman and by Varadi *et al.* (1975) does not clearly distinguish their phenomena from the kinematic waves clearly shown and explained by Kopell and Howard (1973a,b), and from the "trigger waves" first clearly shown by Zaikin and Zhabotinsky (1970), and accounted for in different but not incompatible terms (by Field *et al.*, 1974; Murray, 1976; Stanshine, 1977; Field and Troy, 1977; Troy, 1977; Hastings, 1976) without invoking diffusive instabilities. Ross (1976), Thompson and Hunt (1977), and Defay *et al.* (1977) present no evidence, and ignore both the models just cited, and the experimental fact that the spatially uniform steady state is stable.

Zhabotinsky and Zaikin (1973) suggest that the "chicken-skin" pattern seen in some versions of Belousov–Zhabotinsky reagent (their plate III) might arise by a diffusive instability. But they neglected to exclude the likelihood that oxygen transport in convective Benard cells is the cause.

Despite the richness of geometrical development it has enjoyed in recent years, it is my belief that the linear calculus of diffusive instabilities is not at all what we need to understand the most conspicuous wave phenomena in the Belousov–Zhabotinsky reagent. My reasons for saying this are simply the observations that those wave phenomena occur in the slightly modified version of the reagent that I call "Z reagent" (see Appendix) that has no limit cycle (at least when exposed to air in a thin layer). This reagent's unique stagnation point is a global attractor, and its spatially homogeneous stationary state is stable against small perturbations. So its waves do not arise out of any kind of instability.

Such waves are, nonetheless, "dissipative structures" in the sense that they cannot persist forever in a thermodynamically closed system. The conditions of their existence and stability can presumably be derived using bifurcation theory (Nicolis and Prigogine, 1977).

B. Geometrical Approaches

1. Equal-Diffusion Model

Thus for the third time Z reagent has shot out from under us the sophisticated mathematical technology on which we aspired to arrive at a comprehensive understanding of the observed chemical waves. In this situation, I elect to proceed on foot as follows:

(i) I use the piecewise-linear caricature of an excitable reaction presented above, adding a diffusion term.

(ii) I neglect the differences between diffusion coefficients of small molecules in water at room temperature; the diffusion coefficients of A and of B in the above model will both be set equal to 1. (Changing the diffusion coefficients from 1 to D is equivalent to rescaling space and velocities by a factor $\sqrt{D}$.)

(iii) This model is set out mathematically in Winfree (1974b). Though it has not yet received adequate mathematical attention, I will ask here " What aspects of the geometry and behavior of chemical waves in *Z* reagent can we understand, at least qualitatively, in terms of this model?" Answers come from numerical simulations and geometrical diagrams.

2. *Diffusion in Physical Space = Elasticity in Composition Space*

The geometrical diagram I will use most frequently is a mapping of each volume element of the idealized reacting medium into composition space. This amounts to no more than plotting $A(\mathbf{r})$ and $B(\mathbf{r})$, $\mathbf{r}$ being the volume element's position vector in physical space. Executing the plot as an image of the medium in (A, B) space permits us the following convenient geometrical trick for visualizing the spatial consequences of molecular diffusion.

Consider first the case of a reaction in a capillary tube, with one-dimensional spatial concentration gradients of *A* and *B*. In the limit of small length increments, each volume element along the length of the tube may be regarded as a chemically homogeneous compartment or cell. Except at the two ends, each cell is in contact with two adjacent cells of slightly different instantaneous chemical composition. As each cell reacts, reactants are also diffusing across both cell interfaces according to Fick's Law:

$$d/dt\,{}^{j}C_i = \left({}^{j+1}C_i - {}^{j}C_i\right) + \left({}^{j-1}C_i - {}^{j}C_i\right) + R_i \qquad (3.1)$$

where *C* is concentration, the superscript is the cell index, the subscript is the reactant index, and R_i, a function of composition, is the net rate of synthesis of *i*. The first term in this equation states that cell *j* moves in composition space directly toward cell $j + 1$ at a velocity directly proportional to the separation vector in composition space. (Choosing the proportionality factor as 1 determines our unit of physical space.) The second term says that a similar movement toward the other neighboring cell is to be added linearly. The third term says that the reaction adds a net rate of synthesis of each *i* in compartment *j*, and that this vector is again linearly superimposed. Thus cell *j* moves in composition space as prescribed by the reaction flow (as discussed above in the well-stirred or zero-dimensional case) modified in a simple linear way by molecular fluxes from its two neighboring cells. Those fluxes modify its movement exactly as though it were attached by springs (in composition space!) to the adjacent cells while embedded in a viscous fluid moving as prescribed by *R*.

The image of the capillary tube (Fig. 9) thus moves in composition space exactly as would a massless elastic fiber in a moving viscous fluid. My objective in resorting to this slightly unfamiliar representation is to make the most of our physical intuition, in a form that can be readily translated back to chemistry. This heuristic trick does not yield numerical values for propa-

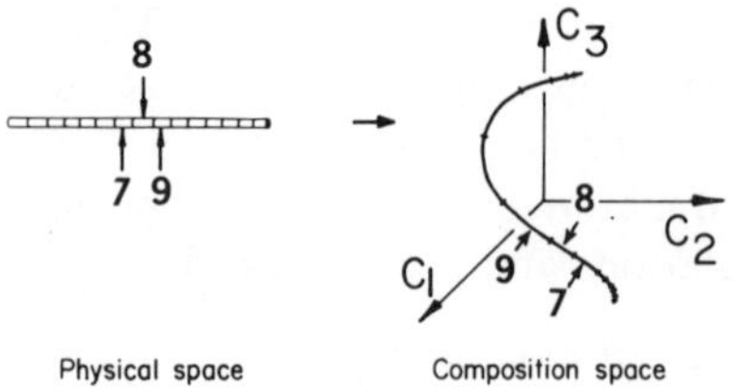

Fig. 9. Composition–space image of a capillary tube full of reacting fluid is obtained by mapping each volume element to its instantaneous position on all N coordinate axes (here 3).

gation velocities, nor does it allow reliable intuition for the *stability* of spatial patterns of reaction. In fact (contra the last sentence of Winfree, 1974c), wave propagation in such models is stable or not according to the values of parameters that do not affect the phase portrait qualitatively. But thinking in this geometrical format does help markedly to understand what initial conditions (i.e., what shape of this medium's image in composition space), what boundary conditions (i.e., what rules for the movement of the image boundary), and what reaction kinetics (i.e., what flow field for the viscous medium) are required to evoke various kinds of solution, and to know what kinds of solution it might be worthwhile to pursue numerically or analytically.

3. Annihilation and Nonreflection

For example, one can see in these terms why colliding chemical waves annihilate each other like grassfires, rather than passing through one another or reflecting as do propagating mechanical deformations of elastic media. To see this we first ask "What is a solitary wave?" It is a temporary excursion of composition from a stagnation point which we may as well put at the origin, as shown in Fig. 10. Suppose we start such an excursion by lifting A concentration above threshold in one compartment, e.g., in the left end of the capillary tube shown in Fig. 10. As such an excursion propagates through the middle, enroute to the right end as shown in Fig. 10, the tube's endpoints lie near the origin while the middle parts, involved in the propagating pulse, flow along a trajectory locally modified only by each cell's attachment to neighbors. By means of a diffusion, compartments already caught up in the pulse "pull" their frontside neighbors out of the attractor at the origin. When the endpoint itself is finally pulled out (when the wave reaches the far end of the tube), it simply follows the local kinetics back around to the origin on very nearly the same path followed by its predecessors. To initiate a reflection, this endpoint would have to lead the image out of the origin for a spatially inverted repeat of the same performance. But its neighbor to the rear preceded it into the origin: the reaction flow into the

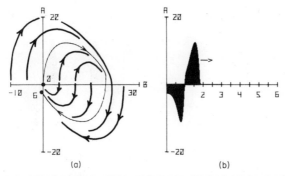

(a)   (b)

Fig. 10. (a) Departing from the tradition of foregoing figures, we henceforth attend to only two concentrations as in the model of Fig. 7. Concentrations are indicated as deviations from the steady state, the origin of this coordinate system. A 6-mm line segment (e.g., a capillary tube) of fluid is reacting along the flow lines of Fig. 7. This image depicts a digitally computed pulse of chemical activity from the tube's left end at 0 mm, which has already relaxed most of the way back to the stagnant composition. The right end at 6 mm still rests there, awaiting excitation. [In (2.1) the time unit is taken as one second and the space unit as $\frac{1}{20}$ mm corresponding to a diffusion coefficient of $2\frac{1}{2} \times 10^{-5}$ cm^2/sec.] The horizontal axis in this and following similar figures is calibrated, not in space units of $\frac{1}{20}$ mm but in millimeters. The heavy arrows repeat the reaction trajectories shown in Fig. 7. (b) Concentration of trigger substance A is plotted against length along the tube. The wave is traveling to the right at $\frac{1}{5}$ mm/sec.

origin is no longer opposed, but now abetted by molecular diffusion and so spatially uniform stationarity prevails.

In the case of a train of several waves propagating along the tube, the tube's image wraps several times around the same path, but when the endpoint at last makes its excursion, activity terminates in the same manner.

Thus waves are absorbed without reflection at boundaries. This case of wave absorption at a boundary is geometrically similar to the case of a head-on collision of two waves. The collision point is a point of mirror symmetry for the concentration pattern in the tubes. This means there is no concentration gradient and no flux through the midpoint. This is similar to the no-flux boundary condition at an endpoint. The tube's image in composition space is thus two-stranded, with the two endpoints leading, and the midpoint last to depart the origin and last to return (Fig. 11). Just as the far endpoint was last in Fig. 10, the midpoint in Fig. 11 joins its predecessors in the origin and no further activity is seen. The two waves have collided at the center and vanished.

In the case of a wave train, with several impulses propagating inward from each side, the diagram is the same except that the two-stranded image of the capillary tube wraps several times around nearly the same path.

The waves mentioned in Sections III,A,1–4 behave in much the same way for much the same geometrical reasons. In contrast, the linear reaction–

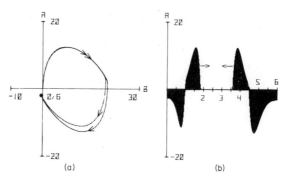

Fig. 11. As in Fig. 10 except that waves were started from slightly different initial conditions (to separate the two strands of the almost mirror-symmetric image) from *both* ends of the tube. The reaction flow trajectories are omitted to keep the diagram uncluttered (see Figs. 7 and 10).

diffusion waves commonly analyzed in connection with dissipative structures (Section III,A,5) admit a superposition principle; they interpenetrate freely. How can that be, in terms of the composition-space images emphasized here? Any concentration pattern in physical space has its corresponding image in composition space, and can be Fourier analyzed into a superposition of periodic waves (each of which is a *winding* of the image around the stagnation point). It turns out that for our case of equal diffusion (prerequisite for the geometrical interpretation of diffusion as elasticity in composition space) all these components grow or decay except one. So superposition obtains only in the trivial sense that this one mode's amplitude is arbitrary.

Reflection *can* occur in reaction–diffusion equations where two pieces of medium are joined together, but with different reaction parameters or boundary conditions on the two sides. Arshavskii *et al.* (1965), Krinskii (1968), Cranefield *et al.* (1971a,b, 1972), Goldstein *et al.* (1974), and Ramon *et al.* (1976) exhibit such reflections in real nerve fibers and in nerve models. It seems likely that something of this sort occurs in the human intestine under pathologically or surgically induced conditions (see references in Section VII). In terms of the medium's image in composition space, reflections can arise only when two attached pieces of the image move under different rules. Examples are diagrammatically complicated, but might provide the reader an amusing exercise.

5. Impermeable Barrier = Scissors Cut

A feeling for the principles of this mapping will be enhanced by considering the consequences of cutting a reacting fiber or inserting an impermeable barrier. In Fig. 12 this is done just ahead of wave arrival, severing a region still near the attractor from the region already excited. The severed piece (right side of barrier X in Fig. 12) never departs the origin, but on the left side of the barrier excitation takes its full clockwise course just as in Figs. 10 and 11.

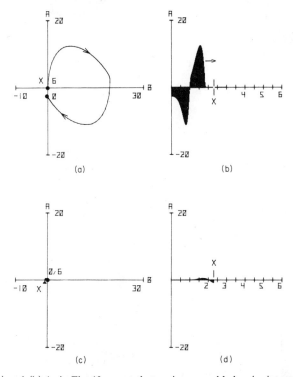

(a) (b)

(c) (d)

Fig. 12. (a) and (b) As in Fig. 10 except that an impermeable barrier interrupts the image ahead of the wave. (c) and (d) As above but 9 sec later: the wave has traveled 1.8 mm, vanishing into the barrier. The severed left end of the image has come full cycle (almost back to the origin). Volume elements beyond the barrier never departed the origin.

If the cut is made at the same point but later (Fig. 13), when that point is already far from the origin, then neither severed endpoint is pulled back to the origin and the wave proceeds almost as though no cut were made. This figure constitutes the justification for calling this kind of wave a "trigger wave" (Zeeman, 1972): its leading edge propagates, triggering a locally controlled excitation and relaxation.

Nerve, heart muscle, and Z reagent behave like these numerical caricatures in that inserting a barrier just ahead of the action potential or the abrupt red-to-blue transition (identified with crossing threshold) does block its conduction, whereas a barrier behind the transition has little effect.

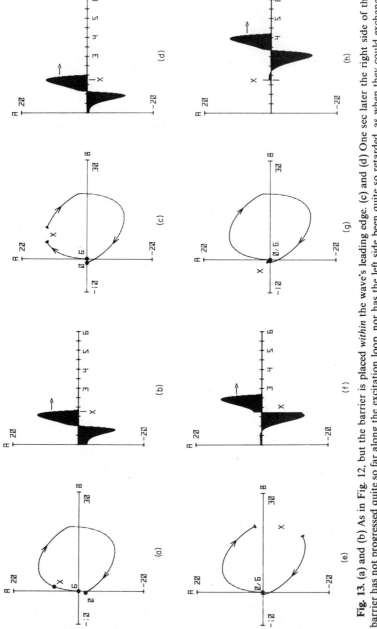

Fig. 13. (a) and (b) As in Fig. 12, but the barrier is placed *within* the wave's leading edge. (c) and (d) One sec later the right side of the barrier has not progressed quite so far along the excitation loop, nor has the left side been quite so retarded, as when they could exchange materials. (e) and (f) At 5 sec the gap is wider, but a wave is definitely advancing beyond the right side of the barrier. On the left side, recovery proceeds much as before, being little affected by diffusion. (g) and (h) After 14 sec the wave is only ⅛ mm or 0.6 sec behind a control experiment unimpeded by barrier implantation.

6. Waves on Rings

a. Size of the Ring: Let us now fashion a *ring* of excitable medium by joining the capillary's endpoints, still using kinetic equations (2.1). This is supposed to caricature a narrow annulus of Millipore material soaked in Z reagent, or Arshavskii's (1965) artificially contrived ring of nerve fiber. Since concentrations vary continuously in space, the ring's image in composition space is necessarily a ring. In the spatially homogeneous case the entire image lies degenerately at the origin. If the ring bears a clockwise or anti-clockwise wave (or *N* of them), then this image (Fig. 14) maps clockwise or

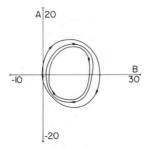

Fig. 14. In the format of previous figures (cf. Fig. 7):

Outermost ring: The composition–space image of a ring of perimeter 1.06 mm, stably conducting a circulating wave. This is equivalent to a train of identical waves 1.06 mm apart on a line.

Middle ring: as above, but 0.75 mm perimeter.

Inner ring: as above, but 0.65 mm perimeter.

In each case, the image of the ring is stationary while cells flow along it, like runners on a racetrack.

anticlockwise around nearly the same path as in Figs. 10–13 (or *N* times around) and circulates clockwise. But the smaller its physical perimeter *P*, the more the ring's image deviates toward the inside of that path. This occurs because molecular diffusion on a sufficiently small ring forms a local average of changing concentrations, thus reducing the range of excursion from the mean. Or in terms of the equivalence of diffusion in physical space with elasticity in composition space, the composition–space image of a physically small ring is stretched tauter against the outward divergence of the reaction flow. And just as molecular diffusion smoothes out local irregularities of concentration in physical space, it also affects a local averaging of velocities in composition space, smoothing over local irregularities of the reaction flow. An overall clockwise rotation thus prevails over the local anticlockwise flow into the origin just below threshold, and the ring's image continues to rotate. If the ring is not too small (too taut), no instabilities arise and a wave circulates forever around the ring.

How do this wave's velocity and period vary with the perimeter P of the rings? Answers are available for periodic wave trains on the open line, which is equivalent to a solitary wave on a ring of that period:

(i) A very small ring cannot stably support a rotating wave. It lapses into a uniform steady state. Kopell and Howard (1973b), Othmer (1977), and Conway *et al.* (1977) have made estimates of size minima for self-sustaining patterns. They vary as the square root of the product of a diffusion coefficient by a term derived from the reaction rates' rates of change with composition.† Edelstein (1972), Winfree (1974b), and Rinzel (1975a) have observed such minima computationally. Line 1 in Fig. 15 indicates this smallest perimeter capable of supporting an endless wave.

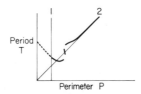

Fig. 15. The period T of a stably circulating wave is plotted against the ring's perimeter P. This is called a dispersion curve. In this hypothetical sketch, wave circulation is not stable on very small rings (boundary 1) and velocity P/T approaches an asymptote on large rings (line 2). The dashed arc depicts a two-dimensional phenomenon discussed below in which P is reinterpreted as the perimeter of a central hole along which a wave circulates in time T.

(ii) On a very large ring, most of which lies near the origin in composition space at any moment, we expect the propagation velocity to approach that of a solitary wave in a very long tube. Call this velocity v_∞. The rotation period is thus P/v_∞, indicated by line 2 in Fig. 15.

Analytical and numerical investigations of such dispersion curves have been carried out by Rinzel and Keller (1973), Rinzel (1975a), and Rinzel (personal communication) for (2.1), but with diffusion only of substance A, Kopell and Howard (1973b) for λ-ω systems with equal diffusion of both substances, Stanshine (1977) for a three-substance model of the Belousov–Zhabotinsky oscillator with equal diffusion, and by Murray (1976) and Hastings (1977) for a two-substance model of the Belousov–Zhabotinsky reaction also with equal diffusion. These studies indicate that dispersion

† Conway's and Othmer's expressions are for Neumann boundary conditions, i.e., impermeable boundaries. Kopell's is for periodic waves on the open line. In contrast, the corresponding expressions for Dirichlet boundary conditions (composition fixed along the boundary or at a point on a ring) vary as the square root of the product of a diffusion coefficient by the period of spatially uniform oscillation according to Goldbeter (1973) and Nazarea (1974, 1977).

curves need not be monotone nor even necessarily continuous or single-valued.

In this one-parameter family of solutions (perimeter P being the parameter) the longer rings can be expected to achieve more nearly the full amplitude of the solitary wave on an open line. Equation (41) of Kopell and Howard (1973b) guarantees this for λ-ω systems if a solitary wave circulating on the ring is to be considered equivalent to a periodic wave train on the open line. (However note that in λ-ω systems, in the piecewise-linear model here emphasized, and in Z reagent, the spatially uniform solution is always a stable alternative.) Figure 14 shows the computed composition–space images of three such rings, stably conducting waves by the piecewise-linear mechanism of (2.1) with $k = \frac{1}{2}$ and $S = 20$.

b. How to Start a Rotating Wave: How is a rotating wave started? These waves were initiated from circular initial conditions, $A(x) = \sin 2\pi x/P$, $B(x) = \frac{1}{2} + \cos 2\pi x/P$, from which the ring's image shrinks down onto a stable cycle. More practically, in laboratory terms one might initiate a wave at a point on the ring and then block its clockwise propagation, leaving only the anticlockwise impulse to circulate forever. This technique is implemented computationally in Fig. 16. One point $(0 = 6)$ on the ring is excited from the origin, and the remaining points begin to follow as mirror-image waves radiate outward. Then a barrier X is inserted as in Fig. 12, cutting only one of the two strands of the ring's image. The barrier is removed after its left side has finished the high-A portion of its cycle. The left side is rejoined to the right, still near the origin. As a result, the previously two-stranded C-shaped image becomes a single-stranded ring with a long double-stranded pigtail. The ring opens wider and continues to rotate stably as the pigtail flows into the origin. This method of initiating a rotating wave has been used by physiologists [Shibata and Bures (1974) and Arshavskii *et al.* (1965)].

IV. Two-Dimensional Situations

I will use the term "rotor" for a rotating wave in a uniform excitable media without holes. Rotors have been reported by experimental physiologists (Gerisch, 1965, 1968; Rosenstraukh *et al.*, 1970; Durston, 1973; Allessie *et al.*, 1973, 1976, 1977), biological theorists using models (Balakhovskii, 1965; Farley, 1965), and physical chemists (Zhabotinsky, 1970, 1971, 1973; Winfree, 1972).

A. FILLING IN THE RING TO MAKE A DISK

We now attempt transition to two-dimensional media by imagining that our ring of medium is only the border of a disk. How does the disk map into

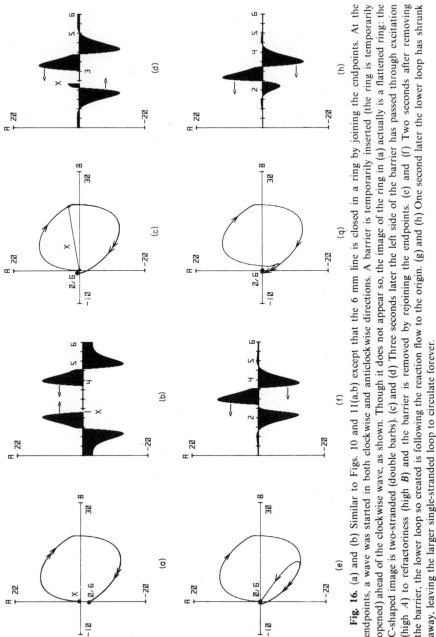

Fig. 16. (a) and (b) Similar to Figs. 10 and 11(a,b) except that the 6 mm line is closed in a ring by joining the endpoints. At the endpoints, a wave was started in both clockwise and anticlockwise directions. A barrier is temporarily inserted (the ring is temporarily opened) ahead of the clockwise wave, as shown. Though it does not appear so, the image of the ring in (a) actually is a flattened ring: the C-shaped image is two-stranded (double barbs). (c) and (d) Three seconds later the barrier has passed through excitation (high *A*) to refractoriness (high *B*) and the barrier is removed by rejoining the endpoints. (e) and (f) Two seconds after removing the barrier, the lower loop so created is following the reaction flow to the origin. (g) and (h) One second later the lower loop has shrunk away, leaving the larger single-stranded loop to circulate forever.

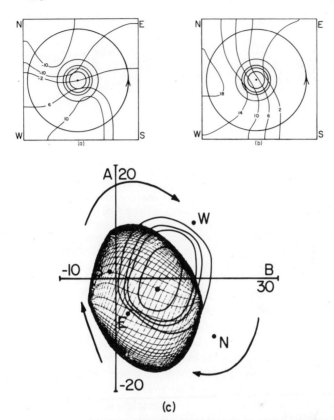

Fig. 17. (a) A 1 mm square, simulated by 61 × 61 equispaced cells in which (2.1) was simultaneously computed. Initial conditions 164 seconds ago established a rotor. Contour lines of *A* concentration now pivot stably around the dot. Circles concentric to the dot are indicated at perimeters 2.8, 1.15, 0.8, 0.65, and 0.35 mm. The four corners of this square are labeled N, E, S, W. (b) As in (a), but contour lines of *B* concentration (cf. Figs. 18 and 19). The pivotal cell has composition $A = -2, B = 7$. (c) The one mm square's image in composition space. The circle of perimeter 2.8 mm (and every larger circle) maps onto the border of the image. The smaller circles lie concentrically inside (cf. Fig. 14). The pivotal cell maps to $(-2, 7)$. The four corners are indicated on the border. This is the anticlockwise rotor of Winfree (1974b), Fig. 7, at $T = 164$ sec.

composition space? Figure 17 answers this question by a numerical experiment (actually using a square " disk ") using (2.1) with $k = \frac{1}{2}$ and $S = 20$. The wave circulating outward along the disk's border extends continuously inward, ultimately pivoting about a cell of unvarying composition. Annuli concentric to this point map into composition space very nearly as do the isolated rings of Fig. 14, except that rings too small to sustain a wave independently are here supported by their outer neighbors in the disk.

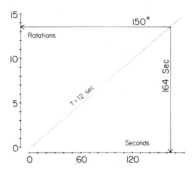

Fig. 18. The rotor of Fig. 17 does not quite turn smoothly because it was confined, not quite symmetrically, within a box with four points (in order to constantly challenge its stability). But near its center, it does turn at nearly constant angular velocity as shown here by plotting the angular orientation of the $B = 7$ contour line at the pivot at 1 sec intervals. Starting soon after initial conditions, the angle was estimated by eye to within $10°$ using a protector on computer printouts, then plotted in this figure. The rotation period is constant at 12 sec during 14 rotations.

This situation is unstable on disks of a diameter less than about the minimum stable length of one-dimensional waves (about 0.65 mm). But the wave seems stable in a larger square:

(i) The initial conditions consisted of perpendicular linear concentration gradients that gradually mold themselves in the patterns shown; other initial conditions lead to the same rotor structure (see below and Fig. 22).

(ii) The numerical experiment of Fig. 17, with a square disk of edge $L = 1$ mm, was continued through 14 complete rotations without detectable change in the shape of the rotating wave, except for its angular position (Fig. 18), after the first few rotations. Similar runs using different parameters, initial conditions, or box sizes have similar results, though the period of rotation depends somewhat on these conditions.

(iii) In some computer runs, errors in the magnetic tape frequently offset random cells by random large distances in composition space, but such disturbances soon damped away rather than growing.

Experiments with $k = \frac{1}{5}$ produce slower rotors of the same geometry, verifying the remark made above that the stagnation point's eigenvalue need not have an imaginary part in order to support this persistent oscillation. Note that the pivotal cell is nowhere near the stagnant composition of the kinetics without diffusion, in contrast to the rotors obtained by Yamada and Kuramoto (1976) and Erneux and Herschkowitz-Kaufman (1977).

Note that we see here, as in the one-dimensional case on a ring, three coexisting alternative solutions to the same reaction–diffusion equation. Each one is locally stable against both uniform and patterned perturbations. The three are a uniform steady state at the stagnation point, a clockwise rotor, and a counterclockwise rotor.

The anatomy of this two-dimensional rotor naturally falls into two parts: a central disk within which the wave is generated and everything outside that disk. The outside of the disk will be considered first.

B. The Wave Radiating from a Rotor

1. *Period and Wavelength*

Far enough from the center of a rotor all cells traverse the same one-dimensional loop in composition space. Therefore, this arbitrarily large two-dimensional region maps degenerately onto the one-dimensional border of the image in composition space (the border of the image includes the image of the border). In this outer two-dimensional region we find only a periodic wavetrain recurrently exciting each cell in the same way. Its wavelength and velocity must lie on the one-dimensional dispersion curve. But where on that curve does it lie? Greenberg (1976) anticipates a one-parameter family of choices, but does not present a stability criterion that might select a unique wavelength. Numerically, the stable answer seems independent of initial conditions and independent of the size of the disk, so long as it is large enough to accommodate a stable wave. The rotor's period slightly exceeds the minimum on the dispersion curve and its wavelength (about 2.8 mm by fitting an involute spiral; see below and Fig. 19) is about the width of the solitary traveling pulse (see Fig. 16d) and substantially greater than the minimum stable wavelength (about 0.65 mm). (In *Z* reagent the same is true qualitatively.)

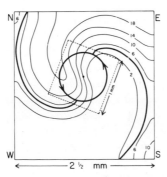

Fig. 19. As in Fig. 17b except that this square is $2\frac{1}{2}$ mm, not 1 mm on edge. The dashed inner box is 1 mm on edge. Even though it is not a boundary here, the portion of *B* contours within it is scarcely distinguishable from Fig. 17b. The nominal boundary between "inside" and "outside" solutions is here a circle of 2.8 mm perimeter, chosen concentric to the observed pivot point. The heavy curves are two sample involutes of that circle. Its circumference was chosen to optimize the eye-fit between the involutes and the concentration level contours (thin curves). This snapshot is at $T = 250$ sec (twentieth rotation) of the anticlockwise rotor shown at $T = 126$ sec in Winfree (1974b). Apart from rotation, there has been no detectable change during the intervening ten rotations.

Imagine a wave circulating around a hole of perimeter P, as allowed in Wiener and Rosenblueth's (1946) analysis. So long as P exceeds the minimum for stability on the one-dimensional dispersion curve [simply the refractory period in Wiener's analysis or Durston's (1973), but more difficult to determine in the more realistic analysis of Rinzel and Keller (1973) or Allessie *et al.* (1977)], it seems plausible to expect circulation time as in the one-dimensional case: the inner edge of the wave circulates along a boundary of perimeter P locally like a one-dimensional wave, and this determines the period of the wave propagating outward in two dimensions. But as the hole is progressively filled in, each volume element of the innermost edge is put in closer contact with the opposite side of the wave. Thus diffusion "short-circuits" the wave across a diameter, increasing the minimum perimeter required for stable maintenance of inhomogeneity. In this range, filling in the hole seems to make it effectively larger! Thus in this two-dimensional situation we expect to see the one-dimensional dispersion curve down to the minimum perimeter at which the one-dimensional wave lapses into uniformity. But in two dimensions around a hole, the curve can be extended to $P = 0$ (suggested by the dashed segment in Fig. 15), presumably increasing the period again to that characteristic of a two-dimensional rotor. A simple and potentially precise measurement recommends itself: a sprinkling of tiny oil droplets onto Millipore material creates unwettable disks of all diameters. After floating such a disk on Z reagent (see Appendix), waves around these empty holes can be clocked through many rotations to establish T at each P.

In contrast to the above numerical calculations, the simplest possible models, with a one-dimensional state space and a unique propagation velocity, require the rotor to emit a wave of the minimum stable wavelength and period (Wiener and Rosenblueth, 1946; Selfridge, 1948; Balakovskii, 1965; Krinskii, 1968; Winfree, 1972). This would make the wave emitted by a rotor only marginally stable, a fact often noted as a difficulty by many physiologists who tend to think in terms of such models (Selfridge, 1948; Gulko and Petrov, 1972; Durston, 1973; Allessie *et al.*, 1973, 1977; Reshodko and Bures, 1975). I believe this may be a deficiency in the simplest possible models, and a reason to provisionally entertain some such picture of a rotor as presented here.

2. Shape of the Outer Wave

Here we can come closer to giving an answer. Wiener and Rosenblueth (1946) showed that a wave of fixed velocity propagating according to Huyghen's construction and circulating around an obstacle takes up a shape described geometrically as the "involute" spiral determined by the obstacle's boundary. Here we have no obstacle, but symmetry considerations would suggest that whatever the equivalent locus might be, it is circular, so the wave should be a circle's involute spiral. Alternatively, Winfree (1972) replaced the " obstacle " by the requirement that the wave rotate uniformly; the resulting differential equation describes a circle's involute. But in both cases the wavelength remains indeterminate. Greenberg (1976), Guckenheimer (1976), and Hastings (personal communication) showed analytically by three different methods that the involute solution cannot be exact, and is worse closer to the center. By any reckoning, the wave should be a spiral of approximately uniform spacing, and this is consistently observed in photographs of Z reagent (Winfree, 1978) and in Kuramoto and Yamada's (1976) analytical solutions using a simplified model. Figure 19 fits an involute spiral to a numerical execution of (2.1) spatially coupled by equal diffusion of A and B as in (3.1).

The involute wave terminates on a circle of circumference equal to the wavelength, which is determined by the rotor's period and the one-dimensional dispersion curve. This is also the envelope of perpendiculars to the outer rotating wave (Winfree, 1978), the effective locus of its origin. It is here that the composition first begins to deviate noticeably toward the inside of the disk's image in composition space. We take this locus as the nominal boundary between outside and inside. Karfunkel (1975) prefers instead to define the boundary in terms of stability as the smallest disk capable of supporting a rotor.

C. THE ROTOR'S CORE

The inside, unlike the outside, harbors steep criss-crossing concentration gradients. Figure 20 shows a contour map of the cross product of the spatial gradients of A and of B: its magnitude, $|(\partial A/\partial x)(\partial B/\partial y) - (\partial B/\partial x)(\partial A/\partial y)|$, decreases tenfold with every $\frac{1}{7}$ mm, or by a factor of 1000 from the pivotal maximum† to the nominal boundary of the inside. It decreases to zero at the

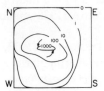

Fig. 20. Contour maps a and b of Fig. 17 were used together to compute the vector product of the spatial gradients of A and B concentrations. This figure shows level contours of $//\nabla A \times \nabla B//$ in arbitrary units on the 1 mm square of Fig. 17. The arrows indicate the direction of rotation of the whole pattern. Note that as is most plainly seen in Fig. 19, the pattern emanates from the unchanging center, outer annuli lagging behind inner annuli.

† No mathematical proof has been given that the pivot coincides with the maximum of $//\nabla A \times \nabla B//$ under conditions of steady rotation. Note that this quantity is defined by vector operators in real physical space, not in composition space, which lacks a metric.

impermeable boundary because gradients have no normal components at the boundary. The inside or "core" is the region that so conspicuously fills the interior of the disk's image in composition space. The geometry here is determined by a balancing of the image's elasticity (corresponding to molecular diffusion) and the divergence of the flow (corresponding to reaction).

Curiously, and quite unlike linearized reaction–diffusion equations (Gmitro and Scriven, 1966), the boundary conditions play little role in determining the core's rotation period or diameter or the wavelength of the surrounding involute spiral. So long as the box is big enough (in real physical space) to allow a stable alternative to uniform stationarity, only left- or right-handed rotors of standard geometry and angular velocity appear. Figure 19 underscores this point by outlining a central 1 mm square of a $2\frac{1}{2}$ mm square computation: the contour lines within this unconfined region are quite similar to those within the 1 mm square of Fig. 17b. The no-flux boundary condition forces contour lines to turn perpendicular to walls, but this happens only in a boundary layer about $\frac{1}{10}$ mm thick.

At the center of the rotor lies a cell in which the composition stays fixed while steep transverse concentration gradients of A and of B pivot around it. Erneux and Herschkowitz-Kaufman (1977) observed the same in solving limit cycle kinetics with diffusion on a circular disk, as did Yamada and Kuramoto (1976).

But here the present model fails badly in representing experimental observations: at least three of the four excitable media known to support rotors do not have fixed pivots:

(i) The visible endpoint of the rotating wave in Z reagent meanders in irregular loops through a central region of circumference approximately equal to the outer wavelength. It does not follow a regular circular path and it often crosses right through the center (Winfree, unpublished data).

(ii) In numerical experiments with a neurophysiologically motivated excitable medium, Gulko and Petrov (1972) made much the same observation: there is no fixed pivot point.

(iii) The experiments of Allessie *et al.* (1973, 1976, 1977) with such a rotor in heart muscle suggest a similar irregularity of core dynamics.

(iv) I do not know whether *Dictyostelium*'s cAMP rotor (Gerisch, 1965, 1968; Durston, 1974; Gross *et al.*, 1976) has a fixed pivot.

In contrast, the principal deficiency of simpler models is that they permit *no* detailed description of the core. Under the name "reverberators," Balakhovskii (1965) and Krinskii (1968) anticipated rotors in excitable media before they were observed in slime molds, heart muscle, or chemical solutions. But their models leave no room for a core: the source of rotating wave is a discontinuity, which in the chemical case, would have to be interpreted as an infinitely steep concentration gradient. The same artefact lingers in the more recent limit-cycle model of Smoes and Dreitlein (1973), in which the reactants are not allowed to diffuse. The caricature I present here at least allows us to form some picture of the rotor's source in terms compatible with physical chemistry. But it needs refinements that only mathematical analysis can bring.

D. How Rotors Are Born

How does one deliberately create a chemical rotor? One way is to obstruct part of a wave with an impermeable barrier, allowing the unobstructed part to propagate around to the backside of the barrier, then remove the barrier to confront the edge of the wave with unexcited medium on the other side. This method was used by Gulko and Petrov (1972) and Winfree (1977). Alternatively, one can abut a solitary wave against reagent at equilibrium. This is achieved by pushing together two blocks of gelled reagent, one of which contains a wave propagating along the interface (Winfree, 1974c). Figure 21 shows a numerical simulation of this experiment. It works as follows:

(i) in physical space, the concentration level contours are obliged to crisscross in the way characteristic of the core of a rotor, thereby initiating a rotor; or stated another way;

(ii) in composition space, the image of the medium is stretched between the wave trajectory and the stagnation point, thus filling out a disk like Fig. 17. This is close enough to the stable image of a rotor so that it soon becomes closer.

The abrupt confrontation between excited and nonexcited regions creates steep concentration gradients near the boundary line. The same geometry can be arranged in a *continuous* way by shearing the reacting fluid to homogenize it at one end of a wave while leaving the other end unsheared. This method was used by Winfree (1972). Both of these techniques work as well with periodic wave trains as with a solitary wave.

The sketch of Fig. 22 anticipates a method of rotor creation that has *not* yet been attempted in the laboratory. The objective is to physically induce a (two-layered) map across the critical central region of composition space by placing point sources of two reactants close together in a solution of all the other constituents of Z reagent. Using cold to inhibit reaction while the necessary diffusion gradients are established, then returning to room temperature should conjure out of the liquid a mirror pair of rotors between the two sources, as sketched. This may provide a counterexample to the assertion that spiral formation depends on preexisting wave propagation. Another is provided by mutants of *Dictyostelium discoideum* (Durston, 1974).

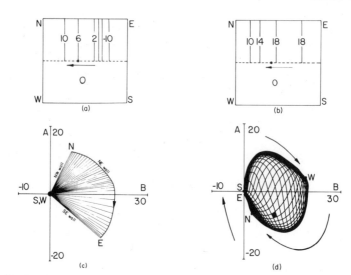

Fig. 21. Creating a rotor by putting a plane wave in lateral contact with inert medium. This is a digital computation using 43 × 43 cells to simulate a 1 mm square. (a) A plane wave propagates from right to left. The full pulse being 2–3 mm wide (see Fig. 10), only its middle part fits in this 1 mm box. Level contours of A are shown in the top half of the box. The bottom half is at $A = 0$, $B = 0$. (b) As in (a), but B level contours. (c) A plotted against B: the composition–space image of the box immediately after the barrier is removed. The whole bottom half of the box in real space initially lies at the origin in composition space. But molecular diffusion along the midline in real space quickly moves nearby cells along rays that join the origin to points along the wave. (d) As in (c) but two seconds later: the initial discontinuity of concentrations is smoothed out. The shaded region [the dot in (a) and (b)] will move up to $(-2, 7)$ to become the pivot of an anticlockwise rotor morphologically indistinguishable from Fig. 17. (However, its rotation period turned out to be 10.25 sec rather than 12.0 sec, possibly on account of the coarser mesh: 43 × 43 compartments rather than 61 × 61 as in Fig. 17.)

E. How Rotors Die

In short, a rotor is created, both in the laboratory and the computer, by contriving concentration gradients that cross transversely through the middle range of A and B concentrations realized in a traveling pulse. So a rotor can be destroyed only by destroying this configuration of gradients:

(i) By increasing or decreasing the concentration of A or B (or both) by a fixed amount in every cell, one moves the physical medium's image by rigid translation in composition space. If it is moved entirely to one side of the line segment ($A = 1$, $B = 0$ to S), then the reaction flow no longer diverges from the middle of the image, no longer opposes the "elasticity" of the image, so it contracts while following the arrows to the origin (Fig. 23). Note that the incrementation of A or B must be effected uniformly throughout the entire medium, not only near the spiral wave's core. Local displacement of A or B

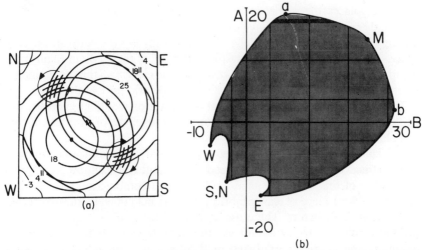

Fig. 22. (a) Sources of *A* and of *B* are placed at points *a*, *b* in a 3 mm square. The level contours of concentration, sketched after about five minutes, are circles concentric to the sources, except near the walls. In the cross-hatched regions, *A* and *B* concentrations and gradients are as in a rotor's core, with directions of rotation as indicated. (b) The image of (a). The four corners and points *M*, *a*, and *b* are all extrema of one sort or another, and the rest of the square lies between them in two layers, like a flattened hot-air balloon opening at the lower left. In this *A* × *B* plane the circles of uniform concentration are level lines parallel to *A* or *B* axes. For example, the heavy circle in (a) is the heavy line in (b). The middle region of this figure (in two sheets) presents initial conditions suitable for establishment of counter-rotating rotors.

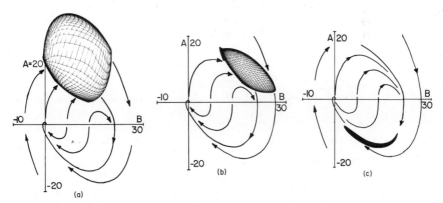

Fig. 23. (a) To the anticlockwise rotor of Fig. 21 we uniformly add $\Delta A = 20$ at $T = 60$ (end of the fifth rotation). Its composition–space image is now displaced into a region of nearly parallel trajectories. (b) One second later, its image has markedly shrunk as all cells react in roughly the same direction and diffusion further homogenizes the square. (c) Several seconds later the rotor, uniformly excited above threshold in (a), is uniformly refractory and flowing into the stagnation point. The sequence provides a simple caricature of defibrillating a heart by elevating voltage (*A*) everywhere, to synchronize a patch of high-frequency wave initiation.

merely displaces the image of that region, and so it only *moves* the rotor within the boundaries (if any) of the medium. Note also that uniformly adding or removing just a little A or B also displaces the rotor in physical space. To displace it all the way to the boundary of the medium (if any), the composition–space image must be displaced *far enough* so that its boundary no longer encloses the rotationally diverging central part of the flow.

The most convenient laboratory test of this notion would be exposure of a Millipore filter bearing spiral waves to an atmosphere of bromine, a consumable reactant.

(ii) The reaction could be stopped (e.g., by cold) while diffusion continues to level the existing concentration gradients. Figures 24–26 show a computer simulation of this kind: the medium's image in composition space swiftly contracts, like a piece of elastic fabric, no longer held open by the divergence of reaction flow vectors. The cross product of the gradients of A and of B, normally confined to a tiny vortex core, rapidly expands to fill the whole medium. The gradients of A and of B relax and ultimately vanish as the medium turns homogeneous.

(iii) Like any living organism, a rotor can be killed by stopping its reactions, as above. Similarly, it can be killed by mechanical trauma: with a sword. Inserting an impermeable barrier from the boundary all the way into

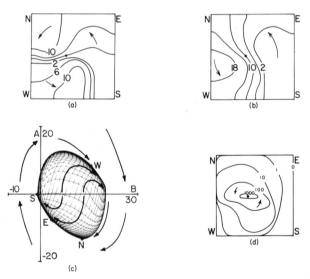

Fig. 24. Anticlockwise rotor created in Fig. 21 is continued up to 60 seconds (five revolutions) in its 1 mm square box. (a) Its A level contours (similar to Fig. 17a), (b) B level contours (similar to Fig. 17b). The dot in (a) and (b) is the pivot point. (c) Its image (A vs. B) (similar to Fig. 17c) and (d) the level contours of $//\nabla A \times \nabla B//$ as in Fig. 20. At this point the reaction is artificially stopped, but diffusion is allowed to continue.

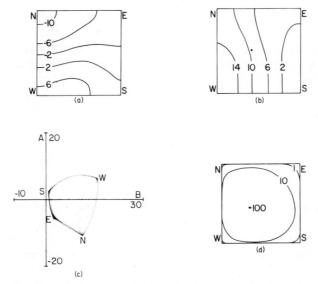

Fig. 25. As in Fig. 24, but $12\frac{1}{2}$ seconds after the reaction stopped.

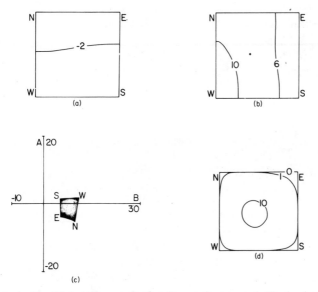

Fig. 26. As in Fig. 24, but 50 seconds after the reaction stopped. The level contours are severely dilated, the image is severely shrunken, and the cross product of gradients is orders of magnitude smaller.

the core of a rotor is equivalent to cutting its elastic image in composition space. This breaks the symmetry of stabilizing tensions, with the result that the image opens along the cut, slides off the central rotationally diverging part of the reaction flow, and follows the arrows to homogeneous equilibrium.

(iv) A rotor too close to an impermeable barrier (e.g., the edge of the medium) is attracted by the barrier, drifts into it, and vanishes. In composition space this process corresponds to that boundary's image being pulled across the central, rotationally diverging part of the reaction flow. This process grades into the mechanism (i) above: putting a boundary too close to the rotor's pivot is the same as displacing the rotor's image so that its margin (that includes the boundary's image) is too close to the pivot composition. An instability ensues in which the image slides by itself the rest of the way off the pivotal composition. In Z reagent the same thing happens near vessel walls, and where two counterrotating rotors lie too close together: the plane of mirror symmetry separating them serves the role of an impermeable barrier, inasmuch as there is no net transport across it in either direction. Mutual annihilation of adjacent counterrotors is a common sight in thin layers of Z reagent. This instability corresponds to the observation (Winfree, 1974b, Fig. 6) that nonuniform solutions are unstable in too small a box. For estimates of the minimal size of a box, see Stibitz and Rytand (1968), Krinskii (1968), Allessie et al. (1973, 1976, and 1977), and Conway et al. (1977).

(v) A rotor can presumably be killed in the same manner as pacemakers (Zhabotinsky and Zaikin, 1973), by encroachment of shorter period waves. This remains to be tested, either computationally or experimentally (perhaps using a periodically heated tungsten filament as the source of waves in Z reagent). My earlier belief that the rotor emits waves at the shortest stable period (Winfree, 1972) requires some such test.

V. Other Structures

A. Multiarm Rotors

The rotor was conceived of by mapping a disk of excitable reagent into composition space in such a way that its elastic coherence would stabilize a perpetual rotation driven by the reaction flow.

From a qualitatively different initial mapping can there arise anything qualitatively different from uniform stationarity or a rotor? A mapping whose border winds more than once around the region ($A \sim -2$, $B \sim 7$) describes a rotor with more than one arm. An effort to depict this image in composition space suggests physical reasons for its (presumable) instability.

In Z reagent multiarm rotors commonly occur only in the sense that a region can harbor several rotors of the same handedness, so that a ring enclosing the region cuts several distinct spiral arms. However, the version in which all rotate about a common pivot seems not to be stable. What this seems to amount to is that rotors too close together repel each other, just as counter-rotors attract each other. So far as I know this possibility has not been studied, either numerically or analytically. Karfunkel (1975) computed a two-armed rotor in an excited enzyme reaction, but did not continue long enough to check its stability.

How does *Dictyostelium* produce stable multiarmed rotors of cAMP excitation (Durston, 1974; Gerisch, 1965, 1968)?

B. REACTION–DIFFUSION "MECHANICS"

Topologically complex two-dimensional media (with holes, twisted bands as handles, etc.) might support exotic wave patterns. It is conceivable that the pieces of such a medium, as they follow local reaction flows in composition space, might tug and push each other in rhythmical patterns that prevent uniform stationarity. But up to now no example of such a "chemical machine" has been invented.

C. TWISTED AND KNOTTED SCROLL RINGS

Even in simply connected media, adding a third dimension permits contemplation of a rich variety of topologically distinct patterns of reaction and diffusion. Only one of these (the scroll ring: Winfree, 1974c) has been demonstrated in Z reagent. None have been computed. The alternatives include scroll rings in which:

(i) the scroll's cross section (a rotor) rotates through a nonzero multiple of 360° along the scroll's closed-ring axis;
(ii) the scroll's axis is tied in a knot;
(iii) both (i) and (ii) occur together.

D. STRUCTURES OBTAINED FROM UNEQUAL-DIFFUSION INSTABILITIES

The beginnings of a bestiary of stable reaction modes in time and space have been collected by Nicolis (1974), Herschkowitz-Kaufman (1975), and Erneux and Herschkowitz-Kaufman (1975) using the "Brusselator" oscillating reaction and unequal diffusion. Nicolis and Prigogine (1977) present a summary. They have discovered several bifurcations of instabilities, resulting in persistent patterns determined by initial conditions and boundary conditions, not unlike the harmonic modes of organ pipes and drum heads.

VI. Other Boundary Conditions

A. DIRICHLET CONDITIONS ON A DISK

Up to now we have only considered finite media with impermeable boundaries; that is, we have used the so-called Neumann boundary conditions. But much work in the theory of reaction and diffusion has used the so-called Dirichlet boundary conditions in which the medium's composition is fixed along the boundary as though it were in contact with an infinite well-stirred reservoir of that composition. In terms of the image of the physical medium in composition space, this corresponds to mapping the boundaries to a fixed point. In the case of a disk of medium, mapping its boundary to a fixed point amounts to reforming the disk into a sphere before mapping it into composition space. Its image is thus two-layered, so whatever wavelike structures may arise, they necessarily arise in complementary pairs. An unpaired rotor is impossible, given Dirichlet boundary conditions. (This is necessarily so only if the *whole* boundary is fixed at the *same* composition.)

B. MEDIA NOT SHAPED LIKE A DISK

Media without boundaries can be mapped into composition space in the same way, with the same results: a spherical or toroidal surface of reacting medium necessarily has a two-layered image, so all its structures come in pairs.

The mapping of three-dimensional media into composition space follows the same principles. The geometric possibilities, of course, depend on whether composition space has fewer or more dimensions than three. This is an area in which collaboration between chemists and topologists seems indispensable. In the case of Z reagent, the first question is "What is the 3-D equivalent of the rotating spiral wave?" One answer is "A vortex ring," as demonstrated experimentally by Winfree (1974c). But, as noted in Section V,C, there seem to be other answers that collectively constitute a taxonomy of topologically distinct, stable modes of spatial and temporal organization of this reaction. To me this looks like a promising area for interaction between theory and experiment, to uncover organizational principles that stand somewhere between physical chemistry and developmental biology.

C. LEADING CENTERS

A conspicuously unresolved question of boundary conditions surrounds the phenomenon of "target patterns" in Z reagent. These are concentric ring-shaped waves (see Appendix Fig. 28 and Kopell and Howard, 1974, Fig. 6) propagating at diverse regular intervals from a point called a "leading center"† (Zhabotinsky and Zaikin, 1973; Zaikin and Kawczynski, 1977) or a "pacemaker" (Winfree, 1972; Durston, 1973; Field, 1973). It appears that there are four kinds of leading centers or pacemakers; the mechanisms of all four stand in need of both experimental and theoretical clarification.

1. Three-Dimensional

One kind of pacemaker occurs only in sufficiently thick layers of reagent: it is an intrinsically three-dimensional wave structure in the physically homogeneous reagent. This is the "scroll ring" demonstrated by Winfree (1974c). All scroll rings emit rings at the same period, the period of their two-dimensional analog, the rotating spiral wave. The remaining types of pacemakers occur in two-dimensional (thin-layer) situations.

2. Pseudowaves

The first is a "pseudowave" (Winfree, 1972) consisting of spontaneous repeats of an earlier circular wave, recurring at the period of the bulk oscillation of unstimulated reaction. As the observed pacemakers have diverse periods, they cannot all be accounted for in this way. Also, "pseudowaves" do not occur in nonoscillating reagent, whereas pacemakers do.

3. Homogeneous Models

A second kind of two-dimensional wave source is proposed in models by Krinskii *et al.* (1971), Shcherbunov *et al.* (1972), Zhabotinsky and Zaikin (1973), Yakhno (1975), and Zaikin and Kawczynski (1977). In these simulations an interaction of reaction with molecular diffusion of *one* of the reactants is invoked together with appropriate initial concentration patterns in space to stabilize a slight acceleration at a point that, as a result, becomes the center of concentric ring waves. Though similar to the rotor in that it consists of a local self-sustaining concentration pattern, this mechanism greatly surprises my physical intuition. If it is not a computational artefact, it seems to me to depend critically on gross inequality of diffusion coefficients, and therefore to be implausible in Z reagent. But the question has not been explored mathematically. As John Tyson has suggested (personal communication), it could be explored experimentally by overwhelming the leading center with shorter period waves from a periodically heated tungsten filament, then stopping. Does the leading center reappear? If so, it was probably not a local self-sustaining concentration pattern. If not, it was probably not a heterogeneous nucleus.

† Zhabotinsky (1970 *et seq.*) introduced the term "leading center" to distinguish ring-emitting sources from rotating spiral sources, which he calls "reverberations." Nicolis and Prigogine (1977) now use "leading centers" for rotating spiral sources. I avoid this usage.

4. Heterogeneous Nuclei

The last type of pacemaker, of greatest interest in terms of boundary conditions and alone capable of accounting for the diversity of periods observed (both in oscillating and nonoscillating reagent), is a heterogeneous nucleus. By this I mean some kind of catalytic particle, perhaps dust or a tiny crystal. Belief in such nuclei is encouraged by the following:

(i) a dark speck is often visible to the naked eye or with a low-powered microscope near the center of a target pattern (but not always);

(ii) pacemakers can often (but not always) be pushed around with a hair, suggesting greater coherence and rigidity than expected of a mere concentration gradient;

(iii) filtration of reagent through 0.22 μm Millipore into adequately cleaned containers usually eliminates pacemakers. Those that remain are usually on the interfaces;

(iv) adding floor dust to filtered reagent restores pacemakers in abundance;

(v) clean reagent that has pacemakers in glassware usually has far fewer when transferred to containers lined with "Sylgard" silicone, which has no effect on the reagent's ability to propagate waves or to oscillate spontaneously.

These observations suggest that some pacemakers may be heterogeneous catalysts (Ortoleva and Ross, 1972, 1973) or surface defects in glassware or in particulate contaminants of unknown origin. So far as I know, this latter possibility has not been investigated either experimentally (e.g., by trying pure inert dusts such as Teflon filings or ground glass, or by serially finer filtration, or by sucking up a pacemaker to deposit on a micropore filter for microscopic examination) or theoretically (e.g., by examining the impact of peculiar boundary conditions on the solution of reaction–diffusion equa-

Fig. 27. Panels b and c illustrate the consequences of slight variations in an "excitable" kinetic scheme such as shown in panel a, which is fashioned after Fig. 7. In each panel the zigzag locus (same in a,b,c) runs through states at which $dA/dt = 0$, and the more nearly horizontal locus (varied from a to b to c) runs through states at which $dB/dt = 0$. In (a) there is one attracting stagnation point at which $dA/dt = dB/dt = 0$. In (b) there is one repelling stagnation point and a limit cycle. In (c) there are two attracting stagnation points and an intermediate repellor.

tions). There may be a clue in the observation of Field and Noyes (1974) that suitably cleaned nichrome wire serves as a pacemaker unless electrically charged. The examination of phenomena on peculiar boundary conditions thus grades into examination of phenomena arising from spatially changing kinetic parameters.

VII. Other Kinetics

Having dwelt briefly on alternative boundary conditions for excitable kinetic schemes, we now complete this article with a few moments' attention to alternative kinetics. As indicated in Fig. 27, there are three major alternatives open to this kind of kinetic scheme:

(i) excitable kinetics;
(ii) oscillatory kinetics, i.e., spontaneous self-triggering of the excitation;
(iii) bistability, i.e., threshold kinetics with an attracting stagnation point on each side of a repellor.

These three cases are familiar siblings in the theoretical literature of nerve membrane kinetics and in electronics (Franck, 1973, 1974). Rossler (1972, 1974) has drawn attention to the analogies between electronic and chemical astable flip-flops (multivibrating, oscillating), "monostable flip-flops" (excitable), and "bistable flip-flops" (bistable). Schmitz (1975) reviews the chemical engineering literature of multiple attractors.

The oscillatory case (uniform in space while oscillating in time) was the first case discovered in the Belousov–Zhabotinsky saga, and it initiated an extensive literature of chemical limit cycles and waves built out of them, with or without essential involvement of diffusion. This case has provided stimulating models to mathematicians and physiologists intrigued by biochemical oscillators and "clocks" in living organisms.

In trying to ask whether the oscillation in time is indeed critical to the observed oscillations in space and to other wave phenomena, I discovered the excitable case (Winfree, 1972). In this form, Z reagent has provided stimulating models to mathematicians and physiologists intrigued by excitable membrane and wavelike phenomena, both normal and pathological, in such excitable tissues as muscle, brain, heart, and intestine. Human small intestine in many respects constitutes a one-dimensional continuum of excitable or even spontaneously oscillatory neuromuscular units with a head-to-tail gradient of spontaneous period or of duration of refractoriness (Nelson and Becker, 1968; McCoy and Baker, 1969; Code and Szurszewski, 1970; Diamant *et al.*, 1970; Sarna *et al.*, 1971; Specht and Bortoff, 1972; Sarna *et al.*, 1972a,b; Diamant *et al.*, 1973; Sarna *et al.*, 1973; Conner *et al.*, 1974; Sarna and Daniel, 1974; El-sharkawy and Daniel, 1975; Linkens,

1977). Electrical and muscular waves propagating along this line have much to do with the unidirectional movement of food during digestion. The activity of intestine resembles the relatively unexplored short-wave limit of the Kopell and Howard (1973a,b) and Thoenes (1973) experiments described above, inasmuch as those chemical experiments established a line of oscillators along a monotone frequency gradient, resulting in apparent propagation of kinematic waves toward the slow end. Such waves, interpreted mechanically by smooth-muscle contraction, could move the contents of the intestine. The chemical analogs, as published, were terminated before phase gradients become steep enough to initiate trigger waves, and before these waves pressed themselves too close together for stable coexistence. In the intestine, waves arriving at a high frequency from the fast end are occasionally deleted where the local maximum response frequency is too low to accommodate the incoming wave train. Presumably the same must occur along a chemical frequency gradient after the stage photographed by Thoenes and by Kopell and Howard.

Something of this sort was shown by Arshavskii (1964) in a nerve fiber arranged along a temperature gradient. Waves from a pacemaker at the warm end were deleted at stages along their journey to the cool end, which had a greatly prolonged refractory period.

But no one has yet come up with the recipe for a bistable version of the isothermal unstirred Belousov–Zhabotinsky reaction. (Though Graziani et al., 1977 and Geiseler and Follner, 1977 have achieved bistability in a continuous-flow stirred-tank reactor. The influx and efflux terms seem to play an essential role in the kinetic equations.) This would be of fundamental interest as a chemical analog to the ability of embryonic cells to differentiate into stable alternative steady states of biochemical operation. Theorists have produced a variety of models of tissue regionalization in terms of alternative attracting stagnation points separated by repellors as in Fig. 27 (e.g., Selkov, 1970; Edelstein, 1972; Gilbert, 1973; Wilson, 1973; Kauffman, 1975; Sanglier and Nicholis, 1976; Lewis et al., 1977). Such bistability was discovered by Degn (1968) in a biochemical reaction previously studied for its limit cycle oscillations (Yamakazi and Yokota, 1967; Degn, 1969; Degn and Mayer, 1969). This same reaction, incidentally, has recently been shown to exhibit *chaotic* dynamics in yet a third parameter range [Olsen and Degn (1977)]. Discussion at the Faraday Society Meeting of 1974 [(*Farad. Symp. Chem. Soc.* 9); Rossler, 1974; Othmer, 1975; and Turner, 1976] suggest that this may be a viable option for the Belousov–Zhabotinsky reagent. For example, it might suffice to run the reaction atop a bromine-saturated Sylgard layer, or beneath a bromine-saturated layer of oil. When this possibility is realized in the laboratory, we will have the first opportunity to visually follow the kinetics of spatial differentiation in a reaction of known mechanism spatially coupled by molecular diffusion.

The behavior of such media has already been anticipated mathematically to some extent in context of embryology by Edelstein (1972) and in context of membrane kinetics and physical chemistry by Wilson (1973), Franck (1973), Ortoleva and Ross (1975), Fife (1976, 1977), and Haedt *et al.* (1976). The usual result is that the chemical medium is partitioned into regions, each of which lingers close to one of the two attractors. A narrow transition zone separates adjacent regions. It moves as one region expands at the expense of the other. Only Edelstein (1972) obtained (numerically) a static partitioning of the medium. I have been unable to reproduce this computation: the transition zone always moves toward one attractor, as it also does in all other examples I know of.

VIII. Appendix

This Appendix contains technical directions for playing with waves in the Belousov–Zhabotinsky reaction. The best general reference is Tyson (1976). (See also the chapter by Field, this volume.)

As first encountered by Belousov (1959), it was an aqueous solution in which bromate oxidatively decarboxylated an organic diketone through the catalytic agency of cerium ions. Belousov's seminal discovery came as an accident during his effort to determine citrate quantitatively. Zaikin and Zhabotinsky (1970) replaced cerium by ferroin to serve additionally as a visible redox indicator and modified other proportions to accentuate wave-like color bands. I eliminated spontaneous bulk oscillations by further empirical refinement of the formula and eliminated spontaneous sources of ring-shaped waves of diverse periods (so called "pacemakers" or "leading centers") by Millipore filtration into a silicone-lined dish. In this form, the reagent, which I call Z reagent, is excitable in the sense that it rests in an orange–red steady state with reduced ferroin and high bromide concentration until excited by any of a variety of disturbances. Then it turns blue, with ferroin oxidized and low bromide concentrations, for several seconds, and slowly relaxes back to the same orange–red steady state. This reaction takes place not uniformly throughout the reacting volume, but in *waves* of oxidative activity, typically a couple of millimeters apart, with a razor-sharp blue profile in orange liquid: hence its whimsical pseudonym "1 normal solution of the wave equation."

Z reagent is easy to prepare and functions reliably. The following aqueous solutions are to be made from anhydrous reagents:

To 67 ml of water add 2 ml of sulfuric acid and 5 g of sodium bromate (total 70 ml). To 6 ml of this (in a glass vessel) add $\frac{1}{2}$ ml of sodium bromide solution (1 g/10 ml). Add 1 ml of malonic acid solution (1 g/10 ml) and wait for bromine color to vanish. Add 1 ml of 25 mM (standard) phenanthroline ferrous sulfate and a drop of 1 g/1000 ml Triton X-100 surfactant to facilitate spreading. Mix well, pour into a covered 90 mm Petri dish illuminated from below. Bubbles of CO_2 can be removed every 15 minutes by stirring the liquid; it turns blue, then reverts to red, and ring propagation begins anew. The reagent turns permanently blue after about 45 min at 25°C. Amounts are not critical, except that no more than 4 mM chloride can be tolerated: i.e., use distilled water, and keep your salty fingers off any surfaces that will contact the liquid. The ferroin can be obtained from the G. F. Smith Chemical Company, Columbus, Ohio, or the Fisher Scientific Company, Fairlawn, New Jersey. Avoid European *Merck* ferroin, which contains 75 mM chloride. "Photoflo" from a darkroom serves as well as Triton X-100 (Rohm and Haas, Philadelphia) to reduce surface tension without obstructing the reaction. I have usually found it necessary to recrystallize the sodium bromate, even from A.R. grade stocks.

Variations of the formula give interestingly different results. For example, the reagent oscillates more quickly if the bromide is omitted. The sulfuric acid can be replaced by sodium bisulfate so that spilled reagent dries into a powder, rather than into concentrated H_2SO_4. Replacing some of the 25 mM ferroin an equal volume of 0.1 N ceric sulfate (another standard oxidometric reagent) also enhances oscillations and makes it possible to see into a deeper layer of reagent. Leaving out half of the sulfuric acid (Fig. 28, bottom) suppresses spontaneous oscillation, at least in thin layers exposed to the air; in this condition, the blue, oxidizing phase propagates like a grassfire, once started, but never occurs spontaneously. A variety of other recipes are described by Busse (1968), Hershkowitz-Kaufman (1970), Zaikin and Zhabotinsky (1970), Beck and Varadi (1971), Field (1973), Zhabotinsky and Zaikin (1973), Kopell and Howard (1973a,b), Thoenes (1973), Tatterson and Hudson (1975), and Graziani, Hudson, and Schmitz (1976).

The easiest way to watch these chemical waves is to pour a 1 mm depth of reagent into a very clean dish resting on top of a light box. Disposable plastic tissue culture dishes are good for this purpose: they are optically perfect, and free of dirt and scratches, which otherwise serve as nuclei for CO_2 bubbles and as pacemaker nuclei that engender a discouraging profusion of waves. Setting the dish over a dish of blue $CuSO_4$ solution (to which a few drops of H_2SO_4 are added for clarity) stops heat and stops the consequent deterioration of waves by convection currents, and greatly enhances visual contrast between the vivid blue waves and their orange background. Be sure to *cover* the reagent dish: the slightest air currents disrupt the perfection of the wave fronts, and escaping Br_2 will give you a headache as a reminder. You must be very careful not to bump, vibrate, or tilt the dish, or liquid flow will destroy the waves. Waves may start spontaneously. If not, a touch of a hot needle will help.

Fig. 28. Top: Circular waves emerge periodically from randomly scattered pacemaker nuclei in a 1.5 mm thickness of reagent. (A few pacemakers lie outside the field of view.) These blue waves of malonic acid oxidation cancel each other where they collide. Each successive collision occurs closer to the lower frequency pacemaker, until it is entrained. These four successive snapshots by transmitted light were taken 60 sec apart. The field of view is 66 mm in diameter. Wave velocity is about 8 mm/min. Pacemaker frequencies range from about one to three rings per minute. The tiny circles are growing CO_2 bubbles. Bottom: Everything is as above except that the reagent is less acidic (to minimize pacemaker activity and to slow wave propagation to 4 mm/min). This reagent, when conducting ring-shaped waves, was deformed by gentle stirring. Segments of blue wave revert to red and the surviving pieces begin to pivot around their free ends, winding into paired involute spirals. Each spiral rotates twice each minute. Notice the widening space between the outermost waves: waves move a little faster into virgin territory than into reagent still recovering from the passage of a prior wave.

Two ml of colloidal SiO_2 (Apache Chemical Co., Rockford, Illinois) can be used to gel a ml of reagent to a transparent substance with the consistency of peanut butter, in which waves propagate exactly as in the unbound fluid. This makes it easy to implant barriers to block a wave, to shear waves in a controlled way, to suck up the center of a spiral or ring source, etc. [This silica gel incidentally provides a remarkably reliable and exact demonstration of the interaction of diffusion with a simpler reaction: by abutting a gel of 25 mM ceric sulfate (yellow) against a gel of 25 mM ferroin (red), a blue boundary is formed that moves as an extremely sharp line advancing in proportion to the square root of time.]

Liquid Z reagent is instantly absorbed by capillarity into the fine (0.22 μm) holes of a Millipore filter disk floating on the liquid surface. These filters, type GSWP, are available from the Millipore Corporation, Bedford,

Massachusetts. I have not found any other kind of micropore filter that works well. Not all versions of the reagent propagate in Millipores, but the unmodified recipe given above does perfectly when the filter is either stuck between plastic sheets or stuck to the inside of the lid of a closed plastic dish, or dropped in oil to retard evaporation and exclude oxygen. CO_2 passes off without forming bubbles. Because its own viscosity immobilizes the liquid in the submicroscopic connected tunnels of the Millipore, it can be freely handled without disrupting wave patterns. However, beware of salty finger-tips and iron forceps: use nylon forceps. Waves can be started with a hot needle, or by electrically heating the Millipore through the oil with the stretched-out tungsten filament of a broken flashlight bulb. Periodic wave trains can be set up, in concentric rings, by pulsing periodically. I have measured the dependence of velocity on period and observed instabilities at the minimum supportable period (about 20 seconds in this recipe, at 20°C).

Waves can be blocked and fragmented by briefly touching the filter through the oil film with iron, e.g., a razor edge. This lays down a red "Iron Curtain" that blocks waves for several seconds. Any free endpoints so formed become the centers of spirals. Particularly amusing phenomena ensue when the iron curtain is used to delete a segment of one wave in a train of waves.

Any of these rapidly changing wave patterns can, of course, be captured by photography. But a chemical trick serves as well: immersion in ice cold saturated salt water instantly stops the reaction, and within five minutes almost everything diffuses out *except* the red ferrous phenanthroline. After blotting dry, dip the filter in 1 gm/40 ml sodium iodide solution, blot dry again, and expose to iodine vapor for 10 min (in a bottle containing iodine crystals). This fixes the ferroin. After a final rinse in water, dry the filter in air and clear it by floating on paraffin oil. After blotting dry again, sandwich it between two sheets of clear "Contact" (Comark Plastics, New York) paper. Filters preserved in this way retain intricate wave patterns with striking clarity that has not lessened since 1973.

It is possible to preserve filters dry, without oil or "Contact" paper, with the ferroin precipitated on the glossy side of the Millipore as a brightly reflecting metallic gold film. This better technique requires very thorough rinsing prior to the iodide dip, and exact control of the iodine atmosphere. I have not been able to do this reliably.

A less troublesome procedure requires only dropping the filters into icy 3% perchloric acid solution, but the pattern remains less than a day, or until the filter dries out.

For demonstration purposes, it is sometimes convenient to have a supply of dry Millipores that need only be wetted to start the reaction. These are not suitable for quantitative experiments, since the wetting typically introduces inhomogeneities of catalyst concentration, etc. But they make excellent waves.

Prepare ammonium bromomalonate at 1 molar in water. Mix with an equal volume of standard 25 mM ferrous phenanthroline. Combine this mixture with an equal volume of 4 molar ammonium bisulfate. Float a GSWP Millipore on this solution, remove surface liquid, and allow to dry on a plastic surface. Next, soak Whatman #1 filter paper in $\frac{1}{3}$ M sodium bromate and dry. To produce waves, wet the bromate paper, stick it on to a clean surface and place a bromomalonate Millipore on top. It wets and adheres. Blue dots soon appear and grow into waves propagating at a few millimeters per minute. This continues for about an hour at 20°C unless the papers dry up first. DeSimone, Beil, and Scriven (1973) describe an alternative preparation involving collodion membranes.

By substituting another substrate for the conventional malonate, and adjusting the proportions of other reactants, one can obtain excitable media that do not accumulate CO_2 bubbles. This is a great technical convenience for experiments in capillary tubes, or between glass slides, or in stacks of Millipore disks. I have had good results with 2 acetoacetamido pyrimidine diethyl and dimethyl malonates. Jessen *et al.* (1976) report excellent results with 2.4 pentanedione (alias acetylacetone) with manganese catalyst instead of ferroin. M. Cohen in my laboratory has prepared an excellent ferroin recipe using trifluoroacetylacetone.

ACKNOWLEDGMENTS

This work has been supported by grants from the National Science Foundation and the National Institutes of Health. The paper was written while enjoying the hospitality of the Biochemistry Department, Medical University of South Carolina. I thank John Rinzel and Richard Field for discussions over the years that have decimated a greater variety of conjectures, Richard Krasnow for editing, and Joan Waybright for typing endless revisions. Endré Koros generously provided a hand-copied translation of Belousov's paper, and Jean Anwyll of Polaroid donated SX-70 film for Fig. 28.

References

Aldridge, J. (1976). *In* " Handbook of Engineering in Medicine and Biology " (D. Fleming, ed.). Chem. Rubber Co. Press, Cleveland, Ohio.

Aldridge, J., and Pavlidis, T. (1976). *Nature (London)* **259**, 343.

Allessie, M. A., Bonke, F. I. M., and Schopman, F. J. G. (1973). *Circ. Res.* **33**, 54.

Allessie, M. A., Bonke, F. I. M., and Schopman, F. J. G. (1976). *Clin. Res.* **39**, 168.

Allessie, M. A., Bonke, F. I. M., and Schopman, F. J. G. (1977). *Circ. Res.* **41**, 9.

Andronov, A. A., Vitt, A. A., and Khaikin, S. E. (1966). *In* " Theory of Oscillators " (W. Fishwick, ed.), pp. 468–478. Pergamon Press, Oxford.

Aris, R. (1975). "The Mathematical Theory of Diffusion and Reaction in Permeable Catalysts," Vols. I and II. Oxford Univ. Press (Clarendon), London and New York.

Arshavskii, Y. I. (1964). *Biofizika* **9**, 365.
Arshavskii, Y. I., Berkenblit, N. B., and Dunin-Barkovskii, V. L. (1965). *Biofizika* **10**, 1048.
Balakhovskii, I. S. (1965). *Biofizika* **10**, 1063.
Balslev, E., and Degn, H. (1975). *J. Theor. Biol.* **49**, 173.
Beck, M., and Varadi, Z. (1971). *Magy. Kem. Foly.* **77**, 167.
Belousov, B. (1959). *Sb. Ref. Radiats. Med.* (Medgis, Moscow). p. 145.
Busse, H. G. (1968). *J. Phys. Chem.* **73**, 750.
Chance, B., Williamson, G., Lee, J. Y., Mela, L., DeVault, D., Ghosh, A., and Pye, E. K. (1973).
    In " Biological and Biochemical Oscillators " (B. Chance, A. K. Ghosh, E. K. Pye, and B.
    Hess, eds.), pp. 285–302. Academic Press, New York.
Clarke, B. L. (1976). *J. Chem. Phys.* **64**, 4165.
Code, C. F., and Szurszewski, J. H. (1970). *J. Physiol.* **207**, 281.
Connor, J. A., Prosser, C. L., and Weems, W. A. (1974). *J. Physiol.* **240**, 671.
Conway, E., Hoff, D., and Smoller, J. (1977). Preprint.
Cranefield, P. F., Klein, H. O., and Hoffman, B. F. (1971a). *Circ. Res.* **28**, 199.
Cranefield, P. F., and Klein, B. F. (1971b). *Circ. Res.* **28**, 220.
Cranefield, P. F., Wit, A. L., and Hoffman, B. F. (1972). *J. Gen. Physiol.* **59**, 227.
Cummings, F. W. (1975). *J. Theor. Biol.* **55**, 455.
Defay, R., Prigogine, I., and Sanfeld, A. (1977). *J. Colloid Interface Sci.* **58**, 498.
Degn, H. (1968). *Nature (London)* **217**, 1047.
Degn, H. (1969). *Biochem. Biophys. Acta* **180**, 271.
Degn, H., and Mayer, D. (1969). *Biochim. Biophys. Acta* **180**, 291.
DeSimone, J. A., Beil, D. L., and Scriven, L. E. (1973). *Science* **180**, 947.
Diamant, N. E., Rose, P. K., and Davison, E. J. (1970). *Am. J. Physiol.* **219**, 1684.
Diamant, N. E., Wong, J., and Chen, L. (1973). *Am. J. Physiol.* **225**, 1497.
Drietlein, J., and Smoes, M. (1974). *J. Theor. Biol.* **46**, 559.
Durston, A. J. (1973). *J. Theor. Biol.* **42**, 483.
Durston, A. J. (1974). *Dev. Biol.* **37**, 225.
Edelstein, B. B. (1972). *J. Theor. Biol.* **37**, 221.
El-Sharkawy, T. Y., and Daniel, E. E. (1975). *Am. J. Physiol.* **229**, 1268.
Erneux, T., and Herschkowitz-Kaufman, M. (1975). *Biophys. Chem.* **3**, 345.
Erneux, T., and Herschkowitz-Kaufman, M. (1977). *J. Chem. Phys.* **66**, 248.
Farley, B. (1965). *In* " Computers in Biomedical Research " (R. W. Stacy and B. Waxman, eds.),
    Vol. 1, p. 265. Academic Press, New York.
Feinn, D., and Ortoleva, P. (1977). *J. Chem. Phys.* **67**, 2119.
Field, R. J. (1973). *Chem. Unserer Zeit* **7**, 171.
Field, R. J., Koros, E., and Noyes, R. M. (1972). *J. Am. Chem. Soc.* **94**, 8649.
Field, R. J., Noyes, R. M., and Koros, E. (1974). *J. Am. Chem. Soc.* **96**, 2001.
Field, R. J., and Troy, W. C. (1977). *Ark. Rat. Mech.* (in press).
Fife, P. (1976). *J. Chem. Phys.* **64**, 554.
Fife, P. (1977). *Proc. SIAM-AMS* **10** (in press).
Fife, P. (1977). Preprint.
Franck, U. F. (1973). *In* " Biological and Biochemical Oscillators " (B. Chance, ed.), pp. 7–30.
    Academic Press, New York.
Franck, U. F. (1974). *Faraday Symp. Chem. Soc.* **8**, 137.
Gerisch, G. (1965). *Wilhelm Roux Arch. Entwicklungsmech. Org.* **116**, 127.
Gerisch, G. (1968). *Curr. Top. Dev. Biol.* **3**, 157.
Geisler, W., and Follner, H. H. (1977). *Biophys. Chem.* **6**, 107.
Ghosh, A. K., Chance, B., and Pye, E. K. (1971). *Arch. Biochem.* **145**, 319.
Gilbert, D. A. (1973). *Biosystems* **5**, 128.

Gilpin, M. (1977). *Am. Nat.* (in press).

Glass, L. (1975). *J. Chem. Phys.* **63**, 1325.

Glass, L., and Perez, R. (1974). *J. Chem. Phys.* **61**, 5242.

Gmitro, J. I., and Scriven, L. E. (1966). *In* "Intracellular Transport" (J. Danielli, ed.), p. 221. Academic Press, New York.

Goldbeter, A. (1973). *Proc. Natl. Acad. Sci. U.S.A.* **70**, 3255.

Goldbeter, A. (1975). *Nature (London)* **253**, 540.

Goldbeter, A., and Caplan, S. R. (1976). *Annu. Rev. Biophys. Bioeng.* **5**, 449.

Goldstein, S. S., and Rall, W. (1974). *Biophys. J.* **14**, 731.

Graziani, K. R., Hudson, J. L., and Schmitz, R. A. (1976). *Chem. Eng. J.* **12**, 9.

Greenberg, J. M. (1976). *SIAM J. Appl. Math.* **30**, 199.

Greenberg, J. M. (1977). *SIAM J. Appl. Math.* (in press).

Gross, J. D., Peacey, M. J., and Trevan, D. J. (1976). *J. Cell. Sci.* **22**, 645.

Gulko, F. B., and Petrov, A. A. (1972). *Biofizika* **17**, 261.

Guckenheimer, J. (1976). *In* "Lecture Notes in Mathematics" (P. Hilton, ed.), Vol. 525. Springer-Verlag, Berlin and New York.

Guckenheimer, J., Oster, G., and Ipaktchi, A. (1977). *J. Math. Biol.* **4**, 101.

Hardt, S., Naspartek, A., Segel, L. A., and Caplan, S. R. (1976). *In* "Analysis and Control of Immobilized Enzyme Systems" (D. Thomas, ed.). North-Holland, Amsterdam.

Hastings, S. P. (1976). *Stud. Appl. Math.* **55**, 327.

Hastings, S. P., and Murray, J. D. (1975). *SIAM J. Appl. Math.* **28**, 678.

Herschkowitz-Kaufman, M. (1970). *C.R. Hebd. Seances Acad. Sci.* **270**, 1049.

Herschkowitz-Kaufman, M. (1975). *Bull. Math. Biol.* **37**, 589.

Hess, B., and Boiteux, A. (1971). *Annu. Rev. Biochem.* **40**, 237.

Higgins, J. (1967). *Ind. Eng. Chem.* **59**, 19.

Jessen, W., Busse, H. G., and Havsteen, B. (1976). *Agnew. Chem.* **15**, 689.

Kalmus, H., and Wigglesworth, L. A. (1960). *Cold Spring Harbor Symp. Quant. Biol.* **25**, 211.

Karfunkel, H. R. (1975). Thesis, University of Tubingen.

Karfunkel, H. R., and Seelig, F. F. (1975). *J. Math. Biol.* **2**, 123.

Kauffman, S. (1975). *Science* **181**, 310.

Kopell, N., and Howard, L. (1973a). *Science* **180**, 1171.

Kopell, N., and Howard, L. (1973b). *Stud. Appl. Math.* **52**, 291.

Krinskii, V. I. (1968). *Probl. Kibern.* **20**, 59.

Krinskii, V. I., Pertsov, A. M., Reshetilov, A. N., and Shcherbunov, A. M. (1971). *In* "Oscillatory Processes in Biological and Chemical Systems," Vol. 2. Puschino-on-oka (in Russian).

Kuramoto, Y., and Tsuzuki, T. (1976). *Prog. Theor. Phys.* **55**, 356.

Kuramoto, Y., and Yamada, T. (1976). *Prog. Theor. Phys.* **56**, 724.

Lewis, J., Slack, J. M. W., and Wolpert, L. (1977). *J. Theor. Biol.* **65**, 579.

Linkens, D. A. (1977). *Bull. Math. Biol.* **39**, 359.

Lorenz, E. N. (1963). *J. Atmos. Sci.* **20**, 130.

May, R. M. (1976). *Nature (London)* **261**, 459.

May, R. M., and Oster, G. F. (1976). *Am. Nat.* **110**, 573.

McCoy, E. J., and Baker, R. D. (1969). *Am. J. Dig. Dis.* **14**, 9.

McKean, H. P. (1970). *Adv. Math.* **4**, 209.

Moore, D. W., and Spiegel, E. A. (1966). *Astro. Phys. J.* **143**, 871.

Murray, J. D. (1976). *J. Theor. Biol.* **56**, 329.

Nazarea, A. D. (1974). *Proc. Natl. Acad. Sci. U.S.A.* **71**, 3751.

Nazarea, A. D. (1977). *J. Theor. Biol.* **67**, 311.

Nelsen, T. S., and Becker, J. C. (1968). *Am. J. Physiol.* **214**, 749.

Nicolis, G. (1974). *Proc. SIAM-AMS* **8**, 33.

Nicholis, G., and Prigogine, I. (1977). " Self Organization in Non-Equilibrium Systems." Wiley, New York.

Nicholis, G., and Portnow, J. (1973). *Chem. Rev.* **73**, 365.

Offner, F., Weinberg, A., and Young, G. (1940). *Bull. Math. Biophys.* **2**, 89.

Olsen, L. F., and Degn, H. (1977). *Nature (London)* (in press).

Ortoleva, P. (1976). *J. Chem. Phys.* **64**, 1395.

Ortoleva, P., and Ross, J. (1972). *J. Chem. Phys.* **56**, 4397.

Ortoleva, P., and Ross, J. (1973). *J. Chem. Phys.* **58**, 5673.

Ortoleva, P., and Ross, J. (1974). *J. Chem. Phys.* **60**, 5090.

Ortoleva, P., and Ross, J. (1975). *J. Chem. Phys.* **63**, 3398.

Oster, G., and Perelson, A. (1974). *Arch. Rat. Mech. Anal.* **55**, 230.

Othmer, H. G. (1975). *Math. Biosci.* **24**, 205.

Othmer, H. G. (1977). *Lec. Math. Life Sci.* **9**, 57.

Othmer, H. G., and Scriven, L. E. (1969). *Ind. Eng. Chem., Fundam.* **8**, 302.

Othmer, H. G., and Scriven, L. E. (1974). *J. Theor. Biol.* **4**, 83.

Ramon, F., Joyner, R. W., and Moore, J. W. (1975). *Fed. Proc., Fed. Am. Soc. Exp. Biol.* **34**, 1357.

Reshodko, L. V., and Bures, J. (1975). *Biol. Cyber.* **18**, 181.

Rinzel, J. (1975a). *J. Math. Biol.* **2**, 205.

Rinzel, J. (1975b). *Biophys. J.* **15**, 975.

Rinzel, J., and Keller, J. (1973). *Biophys. J.* **13**, 1313.

Rosen, G. (1976). *J. Chem. Phys.* **63**, 417.

Rosenshtraukh, L. V., Kholopov, A. V., and Yushamanova, A. V. (1970). *Biofizika* **15**, 690.

Ross, J. (1976). *Ber. Bunsenges. Phys. Chem.* **80**, 1112.

Rossler, O. E. (1972). *J. Theor. Biol.* **36**, 413.

Rossler, O. E. (1974). *In* " Lecture Notes in Biomathematics " (S. Levin, ed.), Vol. 4. Springer-Verlag, Berlin and New York.

Rossler, O. E. (1976a). *Phys. Lett. A* **57**, 397.

Rossler, O. E. (1976b). *Z. Naturforsch.* **31a**, 259.

Rossler, O. E. (1976c). *Z. Naturforsch.* **31a**, 1168.

Rossler, O. E. (1976d). *Z. Naturforsch.* **31a**, 1664.

Rossler, O. E. (1977). *Bull. Math. Biol.* **39**, 275.

Rossler, O. E., and Wegmann, K. (1977). *Nature (London)* **271**, 89.

Rossler, O. E., and Hoffman, D. (1972). " Analysis Simulation." North-Holland Publ., Amsterdam.

Sanglier, M., and Nicholis, G. (1976). *Biophys. Chem.* **4**, 113.

Sarna, S. K., and Daniel, E. E. (1973). *Am. J. Physiol.* **225**, 125.

Sarna, S. K., and Daniel, E. E. (1974). *Am. J. Physiol.* **229**, 1268.

Sarna, S. K., Daniel, E. E., and Kingma, Y. J. (1971). *Am. J. Physiol.* **221**, 166.

Sarna, S. K., Daniel, E. E., and Kingma, Y. J. (1972a). *Am. J. Physiol.* **223**, 332.

Sarna, S. K., Daniel, E. E., and Kingma, Y. J. (1972b). *Dig. Dis.* **17**, 299.

Schmitz, R. A. (1975). *Adv. Chem.* **148**, 156.

Schmitz, R. A., and Garrigan, P. C. (1977). *Science* (in press).

Schmitz, R. A., Graziani, K. R., and Hudson, J. L. (1977). *J. Chem. Phys.* **67**, 3040.

Segal, L., and Jackson, J. (1972). *J. Theor. Biol.* **37**, 545.

Selfridge, O. (1948). *Arch. Inst. Cardiol. Mex.* **10**, 177.

Selkov, E. E. (1970). *Biofizika* **15**, 1065.

Shcherbunov, A. M., Krinsky, V. I., and Pertzov, A. M. (1972). *In* " Mathematical Models of Biological Systems," p. 25. Sci. Publ., Moscow (in Russian).

Shibata, M., and Bures, J. (1974). *J. Neurobiol.* **5**, 107.

Smoes, M., and Dreitlein, J. (1973). *J. Chem. Phys.* **12**, 6277.
Specht, P. C., and Bortoff, A. (1972). *Dig. Dis.* **17**, 311.
Stanshine, J. A. (1977). *Stud. Appl. Math.* **55**, 327.
Stanshine, J. A., and Howard, L. N. (1976). *Stud. Appl. Math.* **55**, 129.
Stibitz, G. R., and Rytand, D. A. (1968). *Circulation* **37**, 75.
Tatterson, D. F., and Hudson, J. L. (1973). *Chem. Eng. Commun.* **1**, 3.
Thoenes, D. (1973). *Nature (London), Phys. Sci.* **243**, 18.
Thompson, J. M. T., and Hunt, G. W. (1977). *Interd. Sci. Rev.* **2**, 240.
Troy, W. C. (1977). *Rocky Mount. J. Math.* **7**, 467.
Troy, W. C., and Field, R. J. (1977). *SIAM J. Appl. Math.* **32**, 306.
Turing, A. (1952). *Philos. Trans. R. Soc. London, Ser. B* **237**, 37.
Turner, J. S. (1976). *Phys. Lett. A* **56**, 155.
Tyson, J. (1976). *In* " Lecture Notes in Mathematics " (S. Levin, ed.), Vol. 10. Springer-Verlag,
    Berlin and New York.
Tyson, J. (1977). *J. Chem. Phys.* **66**, 905.
Tyson, J., and Kauffman, S. (1975). *J. Math. Biol.* **1**, 289.
Varadi, Z. B., and Beck, H. T. (1975). *BioSystems* **7**, 77.
Wei, J. (1962). *J. Chem. Phys.* **36**, 1578.
Wiener, N., and Rosenblueth, A. (1946). *Arch. Inst. Cardiol. Mex.* **16**, 105.
Williams, R. F. (1977). Preprint.
Wilson, H. (1973). *In* " Synergetics " (H. Haken, ed.). Teubner, Stuttgart.
Winfree, A. (1972). *Science* **175**, 634.
Winfree, A. (1974a). *J. Math. Biol.* **1**, 73.
Winfree, A. (1974b). *Proc. SIAM-AMS* **8**, 13.
Winfree, A. (1974c). *Faraday Symp. Chem. Soc.* **8**, 38.
Winfree, A. T. (1974d). *Sci. Am.* **230**, 82.
Winfree, A. (1977). *Adv. Med. Phys.* **16** (in press).
Yakhno, V. G. (1975). *Biofizika* **20**, 669.
Yamada, T., and Kuramoto, Y. (1976). *Prog. Theor. Phys.* **55**, 2035.
Yamakazi, I., and Yokota, K. (1967). *Biochem. Biophys. Acta* **132**, 310.
Zaikin, A. N. (1975). *Biofizika* **20**, 772.
Zaikin, A. N., and Kawczynski, A. L. (1977). *J. Non-Equilib. Thermodyn.* **2**, 39.
Zaikin, A. N., and Zhabotinsky, A. M. (1970). *Nature (London)* **225**, 535.
Zeeman, E. C. (1972). *In* "Towards a Theoretical Biology" (C. H. Waddington, ed.), Vol. 4.
    Aldine, Chicago, Illinois.
Zhabotinsky, A. M. (1970). Thesis, Inst. of Biol. Phys., Acad. Sci. USSR.
Zhabotinsky, A. M., and Zaikin, A. N. (1971). *In* "Oscillatory Processes in Biological and
    Chemical Systems," Vol. 2. Puschino-on-oka (in Russian).
Zhabotinsky, A. M., and Zaikin, A. N. (1973). *J. Theor. Biol.* **40**, 45.

Chemistry of Inorganic Systems Exhibiting Nonmonotonic Behavior

Richard J. Field

Department of Chemistry,
University of Montana, Missoula, Montana

I. Introduction

As a reacting chemical system spontaneously approaches thermodynamic equilibrium, certain species, referred to as reactants, monotonically disappear. Other species, referred to as products, monotonically appear. This monotonic conversion of reactants to products supplies the free energy to drive the spontaneous approach to equilibrium. Intermediates are species that are both produced and consumed in the course of chemical reaction. It was long thought that the concentrations of these intermediate species were constrained to pass through only a limited number of monotonically damped maxima and minima during the approach to equilibrium. However, it has recently been shown (Glansdorff and Prigogine, 1971) theoretically that in systems far enough from equilibrium it is possible for *intermediate concentrations* to undergo a wide range of nonmonotonic behaviors. These include, but are not limited to, large amplitude, only slightly damped temporal oscillations, and traveling waves of high or low intermediate concentrations. Since the development of these ideas, it has been recognized that a number of real chemical reactions exist which among them exhibit the entire gamut of theoretically possible nonmonotonic behavior.

In this chapter we shall begin by discussing fundamental theoretical ideas to a degree sufficient to understand exactly what sorts of nonmonotonic behaviors might be expected in real chemical systems. We shall then attempt to describe the types of dynamic interactions necessary for each type of behavior. The presently known experimental examples of chemical reactions that exhibit nonmonotonic behavior will then be discussed within this theoretical context.

Nonmonotonic behavior is observed in both mainly inorganic systems and in enzyme catalyzed biochemical systems. Because enzyme systems are discussed in the article by Hess and Chance in this volume, the present article will be almost entirely devoted to inorganic systems. Much of the interest in nonmonotonic chemical systems arises because of the similarity of their behavior to many biological systems (Chance *et al.*, 1973; Aldrich, 1976). This topic is also discussed elsewhere in this volume; thus it will be neglected here.

II. Theory-Scope of Nonmonotonic Behavior in Purely Chemical Systems

A. MASS ACTION KINETICS

We will consider here chemical reactions nominally occurring at constant temperature and pressure. Most chemical transformations are the net result of a number of component elementary† reactions. The set of component chemical reactions leading to a net chemical process is referred to as the mechanism of that process. A typical elementary chemical reaction can be represented by Eq. (T1).

$$aA + bB \underset{k_r}{\overset{k_f}{\rightleftharpoons}} cC + + dD. \tag{T1}$$

In Eq. (T1), A and B are reactant species and C and D are product species; a, b, c, and d are the stoichiometric coefficients indicating the number of individuals of each species involved in the reaction. Initially, we will consider systems without diffusion. The law of Mass Action (Benson, 1960) states that the rate of such an elementary reaction in the forward direction v_f by Eq. (T2).

$$v_f = -\frac{1}{a}\frac{d[A]}{dt} = -\frac{1}{b}\frac{d[B]}{dt} = \frac{1}{c}\frac{d[C]}{dt} = \frac{1}{d}\frac{d[D]}{dt} = k_f[A]^a[B]^b \tag{T2}$$

and its rate in the reverse direction v_r is given by Eq. (T3).

$$v_r = -\frac{1}{c}\frac{d[C]}{dt} = -\frac{1}{d}\frac{d[D]}{dt} = \frac{1}{a}\frac{d[A]}{dt} = \frac{1}{b}\frac{d[B]}{dt} = k_r[C]^c[D]^d. \tag{T3}$$

In Eqs. (T2) and (T3), respectively, k_f and k_r are the specific rate constants in the forward and reverse directions. Chemical reactions capable of exhibiting the nonmonotonic behavior of interest here normally involve at least three such elementary processes (Field and Noyes, 1977). The overall time rate of change of the concentration of a species X_i is described by adding the contributions to the overall rate of each elementary reaction producing or consuming X_i. The overall dynamics of a complex chemical reaction is then defined by the set of ordinary differential equations given in Eq. (T4).

$$\frac{d[X_i]}{dt} = F_i([X_1], [X_2], \cdots [X_N], k_1, k_2, k_3 \cdots k_M). \tag{T4}$$

In Eq. (T4), $[X_i]$ is the concentration of the ith chemical species, k_j is the specific rate constant for the jth reaction, and N and M are, respectively, the number of chemical species and the number of elementary reactions involved in the system. Because of the form of the Law of Mass Action, detailed application of it to a set of elementary reactions composing the mechanism of a complex reaction will yield a set of F_i that are polynomials in the $[X_i]$. However, often such a mechanism will have equilibria or steady states involved such that its dynamics may be closely approximated

† Elementary chemical reactions are those that occur in a single step without the formation of any detectable intermediate species.

by a smaller set of differential equations possibly containing nonpolynomial type terms. In either case, terms may be either linear or nonlinear in the $[X_i]$. The nonmonotonic behavior to be discussed here can occur *only* when there is nonlinearity in the dynamic equations (Minorsky, 1962; Cronin, 1977). Thus theoretical investigations in this field involve techniques of treating sets of nonlinear differential equations. Much of the importance and interest in nonmonotonic chemical phenomena results from their use as heuristic models of mathematically similar behavior in physical and biological systems. Normally the mathematics of the chemical systems is somewhat easier to handle than the mathematics of these other systems (Nicolis *et al.*, 1975).

B. Some Ideas from Thermodynamics

As indicated in Eq. (T1), each elementary chemical reaction is accompanied by its reverse, although the rates of forward and reverse processes may differ enormously in nonequilibrium situations. Of course, at thermodynamic equilibrium, the Principle of Detailed Balance (Weston and Schwarz, 1972) holds, and the rates of forward and reverse processes must be equal. In a complex chemical system at thermodynamic equilibrium, the forward and reverse rates of all component elementary reactions are equal and each $d[X_i]/dt = 0$. The equilibrium point is at the minimum of free energy, and in a closed system it is globally, asymptotically stable (Shear, 1967). Furthermore, *final* approach to equilibrium in such a system must be monotonic for all X_i regardless of the dynamic laws of the system (Gray, 1970). This statement implies that *sustained* nonmonotonic behavior is not possible in a *closed* chemical system.

Even in a closed, complex chemical system not at equilibrium, it normally occurs that *some* $d[X_i]/dt$ are very nearly zero. This occurs because of the form of the Law of Mass Action. Certain species, referred to as intermediates, are produced by some elementary reactions, but are then so rapidly consumed by other elementary reactions that their concentrations always remain low compared to reactant species. Reactant concentrations are large and so far from equilibrium that they decrease monotonically in the course of spontaneous chemical reaction. If the concentrations of intermediates always remain small, then the rate of change of their concentrations must also be always small. They are said to be in a pseudosteady state.

In a closed chemical system, the steady state is only an approximation; by setting the appropriate rates in Eq. (T4) equal to zero, algebraic expressions can be derived relating the instantaneous concentrations of intermediates to those of reactants and sometimes of products. However, as the reaction proceeds towards equilibrium, reactant and product concentrations change and the steady-state intermediate concentrations drift. Thus the rates of change of intermediate concentrations are not exactly zero. Normally this drift of intermediate concentrations is either monotonic or passes through a number of strongly damped maxima or minima related to the number of intermediates involved in the overall reaction (Schmitz, 1973).

In the work to be discussed here, the pseudosteady-state problem is often overcome by dealing with *open systems* in which the concentrations of reactants and products are held constant. This can be carried out experimentally by the continuous addition of reactants and removal of products. Under this circumstance, the steady state becomes an exact solution to Eq. (T4). Furthermore, such a system does not decay towards an equilibrium point. The rest of this discussion relates to such open systems.

The characteristics of steady states can be well defined for systems that *do not differ greatly from equilibrium*. Thermodynamically, they are characterized by minimum entropy production (Prigogine, 1967). Equilibrium states are characterized by zero entropy production. It can also be shown (Glansdorff and Prigogine, 1971) that near to equilibrium steady states are, like the equilibrium point itself, globally, asymptotically stable. In the vast majority of cases as the distance from equilibrium is increased† these stable, equilibrium like, steady states persist and nonmonotonic dynamic behavior does not occur in the concentrations of intermediates. However, Glansdorff and Prigogine (1971) have shown that in systems with nonlinear dynamic laws there may be a critical distance from equilibrium beyond which new solutions to the dynamic laws of a chemical system may appear. It is in fact these new solutions that are of interest to us here; they need not show monotonic temporal or spatial behavior. Intermediate concentrations may oscillate and/or undergo other types of nonmonotonic behavior such as excitability or hysteresis. Furthermore, standing or moving concentration inhomogeneities in space may develop in a unstirred reagent.

C. Some Ideas from Bifurcation Theory

Consider an open chemical system at constant temperature and pressure and subject to several elementary chemical reactions among the various species present. The system is initially at equilibrium and this state is globally, asymptotically stable. Of course this state must also be infinitesimally stable as well; the inevitable fluctuations around the equilibrium state must decay. If the system is then moved away from the equilibrium point by the addition of reactants and/or the removal of products, net conversion of reactants to products begins as the concentrations of inter-

† Perhaps as measured by a function like ln $\{(\sum [\text{reactants}])/(\sum [\text{products}])\}$ in a reversible system or by rate constants and reactant concentrations in a system so far from equilibrium that it is essentially irreversible.

mediates move continuously away from the equilibrium point along a line of steady states. Close to equilibrium these steady states are unique and globally, asymptotically stable. As the distance from equilibrium is further increased, however, it is possible in systems with nonlinear dynamic laws for solutions to Eq. (T4) in addition to the equilibrium like steady state to appear. All solutions may be locally, asymptotically stable. New solutions may be merely new steady states or they may be nonmonotonic. With multiple steady states hysteresis (Section II,C,3,d) may be observed. As the distance from equilibrium is increased even further, a *bifurcation point* (Sattinger, 1973) may be reached beyond which the equilibrium like line of steady states becomes unstable infinitesimally as well as globally. Beyond this critical point, fluctuations can cause the system to evolve to one of the other solutions. It may be that the appearance of multiple solutions to the dynamic laws also occurs at the bifurcation point.

Equation (T4) corresponds to a homogeneous, well-stirred chemical system. However, inhomogeneous, unstirred chemical systems in which the concentrations of chemical species are not the same throughout space can be described by adding diffusion terms like that shown in Eq. (T5) to Eq. (T4) to generate a set of partial differential equations

$$\left(\frac{\partial [X_i]}{\partial t}\right)_l = D_{X_i}\left(\frac{\partial^2 [X_i]}{\partial l^2}\right)_t. \tag{T5}$$

The discussions of the previous sections apply equally well to these partial differential equations (Nicolis, 1975). The nonmonotonic behavior observed may include traveling as well as standing areas of inhomogeneous intermediate concentrations. These systems may be very sensitive to boundary conditions.

Knowing of the existence of bifurcation points, three questions become apparent. They are: (1) How can we determine the bifurcation point (as a function of rate constants and concentrations in a chemical reaction) beyond which the equilibrium like steady state becomes unstable? (2) What is the mechanism of the growth of fluctuations around unstable steady states and how is this related to the requirement that the system be beyond a critical distance from equilibrium? (3) What possible dynamic behaviors can be exhibited by chemical systems with multiple solutions to their dynamic equations?

These questions have been the subject of enormous interest by theoreticians and experimentalists. In the following sections we will discuss each of them.

1. Bifurcation Point-Infinitesimal Stability Analysis

By infinitesimal stability of chemical systems we mean the response of the steady state of Eq. (T4) to infinitesimal perturbations in the form of very small concentration fluctuations. If these fluctuations decay back to zero rather than growing, then the steady state is said to be infinitesimally stable. Methods of investigating the stability properties of the zero solutions of sets of nonlinear differential equations are well developed (Minorsky, 1962; Cesari, 1962). They are most easily applied to open systems. Here we are interested in the stability properties of steady states as parameters related to the distance from equilibrium are varied.

Consider concentration fluctuations, δX_i,† around the steady state. These fluctuations are so small compared to the steady state concentrations X_i° that their growth or decay is linear and can be described by Eq. (T6).

$$(X_i^\circ - X_i) = \delta X_i = (\delta X_i^\circ)e^{\lambda t}. \tag{T6}$$

The λ's are evaluated simply by substituting Eq. (T7) into Eq. (T4) and suppressing higher powers of the infinitesimal fluctuations.

$$X_i = X_i^\circ + (\delta X_i^\circ)e^{\lambda t}. \tag{T7}$$

This leads to a secular equation of the form of Eq. (T8).

$$0 = \begin{vmatrix} (a_{11} - \lambda) & \cdots & a_{1j} & \cdots & a_{1n} \\ a_{i1} & \cdots & (a_{ii} - \lambda) & \cdots & a_{in} \\ a_{n1} & \cdots & a_{nj} & \cdots & (a_{nn} - \lambda) \end{vmatrix}. \tag{T8}$$

In Eq. (T8), the a_{ij} are the Jacobian matrix of Eq. (T4), i.e.,

$$a_{ij} = \left(\frac{\partial F_i}{\partial X_j}\right)_{X_i \neq X_j}.$$

Solution of Eq. (T8) yields a set of eigen values λ_i and corresponding eigen vectors δX_i° that describe the infinitesimal stability properties of the steady state. The eigen vectors are normal modes of motion around the steady state, and any fluctuation can be described in terms of its components along these normal modes. If the real part of any λ_i is positive, then inevitably some fluctuation will have a component along the corresponding eigen vector and will grow; the steady state is then infinitesimally unstable. All the λ_i must have negative real parts for any steady state to be infinitesimally stable.

The eigen values and eigen vectors depend upon the same parameters as does the distance from equilibrium. Thus the bifurcation point is easily identified as the point where one or more eigen values with positive real parts appear. However, infinitesimal stability does not tell the whole story. Solutions other than the equilibrium like steady state may exist even if that state is infinitesimally stable, then only larger fluctuations can cause a transi-

---

† We will omit the brackets indicating concentrations of species. Whether we mean $[X_i]$ or X_i should be clear from context.

tion to other states to occur and nonmonotonic behavior to appear. In this case there will be a threshold effect. An infinitesimally stable steady state may also be subject to excitability (Section II,C,3,e).

For many real, complex chemical reactions Eq. (T4) is subject to a numerical instability referred to as "stiffness" (Hirschfelder, 1952). This problem arises from the widely varying time constants possible for various elementary reactions in chemical systems and requires that special methods be used (Gear, 1971) to numerically integrate Eq. (T4). This numerical instability carries over to the secular equation [Eq. (T8)]. Thus determining the signs of the λ_i in not unreasonably complex systems may be difficult both by direct solution of the secular equation or by use of the Hurwitz criterion (Cesari, 1962). Clarke (1974a,b) has developed a "graph theoretic" technique for solving the stability problem that avoids this difficulty.

2. Critical Distance from Equilibrium

The reason for the existence of a critical distance from equilibrium beyond which the steady state becomes unstable is not as well understood as is the technical problem of locating this point. At least in the case of chemical dynamics, however, the cause seems to be (Nicolis *et al.*, 1974) a nucleation phenomenon such as exists in phase transitions. That is, the system must be far enough from equilibrium before local, thermal fluctuations can overcome an activation entropy and grow such that the entire system becomes involved.

3. Possible Nonmonotonic Behavior in Chemical Systems

Nicolis and Auchmuty (1974) have performed a bifurcation analysis of a simple model chemical system and classified the various types of behavior noted beyond a bifurcation point. They also discuss the relationship of their work to the "catastrophe theory" of Thom (1969).

Here we shall use models of varying relationships to real chemical systems in order to illustrate and classify the various nonmonotonic phenomena that can be expected to appear in real chemical systems. The models used will also give the reader some feeling for the sorts of interactions that lead to nonmonotonic behavior in chemical systems.

a. Bifurcation of Periodic Solutions Beyond an Instability. The Brusselator model was invented by Prigogine and Lefever (1968) to illustrate the thermodynamic and kinetic principles discussed in the previous sections. The model is given by Eqs. (B1)–(B4).

$$A \rightleftharpoons X \tag{B1}$$

$$B + X \rightleftharpoons Y + D \tag{B2}$$

$$2X + Y \rightleftharpoons 3X \tag{B3}$$

$$X \rightleftharpoons E. \tag{B4}$$

In this model A and B are reactants, D and E are products, and X and Y are intermediates. Step (B3) has been criticized by chemists as not realistic. The termolecular nature of the reaction is not an overwhelming problem; this could result from two consecutive bimolecular reactions. However, the net reaction corresponds to the conversion of Y to X catalyzed by two molecules of X. No such interaction is presently known in chemical systems, but it seems possible that such an interaction might eventually be found in a highly cooperative biological system, e.g., an enzyme system. Both Tyson and Light (1973) and Hanusse (1972) have shown independently that in a model containing only two intermediate species and obeying mass action kinetics, there must be a termolecular step like (B3) for the steady state to become unstable. Regardless of its realism, the Brusselator has contributed greatly to the development of fundamental theoretical concepts concerning nonmonotonic behavior in chemical systems.

An analysis of the stability of the Brusselator model steady state has been carried out by Nicolis (1971) with the reverse rate constants set equal to zero (in order to assure sufficient distance from equilibrium) and with the forward rate constants set equal to one. The system is presumed to be open so that the concentrations of A and B are constant. Under these conditions, the steady state is given by Eq. (B5).

$$X_0 = A, \quad Y_0 = B/A. \tag{B5}$$

Application of linear stability analysis to this steady state leads to the bifurcation diagram shown in Fig. 1. The bifurcation line divides the $[A, B]$ plane

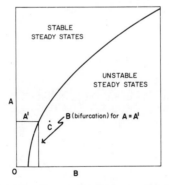

Fig. 1. Bifurcation diagram for the irreversible Brusselator.

into regions of stable and unstable steady states. As B is increased while $A = A'$, the steady state remains stable until $B > B$ (bifurcation). At this point the steady state becomes unstable and the system may evolve to a second state that is oscillatory. The oscillatory solution to the Brusselator dynamic equations beyond the bifurcation point is a limit cycle (Minorsky, 1962). Limit-cycle solutions are orbitally stable such that a solution starting within a certain distance of the orbit will always remain within that distance. Furthermore, all solutions, regardless of initial conditions, asymptotically approach the same limit-cycle oscillatory trajectory. Calculated Brusselator trajectories for A and B equal to point C and for different sets of initial X and Y in Fig. 1 are shown in Fig. 2.

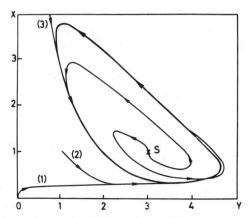

Fig. 2. Approach to a limit cycle orbit in the irreversible Brusselator for $A = 1$, $B = 3$ and for various initial conditions. Arrowheads indicate direction of motion, and the limit cycle is the bold trajectory. [Reproduced by permission from G. Nicolis, *Adv. Chem. Phys.* **19**, 209 (1971).]

Noyes (1976a) has shown that the open system approximation, i.e., ΔA and ΔB approximately equal to zero for one cycle, is only true in the case of the Brusselator model for a small range of rate constants that does not include those used in Figs. 1 and 2.

b. Hard Excitation. In the previous section we discussed the bifurcation of the Brusselator dynamics from an unstable steady state to a stable limit cycle as B exceeds B (bifurcation). In this case the stable limit cycle has a very small amplitude as B critical (B_c) is just exceeded. This amplitude grows as B is made progressively larger than B_c. For some circumstances, non-monotonic solutions to dynamic equations may appear and grow before the equilibrium like steady state becomes infinitesimally unstable. Thus oscillatory solutions with rather large amplitude may appear just at the bifurcation point. Furthermore, when the steady state remains infinitesimally stable after a nonmonotonic solution has appeared, transition to the second solution from the steady state can result from a finite or "hard" perturbation. Nicolis and Auchmuty (1974) have demonstrated just this behavior when diffusion terms are added to the Brusselator dynamic equations as shown in Eq. (B6).

$$\left(\frac{\partial X}{\partial t}\right)_r = D_X \left(\frac{\partial^2 X}{\partial r^2}\right) - (B + 1)X + X^2 Y + A$$

$$\left(\frac{\partial Y}{\partial t}\right)_r = D_Y \left(\frac{\partial^2 Y}{\partial r^2}\right) + BX - X^2 Y.$$

(B6)

Depending upon whether an integer defined as the "critical wave number" and related to A, D_X, and D_Y is even or odd, the bifurcation diagram for Eq. (B6) can look like either Fig. 3 or 4. In Fig. 3, corresponding to an even wave

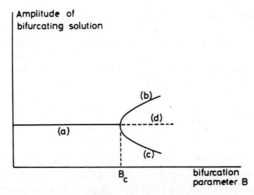

Fig. 3. Bifurcation diagram for Eq. (B6) corresponding to an even critical wave number. (a) stable monotonic solution below the critical point B_c; (d) the same type of solution becoming unstable beyond the bifurcation point; (b) and (c) stable nonmonotonic solution emerging beyond the bifurcation point. [Reproduced by permission from G. Nicolis, I. Prigogine, and P. Glansdorff, *Adv. Chem. Phys.* **32**, 1 (1975).]

number, there are two new solutions that are both stable and appear just beyond the bifurcation point with very small amplitude. This is the same behavior seen in the Brusselator assuming no diffusion except that here there are two new solutions rather than one. Figure 4 shows the situation for odd wave numbers. There is a stable, nonmonotonic solution (b1) that appears exactly at the bifurcation point. However, there is also a stable nonmonotonic solution (c2) that appears at $B < B_c$ where the normal steady state (a) remains stable. Transition from a to c2 while the steady state remains stable requires a finite perturbation or hard excitation.

c. Traveling and Standing Waves of Intermediate Concentrations. Introduction of diffusion terms into the Brusselator dynamics indicates that we are dealing with a system that is not kept spatially uniform by some method like stirring. This allows the circumstance of different concentrations of

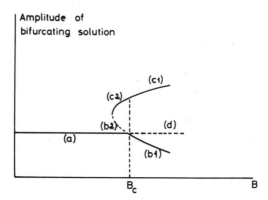

Fig. 4. Bifurcation diagram for Eq. (B6) corresponding to an odd critical wave number. (a) and (d) have the same meaning as in Fig. 3; (b1) supercritical stable nonmonotonic solution arising continuously beyond bifurcation; (b2) subcritical unstable solution; (C1) and (C2) branches extending on both sides of the bifurcation point and separated from the uniform solution by a finite jump. [Reproduced by permission from G. Nicolis, I. Prigogine, and P. Glansdorff. _Adv. Chem. Phys._ **32,** 1 (1975).]

intermediates at different points in the system to arise, and this situation can be described by Eq. (B6). The nonmonotonic solutions indicated in the bifurcation diagrams in Figs. 3 and 4 correspond to standing or traveling waves of intermediate (X, Y) concentrations. The exact characteristics of these waves seem to depend upon the boundary conditions, reactant concentrations and rate constants, and the type of perturbation to the steady state. Nicolis (1975) has solved Eq. (B6) using numerical techniques and exhibited the profiles of the concentration waves.

d. Multiple Steady States and Hysteresis. Another type of nonmonotonic dynamic behavior can be illustrated by the model of an enzyme process illustrated by reactions (C1)–(C4).

$$A + X \rightleftharpoons 2X \tag{C1}$$

$$X + E \rightleftharpoons C \tag{C2}$$

$$C \rightleftharpoons E + B \tag{C3}$$

$$E + C \rightleftharpoons \text{constant.} \tag{C4}$$

In this model A is the reactant, B is the product, X and C are intermediates, and E is an enzyme. This model was suggested by Edelstein (1970) and further investigated by Turner (1975). Calculation of the steady state X as a function of A gives the plot shown in Fig. 5. For A between A_1 and A_2 three steady states are possible. Linear stability analysis shows that all steady states are stable except for those between P and P'. Thus a hysteresis phenomenon could appear in that as A is increased past A_2 a jump from the

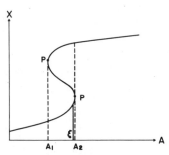

Fig. 5. Steady state $[X]$ for the model (B1)–(B4) as a function of $[A]$. [After I. Prigogine and R. Lefever, *Adv. Chem. Phys.* **29**, 1 (1975).]

lower steady state to the higher should be observed. However, when A is decreased below A_2 from above, the transition from the higher steady state to the lower is not expected until A falls below A_1. This is a hysteresis loop.

The actual behavior of this model is different, however, indicating that infinitesimal stability does not tell the whole story in highly nonlinear systems. Turner (1975) has performed calculations on the model of reactions (C1)–(C4) using fluctuations, some finite, supplied by a stochastic model. He found that transitions between the lower and higher steady states occur at the same A (indicated by ξ in Fig. 5) regardless of which way A is moving. However, Turner (1976) has demonstrated numerically a hysteresis loop in a model of the Belousov–Zhabotinskii reaction suggested by Field (1975).

e. Excitability. Excitability may be one of the most interesting phenomena of the group we have discussed because of its similarity to signal transmission by excitable membranes in biological systems. Nerve impulse conduction (Lefever and Deneubourg, 1975) is an important example of this phenomenon. Troy (in this volume) discusses the similarity between the Field and Noyes (1974a) model of the Belousov–Zhabotinskii reaction and the Hodgkin–Huxley (1967) model of nerve impulse conduction. We will not discuss the model of Field and Noyes in detail here, preferring to wait until the chemistry of the Belousov–Zhabotinskii reaction has been discussed (Section III,A). However, the model of Field and Noyes is a five-step mechanism involving three intermediates. It contains two expendable parameters, and the stability properties of the steady state depend strongly upon these parameters. For some values of these parameters, the model has a large amplitude limit-cycle solution to its dynamic equations. Excitability is exhibited for values of these parameters for which the steady state is globally, asymptotically stable. Small perturbations of this steady-state decay. However, Troy and Field (1977) have demonstrated that for this circumstance the system is still excitable in that a large perturbation to this stable steady state causes the system to take an excursion nearly up to the limit-cycle trajectory before decaying back to the steady state. Troy (1977) has demonstrated the existence of a threshold perturbation necessary to cause the excitability excursion.

f. Chaotic Behavior. Limit-cycle behavior (Section II,C,3,a) is both oscillatory and periodic. This means that all cycles are identical in that they have the same period and amplitude. It has recently been recognized (Rössler, 1976a,b) however, that dynamic equations resulting from chemical systems may sometimes possess solutions that are oscillatory but aperiodic. This phenomenon is called "chaos" because these oscillations have periods and amplitudes of apparently random magnitude. These systems are deterministic, however. A particular set of parameters and initial conditions yields a specific aperiodic solution. The kinetic steady state must still be unstable for chaos to appear.

The mathematics of chaos is abstract and apparently not yet well developed. Chaos may appear in models (Rössler, 1976a) containing a chemical oscillator, such as the Brusselator (Section II,C,3,a), strongly coupled through a common intermediate to a chemical hysteresis system such as the Edelstein model (Section II,C,3,d). It is the perturbation of the oscillator limit cycle by the coupled hysteresis switching that leads to chaos in this system. Tyson (1978) has shown the existence of chaos in a modified version of the Oregonator model (Field and Noyes, 1974a) (Section III,A,4) of the oscillatory Belousov–Zhabotinskii reaction (Section III,A). This model also exhibits bistability and limit-cycle oscillation. Thus chaos in this case may also be related to coupling of limit-cycle oscillation and hysteresis switching. Rössler (1976b) has shown the existence of chaos in a model of a spatially nonuniform system (Section V).

g. Summary. The nonmonotonic behavior that may appear in chemical systems include the following.

A. Spatially uniform systems (diffusion unimportant):
    1. Limit-cycle oscillation of intermediates around an infinitesimally unstable steady state beyond a bifurcation point.
    2. Limit-cycle oscillation of intermediate concentrations around an infinitesimally stable steady state. Hard excitation oscillations.
    3. Multiple steady states and hysteresis phenomena involving intermediates.
    4. Excitability.
    5. Chaos.

B. Spatially nonuniform systems (diffusion important):
    1. Stationary nonuniform distribution of intermediate concentrations in space.

2. Traveling waves of intermediate concentrations in space.
3. Excitability.
4. Chaos.

All of these phenomena are exhibited by one or the other of the real chemical reactions to be discussed in later sections.

III. The Chemistry of Known Systems Exhibiting Temporal Oscillations in Stirred Media

A. BELOUSOV–ZHABOTINSKII (BZ) REACTION

1. Experimental Characterization

Of the presently known purely chemical reactions that exhibit nonmonotonic behavior, the Belousov–Zhabotinskii (BZ) reaction is the most versatile both in terms of the range of phenomena exhibited and in the range of modification of the reagent over which these phenomena appear. Belousov (1959) first reported long-lived oscillations in $[Ce(IV)]/[Ce(III)]$ during the cerium ion catalyzed oxidation of citric acid by bromate ion in a $1\ M$ H_2SO_4 solution. However, in its most general form, the BZ reaction is the metal ion catalyzed oxidation of easily brominated organic materials by bromate ion in aqueous media of pH 0–1. The reaction exhibits easily observed oscillations in the ratio of oxidized and reduced forms of the metal ion catalyst, in the concentration of bromide ion, and in heat evolution. Figure 6 shows potentiometrically determined forms of the $[Ce(IV)]/[Ce(III)]$ and $[Br^-]$ oscillations in a typical stirred Belousov–Zhabotinskii reagent. Note the presence of an induction period followed by the sudden appearance of full blown oscillations whose period and amplitude then change only very slowly. The stability and reproducibility of the oscillations indicate that they correspond to limit-cycle solutions to the BZ

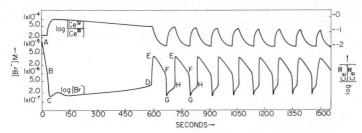

Fig. 6. Potentiometric traces of log $[Br^-]$ and log $[Ce(IV)]/[Ce(III)]$ for a representative BZ reaction. Initial concentrations were $[CH_2(COOH)_2]_0 = 0.032\ M$, $[KBrO_3]_0 = 0.063\ M$, $[KBr]_0 = 1.5 \times 10^{-5}\ M$, $[Ce(NH_4)_2(NO_3)_5]_0 = 0.001\ M$, $[H_2SO_4]_0 = 0.8\ M$. [Reproduced by permission from R. J. Field, E. Körös, and R. M. Noyes, *J. Am. Chem. Soc.* **94**, 8649 (1972).]

dynamic equations. Field (1975) observed oscillations of nearly random period near the end of the period of oscillation in a worn out BZ reagent. Although Field did not interpret these aperiodic oscillations as an example of chaos, they may well be. The metal ion concentration oscillations have been monitored spectrophotometrically (Bowers *et al.*, 1972; Zhabotinskii, 1964a; Kasperek and Bruice, 1971) and potentiometrically (Field *et al.*, 1972; Degn, 1967) while oscillation of the concentration of bromide has been monitored potentiometrically (Field *et al.*, 1972; Körös and Burger, 1973). The concentrations of many other intermediate species surely oscillate, but these are not easily measured. Heat evolution may be measured calorimetrically (Körös *et al.*, 1973). Beyond the temporal oscillations in a stirred medium, such as illustrated in Fig. 6, modifications of the basic BZ reagent exhibit traveling concentration waves in unstirred media (Section V,B) (Field and Noyes, 1974b), multiple steady states, hysteresis, and standing waves in flow systems (Section IV) (Marek and Svobodova, 1975).

All BZ reagents are strongly acidic and contain bromate ion. Oscillations have been observed using a wide range of metal ion catalysts and organic materials, however. Observed oscillatory periods have ranged from several minutes (Field *et al.*, 1972) to tenths of seconds (Dayantis and Sturm, 1975).

Present mechanistic knowledge of the BZ reaction (Section III,A,2,3) rationalizes the absolute requirement for bromate ion and the range of substitutions possible for catalyst, organic material, and inorganic acid.

Zhabotinskii (1964a,b) demonstrated that cerium ion could be replaced by the weak one-electron reductant Mn(II). Zaikin and Zhabotinskii (1970) used $Fe(phen)_3^{2+}$[†] as the catalyst. Demas and Dimente (1973) demonstrated the use of $Ru(II) (bipy)_3^{2+}$[‡] as catalyst. These metal ion catalysts all undergo one-electron redox reactions with a potential between about 1 and 1.5 V. For oscillation to occur, acidic bromate must be able to oxidize the reduced form of the metal ion catalyst, and the oxidized form of the catalyst must be able to oxidize the mixture of brominated and unbrominated organic material with the liberation of Br^-. It is expected that any catalyst that meets these requirements can be used in the BZ reaction.

Zhabotinskii (1964a) and Kasperek and Bruice (1971) found that replacement of malonic acid with several easily brominated organic acids yields oscillatory reagents. These acids include malonic, bromo-malonic, maleic, and malic acids. Bowers *et al.* (1972) reported oscillations in a manganese ion-2,4-pentanedione flow system. The pentanedione system behaves quite differently depending upon whether manganese or cerium ion is used as the

† phen ≡ 1,10,-*o*-phenanthroline.
‡ bipy ≡ 2,2′-bipyridine.

catalyst (Körös *et al.*, 1973). Carbon dioxide is not a final product in this system. Stroot and Janjic (1976) expanded the range of noncarboxylic acid organic materials known to lead to oscillation. Beck and Varadi (1975) found oscillations using acetylene dicarboxylic acid with manganese, cerium, or Fe(phen)$_3$ catalyst. These oscillations were complex showing two oscillatory periods characterized by separate induction periods, frequency, amplitude, and attenuation. Apparently bromination can take place by either an enolization mechanism or by direct addition to an unsaturation. Oxalic and succinic acids do not yield oscillatory reagents (Kasperek and Bruice, 1971). Rastogi *et al.* have carefully investigated the characteristics of the oscillations in the cerium ion-malonic acid (Rastogi and Yadava, 1974), the manganese ion-malonic acid (Rastogi *et al.*, 1974), and the cerium ion-malic acid (Rastogi and Kumar, 1976) systems. In the cerium ion-malic acid system they found that oxaloacetic acid was an important intermediate and that bromooxaloacetic and carbon dioxide were final products.

Bornmann *et al.* (1973) have shown that the products of the oscillatory reaction with cerium ion and malonic acid include carbon dioxide bromomalonic acid and dibromoacetic acid. They specifically did not detect formic acid among the products. Noszticzius (1977) discovered that CO is a minor product in this system and that CO tends to inhibit the oscillations.

The oxidizing power of bromate is strongly dependent upon acidity, and all BZ reagents are strongly acidic. In principle, any strong mineral acid could be used, but in practice effects of the accompanying anion are important. H_2SO_4 is the most commonly used acid; its use is necessary in cerium ion catalyzed systems since bromate is thermodynamically unable to oxidize Ce(III) to Ce(IV) without the added Ce(IV) stability resulting from complexation by sulfate ion. Perchloric acid should be suitable for use in systems where the anion need only be inert, but we know of no instance of its use. Chloride and nitrate ions are both chemically reactive in a BZ reagent. Chloride ion inhibits the oscillations (Jacobs and Epstein, 1976) and nitrate ion increases the frequency of oscillation (Körös, 1977). Thus HCl and HNO_3 are not normally used in BZ reagents.

The Belousov–Zhabotinskii reaction is strongly exothermic. While the temperature of a reasonably well-insulated system does not oscillate, the *rate* of temperature increase does because the rate of heat evolution is not equal throughout the oscillatory cycle. Calorimetric investigations of the cerium ion-malonic acid system have been carried out by Franck and Geiseler (1970), Busse (1971), and Körös *et al.* (1973). The last authors have also investigated the manganese ion-malonic acid and manganese ion-2,4-pentonedione systems and compared these systems with each other and with the cerium ion-malonic acid one.

2. The Field–Körös–Noyes (FKN) Mechanism of the BZ Reaction

Field et al. (1972) (FKN) have proposed an elaborate mechanism for the BZ reaction. Their mechanism is based upon extensive kinetic and thermodynamic considerations, and it deals in particular with the temporal oscillations observed in the stirred cerium ion system with malonic acid (MA). However, it is generally assumed that the basic features are valid in reagents utilizing different catalysts and organic acids and in both unstirred and flow physical configurations. The oscillations are intuitively easily understood to arise from a delayed feedback loop contained in the FKN mechanism.

While there is little controversy concerning the basic validity of the sorts of interactions involved in the FKN mechanism, there are a number of problems concerning the details of some parts of the mechanism. These problems are becoming apparent as increasingly sophisticated (Clarke, 1976a,b) theoretical and computational (Edelson and Noyes, 1978) methods are used to investigate the properties of the FKN mechanism.

Table I summarizes the most important steps in the FKN mechanism. A detailed description must include the effects of reversibility and side reactions. The mechanism can be divided into three overall processes: A, B, and C. The oscillations occur as control of the system is passed back and forth between Processes A and B, partially through mediation by Process C. It is assumed to a first approximation that there is no "cross-talk" among the processes, i.e., intermediates from one process do not interact with those from another. This is certainly not entirely correct.

The competition between Br^- and BrO_3^- ions for $HBrO_2$ embodied in reactions (R2) and (R5), respectively, is a principal feature of the mechanism. When $[Br^-]$ is high, reaction (R2) and Process A are dominant and the reactions occurring are a series of oxygen atom transfer reactions (two-electron redox processes) among the singlet oxybromine compounds (BrO_3^-, $HBrO_2$, $HOBr$, etc.). Process A is dominant during section AB in Fig. 6. Apparently, combination of the thermodynamic preference of these oxybromine compounds for two-electron redox processes coupled with the metal ion requirement of one-electron redox processes inhibits the direct oxidation of M^n to M^{n+1} by bromate ion during Process A (see Fig. 6). But there is always a net consumption of Br^- during Process A, and eventually (R5) must become an important fate of $HBrO_2$. This leads to the production of $BrO_2 \cdot$ and thus an entry into the radical (one-electron redox process) regime represented by Process B. This series of reactions generates $HBrO_2$ autocatalytically according to net reaction G. Thus when step (R5) does become dominant over (R2), Process B gains control of the system almost discontinuously as $[HBrO_2]$ grows exponentially until its disproportionation [reac-

Table I^a

Process A

$Br^- + BrO_3^- + 2H^+ \rightarrow HBrO_2 + HOBr$	(R3)
$Br^- + HBrO_2 + H^+ \rightarrow 2HOBr$	(R2)
$3(Br^- + HOBr + H^+ \rightarrow Br_2 + H_2O)$	(R1)
$3(Br_2 + MA \rightarrow BrMA + Br^- + H^+)$	(R8)

$$2Br^- + BrO_3^- + 3H^+ + 3MA \rightarrow 3BrMA + 3H_2O \qquad (A)$$

Process B

$BrO_3^- + HBrO_2 + H^+ \rightarrow 2BrO_2\cdot + H_2O$	(R5)
$2(M^n + BrO_2\cdot + H^+ \rightarrow M^{n+1} + HBrO_2)$	(R6)

$$2M^n + BrO_3^- + HBrO_2 + 3H^+ \rightarrow 2M^{n+1} + H_2O + 2HBrO_2 \qquad (G)$$

$$2HBrO_2 \rightarrow BrO_3^- + HOBr + H^+ \qquad (R4)$$

followed by (R1) and (R8)

$$BrO_3^- + 4M^n + MA + 5H^+ \rightarrow 4M^{n+1} + BrMA + 3H_2O \qquad (B)$$

Process C

$$M^{n+1} + MA + BrMA + H_2O \rightarrow M^n + FBr^- + \text{other products} \qquad (D)$$

^{a} M = metal ion, MA = malonic acid $[CH_2(COOH)_2]$, BrMA = bromomalonic acid $[BrCH(COOH)_2]$. Numbering is as in R. J. Field, E. Körös, and R. M. Noyes, *J. Am. Chem. Soc.* **94**, 8649 (1972).

tion (R4)] leads to a steady state. Process B actually gains control as $[Br^-]$ passes through a critical value where the rates of reactions (R2) and (R5) are equal. At this point $[HBrO_2]$ shifts suddenly from a lower *partial* steady-state value (partial because $[Br^-]$ is rapidly changing) maintained by reactions (R2) and (R3), to a higher *partial* steady-state value (partial because $[M^{n+1}]/[M^n]$ is rapidly changing) maintained by reactions (R5) and (R4). It is assumed that (R5) is rate determining for net reaction (G). The *overall* steady state with the rates of change of *all* intermediates approximately zero lies between the Process A and Process B *partial* steady states and is normally unstable when oscillations are observed. Point B in Fig. 6 indicates this critical $[Br^-]$ and Process B gains control through Section BC. While Process B is dominant, M^n is rapidly oxidized to M^{n+1} by the radical species $BrO_2\cdot$ [reaction (R6)] and most remaining Br^- is scavenged by $HBrO_2$ in reaction (R2). But M^{n+1}, a product of Process B, reacts in Process C and is reduced back to M^n. Process C has both MA and bromomalonic (BrMA) acids as reactants and includes Br^- as a product. The amount of Br^- produced per M^{n+1} reduced in Process C is given by a stoichiometric factor F. If F is large enough, then sufficient Br^- is produced to pass control of the system back to Process A. However, F is roughly proportional to $[BrMA]/[MA]$ and is small early in the reaction before substantial amounts

Richard J. Field

of BrMA have been synthesized by Processes A and B. Thus during Section CD in Fig. 6, Process B retains control and a pseudosteady state $[M^{n+1}]/[M^n]$ is maintained while bromate ion both oxidizes and brominates MA. Through Section CD BrMA accumulates and, as F becomes larger, $[Br^-]$ increases. Eventually F increases to the point where sufficient Br^- is produced to shift control back to Process A. Then accumulated M^{n+1} drives $[Br^-]$ upwards rapidly through Section DE. Through Section EF $[Br^-]$ is consumed by Process A (although at a slower rate than in Section AB) until the critical $[Br^-]$ is again reached and Process B regains control as the system passes through Section FG. However, in this instance $[BrMA]$ has increased to the point where dominance of Process B is only fleeting as the oscillatory cycle repeats itself. F apparently can increase to a point where Process C produces Br^- so rapidly that Process B never is able to achieve dominance and a Process A like pseudosteady state is maintained.

The FKN mechanism clearly contains a delayed negative feedback loop. Process B produces M^{n+1}, which interacts in Process C to produce Br^-, which inhibits Process B. Because of the autocatalytic nature of Process B, even a small delay as M^{n+1} accumulates and gets involved in Process C is sufficient to allow Process B to overshoot the overall steady state. Furthermore, the reservoir of Br_2 and M^{n+1} produced during the dominance of Process B drives the system far to the other side of the overall steady state after Process A regains dominance.

3. Detailed Consideration of Processes A, B, and C of the FKN Mechanism

In this section we shall discuss the detailed chemistry of each Process under the assumption that there is no "cross-talk" among them.

a. Process A. Free energies of formation of all the singlet oxybromine species involved in reactions (R3), (R2), and (R1) were tabulated by FKN. They were also able to extract from literature data forward and reverse rate constants for each of these reactions. They observed kinetics during Section AB of Fig. 6 that were independent of $[Ce(III)]$ and $[MA]$ and, after adjustment for stoichiometry, quantitatively identical to reported results on the reaction of Br^- with BrO_3^-. FKN also demonstrated that the slower rate of Br^- consumption during Section EF of Fig. 6 can be understood as the result of the simultaneous occurrence of Processes A and C.

Note that the sum of (R1) and (R8) is (R0).

$$HOBr + CH_2(COOH)_2 \rightarrow BrCH(COOH)_2 + H_2O. \qquad (R0)$$

It is not known whether malonic acid is brominated by (R0) or the sequence (R1) plus (R8), but this uncertainty is of little kinetic significance.

b. Process B. Process B is essentially the oxidation of M^n to M^{n+1} by bromate ion. The basic form of the mechanism of such oxidations was devised by Noyes *et al.* (1971) in an interpretation of Thompson's (1971) data on the reaction of the weak one-electron reductants Ce(III), Mn(II), and Np(V) with bromate ion in 3 M H_2SO_4. Thompson (1971) found that the kinetics of the reaction with $[M^n]/[BrO_3^-]$ very large are invariant with metal ion and given by

$$\frac{-d[BrO_3^-]}{dt} = k_{ex}[BrO_3^-]^2$$

where k_{ex} is numerically the same for all three reductants. In the manganese and cerium cases, if $[M^n]/[BrO_3^-]$ is not kept very large, then the kinetic situation becomes very complex. Reactions (R7), $(-R6)$, and $(-R5)$ become important in the cerium case, and equivalent reactions do in the manganese case.

$$Ce^{4+} + BrO_2^- + H_2O \rightarrow BrO_3^- + Ce^{3+} + 2H^+ \qquad (R7)$$

$$Ce^{4+} + HBrO_2 \rightarrow Ce^{3+} + H^+ + BrO_2^- \qquad (-R6)$$

$$H_2O + BrO_2^- + BrO_2^- \rightarrow BrO_3^- + HBrO_2. \qquad (-R5)$$

The reaction involving neptunium is always second order in bromate ion.

The evidence in support of Process B is extensive, albeit mainly indirect. The observation (Yatsimirskii *et al.*, 1976) of CIDNP in reacting BZ reagents implicates radical mechanisms such as Processes B and C. All observed kinetic behavior of the Ce(III)–BrO_3^- system can be simulated numerically using a Process B like mechanism (Barkin *et al.*, 1977). One of the more critical steps in Process B is (R5).

$$HBrO_2 + HBrO_3 \rightarrow 2BrO_2^- + H_2O. \qquad (R5)$$

Because of the rapidity of the disproportionation of $HBrO_2$ [step (R4)], this reaction is difficult to observe directly. However, in a strongly basic medium, Buxton and Dainton (1968) have directly observed BrO_2^- and the reaction

$$2HO^- + BrO_2^- + BrO_2^- \rightleftharpoons Br_2O_4 + 2OH^- \rightarrow BrO_3^- + BrO_2^- + H_2O.$$

Thompson (1973) has shown that the analogous reaction of $HClO_2$ with BrO_3^-, shown below, must be invoked to explain the kinetics of the oxidation of $HClO_2$ by bromate ion.

$$HClO_2 + BrO_3^- + H^+ \rightleftharpoons BrO_2^- + ClO_2^- + H_2O.$$

FKN showed that Betts and McKenzie's (1951) data on the exchange of radioactive bromine atoms between Br_2 and BrO_3^- can be interpreted in terms of reaction (R5) being rate determining. On the basis of their interpretation, FKN assigned k_5 a value of 1.0×10^4 M^{-2} sec^{-1}.

According to the FKN mechanism, the critical $[Br^-]$ occurring at Point B of Fig. 6 can be calculated by setting the rates of reactions (R2) and (R5) equal to each other. This leads to Eq. (BZ1).

$$[Br^-]_B = \frac{k_5}{k_2}[BrO_3^-].$$ \hfill (BZ1)

FKN used the value of k_5 calculated from Betts and MacKenzie's (1951) data, along with a value of k_2 also extracted by them from literature data, to calculate the proportionality constant in equation (BZ1) to be 5×10^{-6} M^{-2}. FKN found experimentally that the critical $[Br^-]$ at Point B in Fig. 6 is indeed given by Eq. (BZ1) with a proportionality constant of 18×10^{-6} M^{-2}. This is in excellent agreement with the calculated value considering the indirect methods used to evaluate the rate constants involved. For example, FKN used a value of $pK_a(HBrO_2)$ of 10^{-2} derived from Pauling's (1970) Rules in their calculation of k_2. Massagli, Indelli, and Pergola (1970) infer a considerably lower value from their experiments on the decomposition of bromite ion, and this will tend to reduce the discrepancy between calculated and experimental critical bromide ion concentrations.

Herbo, Schmitz, and van Glabbeke (1976) have recently reported an extensive series of experiments on the kinetics of the $Ce(III)-BrO_3^-$ system. These authors claim that their data cannot be rationalized in terms of a Process B-type mechanism. This claim is contrary to the results of Barkin *et al.* (1977), and probably results from their neglect of reaction (R7). They propose an alternate mechanism in which they delete reactions (R3)–(R5), and (R7) from the FKN mechanism and insert reactions (H1)–(H3).

$$BrO_3^- + Br^- + 2H^+ \rightleftharpoons Br_2O_2 + H_2O$$ \hfill (H1)

$$Br_2O_2 + H_2O \rightleftharpoons HBrO_2 + HOBr$$ \hfill (H2)

$$HBrO_2 + Br_2O_2 \rightleftharpoons 2BrO_2\cdot + Br^- + H^+.$$ \hfill (H3)

Herbo *et al.* (1976) omit (R5) even though this reaction has been directly observed by Buxton and Dainton (1968). The net of (H1) + (H2) is (R3), but there is little support for the existence of Br_2O_2. Herbo *et al.* (1976) do not specify whether this species is

$$Br-Br\begin{smallmatrix}O\\ \diagup\\ \diagdown\\ O\end{smallmatrix} \quad \text{or} \quad O-Br-Br-O.$$

Noyes and Bar-Eli (1977) have criticized the Herbo mechanism using several arguments. The most convincing of their arguments concerns the concentration of bromide ion, which is a catalytic intermediate in the Herbo mechanism. Noyes and Bar-Eli showed that the steady state [Br⁻] in a reacting, acidic Ce(III)–BrO₃⁻ system is much too small to drive the Herbo mechanism at the experimentally observed rate. Furthermore, the Herbo mechanism predicts a rate dependence on [HOBr], but Knight and Thompson (1973) find no effect on rate upon addition of HOBr in the Np(V)–BrO₃⁻ system, which apparently (Noyes, Field, and Thompson, 1971) proceeds by essentially the same mechanism as the Ce(III)–BrO₃⁻ reaction. Thus while the mechanism would certainly benefit from a more directly determined knowledge of the oxybromine chemistry involved, there seems little substantive reason to doubt the FKN Process B mechanism.

c. Process C. Field, Körös, and Noyes (1972) said relatively little about the details of Process C. They considered its function to be simply the reduction of Ce(IV) accompanied by the liberation of bromide ion. They suggested that it could·be regarded as some linear combination of (R9) and (R10). The value of F in reaction D of Table I is determined by the relative mix of (R9) and (R10).

$$6Ce(IV) + CH_2(COOH)_2 + 2H_2O \rightarrow 6Ce(III) + HCOOH + 2CO_2 + 6H^+ \tag{R9}$$

$$4Ce(IV) + BrCH(COOH)_2 + 2H_2O \rightarrow Br^- + 4Ce(III) + HCOOH + 2CO_2 + 5H^+. \tag{R10}$$

After proposal of the FKN mechanism Field and Noyes (1974a) proposed a simple model of the BZ reaction based upon the simplified FKN mechanism of Table I. We will discuss this model in detail in the next section, however, the model indicates that the FKN mechanism is most susceptible to oscillation if $F \sim \frac{1}{2}$.

Jwo and Noyes (1975) investigated the reaction of Ce(IV) with mixtures of MA and BrMA. The reaction obeys Michaelis–Menton kinetics and is always first order in Ce(IV). If [MA] ≫ [BrO₃⁻], then MA will be the dominant organic material in the system, and Jwo and Noyes' results can be explained by the sequence below.

$$CH_2(COOH)_2 + Ce(IV) \rightleftharpoons [CH_2(COOH)_2 - Ce(IV)]$$
$$\rightarrow Ce(III) + H^+ + \cdot CH(COOH)_2 \tag{BZ2}$$

$$\cdot CH(COOH)_2 + BrCH(COOH)_2 \rightarrow CH_2(COOH)_2 + Br\dot{C}(COOH)_2 \tag{BZ3}$$

$$Br\dot{C}(COOH)_2 + H_2O \rightarrow HO\dot{C}(COOH)_2 + Br^- + H^+ \tag{BZ4}$$

$$HO\dot{C}(COOH)_2 \rightarrow \tfrac{1}{2}HOC(COOH)_2 + \tfrac{1}{2}0{=}C(COOH)_2. \tag{BZ5}$$
$$\overset{|}{H}$$

Jwo and Noyes further found that O=C(COOH)₂ is rapidly oxidized by Ce(IV), but that

$$HO{-}C(COOH)_2$$
$$\overset{|}{H}$$

is oxidized by Ce(IV) somewhat more slowly than is malonic acid. Thus the overall stoichiometry of Process C under the $[MA] \gg [BrO_3^-]$ condition becomes

$$BrCH(CO_2H)_2 + 2Ce(IV) + \tfrac{3}{2}H_2O \rightarrow Br^- + \tfrac{1}{2}HOCH(CO_2H)_2$$
$$+ CO_2 + \tfrac{1}{2}HCO_2H + 2Ce(III) + 3H^+. \qquad (BZ6)$$

This stoichiometry yields $F = \tfrac{1}{2}$, but F is decreased as partially oxidized malonic acid derivatives do react with Ce(IV). No tartronic acid $[HOCH(CO_2H)_2]$ was found among the products of BZ reaction mixtures by Bornmann et al. (1973). However, we have found that tartronic acid is fairly quickly oxidized to carbon dioxide in acidified bromate solutions. This is probably true of other partially oxidized malonic acid derivatives as well, and indicates the basic fallacy of assuming that there is no interaction between Process A, B, and C intermediates in the FKN mechanism.

If $[MA] \ll [BrO_3^-]$, then most malonic acid becomes brominated and the stoichiometry of Process C can be represented by (R10).

Reaction (R10) has an F of $\tfrac{1}{4}$, which is too far from $\tfrac{1}{2}$ for oscillation to occur. Noyes and Jwo (1975) added reaction (R11) to give the overall stoichiometry of (BZ7).

$$HOBr + HCO_2H \rightarrow Br^- + CO_2 + H^+ + H_2O \qquad (R11)$$

$$BrCH(CO_2H)_2 + 4Ce(IV) + H_2O + HOBr \rightarrow 2Br^- + 3CO_2 + 4Ce(III) + 6H^+. \qquad (BZ7)$$

Regardless of whether (R11) is actually the reaction responsible for the further oxidation of partially oxidized malonic acid derivatives by bromine containing species, it is clear from the stability properties of the FKN mechanism that some such process(es) must occur.

4. The Oregonator

Even though the FKN mechanism as described in Table I leaves out reversibility and side reactions, as well as treating the very complex Process C as a single step, the mass action dynamic equations resulting from it are too complex to be handled other than by numerical integration. Field and Noyes (1974a) simplified the FKN mechanism even further and arrived at a model that still seems to retain the basic features of the full FKN mechanism. They chauvinistically dubbed the model the Oregonator. The properties of the Oregonator have been extensively investigated (Tyson, 1976), both because of its clear relationship to a real chemical system and because it combines a relatively tractable level of mathematical difficulty with a great complexity of dynamic behavior. These results are reviewed from a mathematical point of view in this volume by Troy. However, it is useful to introduce the Oregonator here both to support the validity of the FKN mechanism and to relate some behaviors of the BZ reaction to the FKN mechanism.

The Oregonator model consists of the five steps, (O1) to (O5).

$$A + Y \rightarrow X \tag{O1}$$

$$X + Y \rightarrow P \tag{O2}$$

$$B + X \rightarrow 2X + Z \tag{O3}$$

$$X + X \rightarrow Q \tag{O4}$$

$$Z \rightarrow fY. \tag{O5}$$

Identities are: $A \equiv B \equiv BrO_3^-$, $X \equiv HBrO_2$, $Y \equiv Br^-$, $Z \equiv M^{n+1}$ and P and $Q \equiv$ inert products. Normally the open system approximation (Noyes, 1976a) of A and B constant on the time scale of nonmonotonic behavior is valid. The correspondence between individual steps of the Oregonator and the FKN mechanism (Table I) is readily apparent: $(O1) \equiv (R3)$, $(O2) \equiv (R3)$, $(O3) \equiv (R5)$, $(O4) \equiv (R4)$, and $(O5) \equiv$ Process C. In FKN Process C, F is related to the Oregonator f by $F = f/2$. The values estimated by FKN for the rate constants of reactions (R2)–(R5) are carried directly over to steps (O1)–(O4) of the Oregonator but, because of the uncertainties concerning the details of Process C, the rate constant for step (O5) and f of the Oregonator are treated as expendable parameters.

The mass action dynamic law for the Oregonator is most conveniently written in terms of scaled dimensionless variables. These equations are

$$d\alpha/d\tau = s(\eta - \eta\alpha + \alpha - q\alpha^2) \tag{O6}$$

$$d\eta/d\tau = s^{-1}(-\eta - \eta\alpha + f\rho) \tag{O7}$$

$$d\rho/d\tau = w(\alpha - \rho) \tag{O8}$$

where for $k_{(O5)} = 1$

$$[HBrO_2] \equiv X = \frac{k_{(O1)}A}{k_{(O2)}}\alpha = 5.025 \times 10^{-11}\alpha \tag{O9}$$

$$[Br^-] \equiv Y = k_{(O3)}B/k_{(O2)}\eta = 3.00 \times 10^{-7}\eta \tag{O10}$$

$$[Ce(IV)] \equiv Z = (k_{(O1)}k_{(O3)}/k_{(O2)}k_{(O5)})AB\rho = 2.412 \times 10^{-8}\rho \tag{O11}$$

$$\text{Time} \equiv t = \tau/(k_{(O1)}k_{(O3)}AB)^{1/2} = 0.1610\tau \tag{O12}$$

$$S = (k_{(O3)}B/k_{(O1)}A)^{1/2} = 77.27 \tag{O13}$$

$$w = k_{(O5)}/(k_{(O1)}k_{(O3)}AB)^{1/2} = 0.1610 \tag{O14}$$

$$q = \frac{2k_{(O1)}k_{(O4)}A}{k_{(O2)}k_{(O3)}B} = 8.375 \times 10^{-6}. \tag{O15}$$

The overall steady state is given by

$$\alpha_0 = \rho_0 = \{(1 - f - q) + [(q + f - 1)^2 + 4(f + 1)q]^{1/2}\}/2q$$

$$\eta_0 = f\alpha_0/(1 + \alpha_0).$$

Linear stability analysis (Section II,C,1) of this steady state as a function of the expendable parameters $k_{(O5)}$ and f yields the result shown in Fig. 7. The steady state is infinitesimally unstable in region III. Numerical integration of Eqs. (O6)–(O8) with $f = k_{(O5)} = 1$, a point contained in region III, yields the oscillatory trajectory shown in Fig. 8. The phase plane plot in Fig. 9 indi-

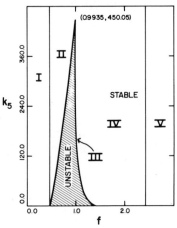

Fig. 7. Regions of stable and unstable Oregonator steady states for $q = 8.375 \times 10^{-6}$, $s = 77.27$, and $w = 0.1610 \, k_{O5}$. The steady state is stable to oscillation except for sufficiently small values of k_{O5} with f in the neighborhood of unity. [Reproduced by permission from R. J. Field and R. M. Noyes, *Faraday Symp. Chem. Soc.* **9**, 21 (1974).]

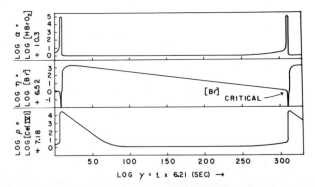

Fig. 8. Traces of log [Ce(IV)] (ρ), log [Br$^-$] (η), and log [HBrO$_2$] (α) vs time (τ) obtained by numerical integration of Eqs. (O6)–(O8) with $f = k_{O5} = 1$. [Reproduced by permission from R. J. Field and R. M. Noyes, *J. Chem. Phys.* **60**, 1877 (1974).]

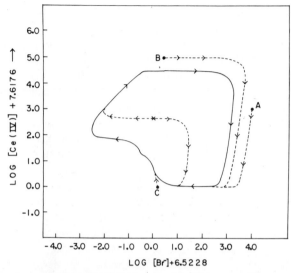

Fig. 9. Phase plane plot of log [Ce(IV)] (ρ) vs log [Br$^-$] (η) obtained by numerical integration of Eqs. (O6)–(O8) with $k_{O5} = f = 1$. The solid line indicates the unique limit-cycle solution for the kinetic constants used and the dashed lines indicate the path of approach to the limit cycle of systems not originally on it. The cross indicates the steady state, and the arrowheads indicate the direction of time evolution. The numbers indicate times in the cycle. $\tau = 302.9$ for the entire cycle. [Reproduced by permission from R. J. Field and R. M. Noyes, *J. Phys. Chem.* **60**, 1877 (1974).]

cates that the oscillatory trajectory is a limit cycle, which is asymptotically attained regardless of the initial values of X, Y, and Z chosen. Hastings and Murray (1975) have proved analytically the existence of finite, oscillatory solutions of Eqs. (O6)–(O8).

The behavior of the Oregonator model as f and $k_{(O5)}$ are varied is considerably more complex than is indicated by simple infinitesimal stability analysis. The similarity of $k_{(O5)}$ and f in the Oregonator model to the bifurcation parameters discussed in Section II,C is obvious. The parameter f varies considerably in the course of the BZ reaction because it is related to [BrMA]/[MA]. Troy and Field (1977) have shown that the steady state is globally as well as locally stable for sufficient small or large values of f regardless of the value of $k_{(O5)}$. The regions of such stability probably correspond to regions I and V of Fig. 7. For $0.5 < f < 1 + \sqrt{2}$, behavior is more complex. Hsu and Kazarinoff (1976) used Hopf (1942) bifurcation ideas to find small amplitude periodic solutions near the bifurcation line defining region III. Tyson (1977) reduced the Oregonator dynamic equations from order 3 to 2 by setting $d\alpha/d\tau = 0$ at all times. This chemically reasonable approximation was suggested by Field and Noyes (1974a). The second-order system is less difficult to handle analytically. Tyson (1977) then showed that within regions III and IV there exist stable, large amplitude periodic solutions even though the steady state is infinitesimally stable. This situation is analogous to that in Fig. 4, and hysteresis and "hard oscillation" phenomena may be possible. Thus moving along the line at $k_{(O5)} = 120$ in Fig. 7 from left to right we move from region I where the steady state is globally, asymptotically stable into region II where the steady state is only locally stable, and there exists a stable oscillatory solution. A large perturbation could move the system from the steady state to the limit cycle, but without such a perturbation the system remains on the steady state until the bifurcation point between regions II and III is reached and the small amplitude oscillations of Hsu and Kazarinoff (1976) are expected to appear and quickly grow to the large amplitude limit cycle of Tyson (1977). If f is now reduced the system will stay on the limit-cycle oscillation as the bifurcation line between regions III and II is recrossed. Oscillation is not expected to die out until region I is entered and the steady state becomes globally, asymptotically stable again. This constitutes a hysteresis loop. Turner (1976) has observed this sort of hysteresis behavior in a reversible version of the Oregonator due to Field (1975). He numerically integrated the dynamic equations of this model while slowly varying a bifurcation parameter and found that the point at which the oscillatory–nonoscillatory transition occurs depends upon when the bifurcation parameter is increasing or decreasing.

The actual behavior of the BZ reaction can be related to the above discussion. Hard oscillations have not been observed, although both Tyson (1976) and Turner and Field have looked for them. However, the transition from a locally stable steady state to an already existing, large amplitude limit cycle is apparently the cause of the sudden appearance of full blown oscillations (Fig. 6) at the end of the induction period. During the induction period the parameter f is growing towards the bifurcation point.

The simple bifurcation of small amplitude oscillations which then grow greatly in amplitude (Fig. 4) has been observed by Turner *et al.* (1977) in the manganese acetyl acetonate system with malonic acid. Their reagent also contained added formic acid. Manganese catalyzed systems containing diones seem to behave quite differently than other BZ systems (Körös *et al.*, 1973).

Tyson (1975) identified the feedback cycles that may destabilize the Oregonator steady state. The stability matrix (Section II,C,1) has the sign structure shown below.

$$\begin{pmatrix} 0 & - & - \\ - & - & + \\ + & 0 & + \end{pmatrix} \qquad (O16)$$

There are two potentially destabilizing cycles apparent in the matrix. Because all a_{ii} are negative, there is no direct autocatalysis interaction. There is a competition interaction between X and Y indicated by a_{12} and $a_{21} < 0$, and there is a negative feedback cycle indicated by $a_{12} a_{23} a_{31} < 0$. The negative feedback cycle can be understood thusly; because a_{12} is negative, a decrease in Y leads to an increase in X, which, because a_{31} is positive, leads to an increase in Z, which, because a_{23} is positive, finally feeds back to cause an increase in Y. This increase in Y directly counteracts the original perturbation in Y, and this cycle is thus a negative feedback cycle. If the feedback current leads to an effect much different than the initial perturbation, this cycle can be destabilizing.

The stability matrix only demonstrates the existence of potentially destabilizing cycles. The actual stability properties of the system depend upon the currents through the loops, which in turn depend upon the magnitude of the rate constants involved. In order to actually determine for what range of rate constants the steady state is actually unstable, the entire stability problem must be solved to find the eigen values of Eq. (O16). The methods of Clarke are useful here. The magnitudes of the currents in the Oregonator have not yet been determined, so it is not known whether the X-Y competition or the negative feedback loop is the principal interaction destabilizing the Oregonator. However, the strong dependence of stability on the step (O5) parameters $k_{(O5)}$ and f, and the participation of step (O5) in the negative feedback loop indicates that this cycle is important. The negative feedback cycle, $a_{12} a_{23} a_{31}$ is the one discussed qualitatively in our description of the FKN mechanism.

As an aside, it should be mentioned here that Zhabotinskii *et al.* (1971) have proposed a three variable model for the BZ reaction. This model is significantly different from the Oregonator because it is based upon the macroscopic form of the BZ oscillations rather than on a chemical mechanism. It reproduces the BZ oscillations quite well. Tyson (1976) has listed the differences between the Zhabotinskii model and the Oregonator. The Zhabotinskii model exhibits a range of dynamic behavior probably wider than does the Oregonator (Othmer, 1975). However, doubts concerning the model's exact relationship to the actual chemistry of the BZ reaction has limited interest in it.

Another model of the BZ reaction, which is very similar to the Oregonator, has recently been proposed (Weisbuch *et al.*, 1975).

5. Stability Properties of the Full FKN Mechanism

FKN argued convincingly, but qualitatively, that the FKN mechanism does indeed possess oscillatory dynamics. However, because of the technical difficulty of the problem, they did not undertake a steady-state stability analysis of their mechanism. The Oregonator was at least partially conceived to quantitatively demonstrate that the sort of interactions present in the FKN mechanism do lead to an unstable steady state and oscillations. Edelson, Field, and Noyes (EFN) (1975) added further evidence for oscillation by numerically integrating the FKN dynamic equations. They used a fairly complete reaction set that included reversibility of many steps, the principal known side reactions, and considerable detail for Process C. They found limit-cycle oscillations that were qualitatively and quantitatively very similar to the BZ oscillations. However, it was subsequently discovered that the set of dynamic equations actually used in the integrations did not correspond exactly to the set of stoichiometric equations presented by EFN. The set of dynamic equations actually integrated contained a discrepancy in Process C that increased the stoichiometry of Br^- production. Without this added source of Br^-, the published EFN model does not oscillate. This observation further emphasizes the message of the Oregonator concerning the importance of the stoichiometry of Process C to the appearance of oscillations in the FKN mechanism.

The complete FKN stability problem is an attractive one not only because of its importance to understanding the BZ reaction, but also because it is an excellent stage for the development of general theoretical ideas on the stability of chemical networks. Thus Clarke undertook a conceptual reorganization (1974a,b) of the problem in which diagrams or " graphs " were used to focus attention on the role of network topology (1975a,b) in stability. The principal result from network topology seems to be that steady-state stability is dependent upon the currents through certain destabilizing feedback cycles in the network. The " graphs " (Clarke, 1974a,b) mentioned previously are used to identify feedback cycles. Recognition of some properties of these cycles leads to the result that a complex network can be simplified by elimination of reactions and intermediates *that do not affect the net topology of cycles.* The reduced network thus obtained retains nearly the same stability properties as the full network. Furthermore, the theory can be extended to calculate any differences that may exist. For example, reverse reactions may be eliminated because they affect the current through a cycle but not its basic topology. Reverse reactions then can only *stabilize* a steady state by diminishing the net current through a destabilizing cycle. Clearly this reflects the fact that as equilibrium is approached, reverse reactions become important and steady states are stabilized.

Flow through intermediates (FTR), which are those produced by a single reaction and consumed by only one other reaction, may also be eliminated without affecting overall network topology. If a FTR is present in large quantity, then the network is reduced by assuming that the concentration of this species is large and constant. If a FTR is present in a small steady-state concentration, then the network is reduced by simply adding stoichiometrically the reactions producing and consuming the FTR. The resulting reduced network will have stability properties essentially identical to those of the full network. As the steady-state concentration of a FTR increases, the importance of a destabilizing cycle containing it decreases and the steady state is stabilized. These ideas on network stability are a very clever and elegant formulation of some ideas that kineticists (Noyes, 1974a) have been using for decades to reduce complex reaction networks, and they should be more widely understood by kineticists. Field and Noyes' reduction of the FKN mechanism to the Oregonator follows the rules of network topology.

Elimination of steps and intermediates using Clarke's topological ideas can grossly simplify a steady-state stability problem with little of the risk of changing drastically the conclusions that may accompany reduction by chemical intuition. The graph theoretical (Clarke, 1974a,b) method of analyzing the reduced network, which may still be quite complex, greatly eases the algebraic and computational difficulty of carrying the problem through to completion.

Clarke (1976a) has applied graph theoretical and network topology ideas to the full FKN model consisting of the reactions in Table I plus (R0), (R7), (R9)–(R11). A topologically similar reduced network consisting of reactions (R1)–(R5), (R9)–(R11) was derived by eliminating reverse reactions and the FTR's BrO_2·, Br_2, $CHBr(COOH)_2$, and $HCOOH$.

If (R11) is excluded at this point, the reduced network becomes essentially, but not exactly, the Oregonator with f constrained to be ≤ 0.5; (R10) yields $f = 0.5$, and f is reduced to the extent that (R9) occurs. Clarke (1976a) found the network without (R11) to be always stable as the Oregonator suggests. The published EFN (Edelson *et al.*, 1975) model is topologically similar to this reduced network without (R11) because it differs from Clarke's reduced network only in that it contains added reverse reactions and FTR's in Processes A and B and by expansion of Process C including only FTR's that leave the overall stoichiometry simply some linear combination of (R9) and (R10).

In the reduced network including (R11), formic acid is essentially a FTR intermediate, but one that could conceivably be present either in large quantities or in steady-state quantities. Further treatment of the reduced network depends upon which situation obtains. Both cases are experimentally possible depending upon whether or not formic acid is initially added as a reactant. Only the analysis of the steady-state formic acid (Clarke, 1976b) case has appeared. This is the situation comparable to the normal experimental conditions used in the BZ reaction. In this analysis Clarke used his methods to show that the steady state of the network is indeed unstable. There are three destabilizing feedback cycles present. They are: (1) a positive feedback two-cycle arising from mutual inhibition (a competition interaction in Tyson's terminology) between Br^- and $HBrO_2$; (2) a negative feedback three-cycle in which $HBrO_2$ activates HOBr, HOBr activates Br^-, and Br^- inhibits $HBrO_2$; and (3) a negative feedback three-cycle in which $HBrO_2$ activates Ce(IV), Ce(IV) activates Br^-, and Br^- inhibits $HBrO_2$. The negative feedback cycles (1) and (3) correspond to the destabilizing cycles in the Oregonator. The Ce(IV) activates Br^- link in cycle (3) is very complex, however. This link corresponds to Process C in the FKN mechanism, which is still the least well-understood part. The resolution of this problem must await exact and extensive experimental kinetic work on the reactions of the oxybromine compounds of Processes A and B with the partially oxidized organic materials of Process C. This work is being undertaken in our laboratory for the cerium ion catalyzed system with malonic acid. Even though experiments (Field, *et al.*, 1977) indicate that reaction (R11) is probably too slow to play the role designated by Clarke (1975a,b) and by Noyes and Jwo (1975), it is likely that some such Br^- producing interaction of an oxybromine compound with an organic material will be important. Glyoxylic acid (OCH—COOH) seems a good candidate because of its stunning effect on the induction period and oscillations in the BZ reaction. Kaner and Epstein (1978) have invoked glyoxylic acid as a very important species mechanistically in their analysis of the results obtained when iodide ion is added to the Belousov–Zhabotinskii reaction. Another possibility is that oxidation of organic material does not proceed to HCOOH and CO_2 on the time scale of the oscillations, thus increasing F. Partially oxidized malonic acid derivatives are not found upon analyses of BZ reaction mixtures, but the time scale of such analyses is very long compared to the time scale of the oscillations. Edelson and Noyes (1977) are using both of these options in their efforts to computationally simulate the BZ reaction. So far no set of stoichiometric reactions and rate constants has proven to be entirely satisfactory, but the situation is clarifying.

B. THE BRAY–LIEBHAFSKY (BL) REACTION

1. Experimental Characterization

In an acidic medium, hydrogen peroxide (HOOH) is thermodynamically capable of both oxidizing I_2 to IO_3^- (BLi)† and reducing IO_3^- to I_2 (BLii). Thus Bray (1921) concluded that IO_3^- should be an ideal catalyst for the decomposition of HOOH. Two net processes are involved, and both are strongly favored thermodynamically.

$$2IO_3^- + 5HOOH + 2H^+ \rightarrow I_2 + 5O_2 + 6H_2O. \qquad \text{(BLi)}$$

$$I_2 + 5HOOH \rightarrow 2IO_3^- + 2H^+ + 4H_2O. \qquad \text{(BLii)}$$

† In the following discussion BL reaction numbers using lower case Roman numerals refer to overall stoichiometries and BL numbers using Arabic numbers refer to elementary reactions.

Overall reactions (BLi) and (BLii) are expected to reach a pseudosteady state at which $d[IO_3^-]/dt \approx d[I_2]/dt \approx 0$. At this point the net reaction is catalysis of the overall thermodynamically favored process (BLiii).

$$2HOOH \rightarrow 2H_2O + O_2. \qquad (BLiii)$$

However, Bray (1921) found that in fact a steady state is sometimes not reached. Instead oxygen production occurs in pulses while $[I_2]$ visibly oscillates. A further curious fact is that pulses of oxygen occur while I_2 is being oxidized to IO_3^-. Oxygen does not occur as a stoichiometric product in (BLii).

After Bray's initial discovery of oscillations in this reaction, little more was done on the reaction until the late 1960's when Liebhafsky, who had been a student with Bray, returned to the problem after a distinguished career in other areas of chemistry. Liebhafsky† applied modern analytical techniques to the experimental characterization of the oscillations. Thus we refer to this oscillator as the Bray–Liebhafsky (BL) reaction. Figs. 10–12 show some oscillatory traces of, respectively, $[O_2]$, $[I_2]$, and $[I^-]$ vs time in typical BL reagents. Oscillations are most readily observed at temperatures near 50°C. The observed behavior of the BL reaction is very complex. The oscillations appear only over a rather narrow range of pH and reactant concentrations. The induction period as well as the shape, frequency, etc., of the oscillations

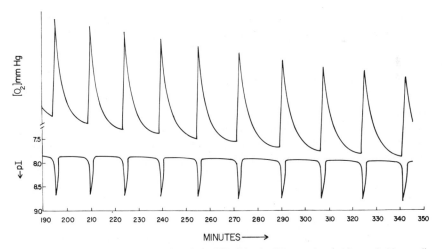

Fig. 10. Potentiometric traces of pI and of $[O_2]$ in the BL reaction (without absolute calibration) vs time at 50°C. Initial concentrations were $[H_2O_2]_0 = 0.098\ M$. $[KIO_3]_0 = 0.105\ M$, $[HClO_4]_0 = 0.059\ M$. Amplitude of O_2 oscillations corresponds to about 300 mm of Hg. Solution was not specifically illuminated. [Reproduced by permission from K. R. Sharma and R. M. Noyes, *J. Am. Chem. Soc.* **97**, 202 (1975).]

† See references 4–23 in Sharma and Noyes (1976).

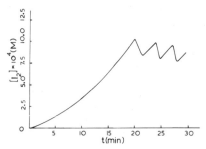

Fig. 11. Spectrophotometric recording at 50°C of iodine concentration in solution in a BL reagent initially containing 0.104 *M* KIO$_3$, 0.490 *M* HOOH, and 0.047 *M* HClO. [Reproduced by permission from K. R. Sharma and R. M. Noyes, *J. Am. Chem. Soc.* **98**, 4345 (1976).]

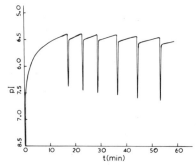

Fig. 12. Potentiometric recording at 50°C with an iodide specific electrode in a BL reagent initially containing 0.432 *M* KIO$_3$, 0.490 *M* HOOH, and 0.055 *M* HClO$_4$. [Reproduced by permission from K. R. Sharma and R. M. Noyes, *J. Am. Chem. Soc.* **98**, 4345 (1976).]

vary greatly as parameters are changed. Furthermore, the occurrence of oscillations is strongly dependent upon visible light and applied oxygen pressure (Sharma and Noyes, 1975). This variability of behavior is to be contrasted with that of the BZ reaction in which oscillations of qualitatively unchanging characteristics appear at room temperature and over a rather wide range of reactant concentrations and pH.

2. The Sharma–Noyes (SN) Mechanism of the BL Reaction

Sharma and Noyes (SN) (1976) have proposed a detailed mechanism for the BL reaction. Their approach to solving the mechanism is classic, and will serve as a model of mechanism elucidation. SN initially wrote down all conceivable chemical reactions that could be components of the mechanism. They then used every piece of kinetic and thermodynamic information available to them, as well as analogies to oxybromine chemistry, to settle upon a set of plausible component reactions that are neither thermodynamically unfavored nor kinetically slow on the time scale of the BL reaction. They then pieced together a mechanism that rationalizes essentially all important known features of the BL reaction. The very diversity of unusual behaviors exhibited by the BL reaction, which made the mechanism so difficult to solve, now gives us confidence that the SN mechanism is unique. This author has little doubt of the fundamental validity of the SN mechanism.

The SN mechanism resembles the FKN mechanism in an important way. It involves several separable overall processes, and oscillation occurs as control of the system is passed from one process to the other through a delayed feedback loop. Furthermore, the first of these processes is nonradical in nature while the second is radical in nature. We shall identify these processes as I and II. Process I is nonradical and comparable to the FKN Process A. Its overall effect is the reduction of IO_3^- to I_2, and it is often slow compared to Process II. The Process II mechanism involves radicals and oxidizes I_2 to IO_3^- with simultaneous catalysis of (BLiii). Process II bears similarity to the FKN Process B. There is nothing in the SN mechanism analogous to the FKN Process C, however.

Process I is the nonradical, two-equivalent oxidation of HOOH. It can be represented by reactions (BL1)–(BL4).

$$2(IO_3^- + I^- + 2H^+ \rightarrow HIO_2 + HOI) \qquad \text{(BL1)}$$

$$2(HIO_2 + I^- + H^+ \rightarrow 2HOI) \qquad \text{(BL2)}$$

$$5(HOI + H_2O_2 \rightarrow I^- + O_2 + H^+ + H_2O) \qquad \text{(BL3)}$$

$$H^+ + HOI + I^- \leftrightharpoons I_2 + H_2O \qquad \text{(BL4)}$$

$$I_2 + 5HOOH \rightarrow I_2 + 5O_2 + 6H_2O. \qquad \text{(BLi)}$$

The net effect of Process I is (BLi). Reaction (BL4) is rapid in both directions and thus is always near the equilibrium condition in which I_2 grossly predominates over HOI and I^-. However, reaction (BL3) maintains $[I^-]/[HOI]$ higher than the hydrolysis of I_2 would. The direct reduction of HOOH by I^- is thermodynamically favored, but it is kinetically slow on the time scale of the BL reaction. However, I^- does serve as catalyst for Process I. As Process I proceeds, $[I_2]$ and $[O_2]$ increase, and it is this increase that is largely responsible for transfer of control from Process I to Process II.

Process II is initiated by reaction (BL5). Reaction (BL5) is analogous to (R5), which initiates the transfer of control from Process A to Process B in the FKN mechanism. Initial interaction of HIO_2 with IO_3^- leads to $HOO\cdot$, as indicated in reactions (BL5)–(BL7). The $HOO\cdot$ produced can either disproportionate by (BL8), or it can enter the radical chain I_2 oxidation sequence (BL9)–(BL13).

$$H^+ + IO_3^- + HIO_2 \rightarrow 2IO_2\cdot + H_2O \qquad \text{(BL5)}$$

$$2(IO_2\cdot + HOOH \rightarrow IO_3^- + H^+ + HO\cdot) \qquad \text{(BL6)}$$

$$2(HO\cdot + HOOH \rightarrow H_2O + HOO\cdot) \tag{BL7}$$

$$HIO_2 + 4HOOH \rightarrow IO_3^- + H^+ + 3H_2O + 2HOO\cdot \tag{BLiv}$$

$$2HOO\cdot \rightarrow HOOH + O_2 \tag{BL8}$$

$$HOO\cdot + I_2 \rightarrow I^- + O_2 + H^+ + I\cdot \tag{BL9}$$

$$I\cdot + O_2 \rightleftharpoons IOO\cdot(I-O-O) \tag{BL10}$$

$$IOO\cdot + I^- + H^+ \rightarrow HOI + IO\cdot \tag{BL11}$$

$$IO\cdot + HOOH \rightarrow HIO_2 + HO\cdot \tag{BL12}$$

$$HO\cdot + HOOH \rightarrow H_2O + HOO\cdot \tag{BL13}$$

$$I_2 + 2HOOH \rightarrow HIO_2 + HOI + H_2O. \tag{BLv}$$

Note that this sequence regenerates the $HOO\cdot$ consumed and is dependent upon $[I_2]$ and $[O_2]$. The sum of (BLiv) and (BLv) can be autocatalytic since the HIO_2 produced by (BLv) may enter (BLiv) to produce more $HOO\cdot$ and thus more HIO_2. The autocatalytic increase in $[HIO_2]$ begins when $[I_2]$ and $[O_2]$ reach high enough levels during Process I for (BL9)–(BL13) to compete with (BL8). It continues until disproportionation of HIO_2 by BL14 leads to a steady state.

$$2HIO_2 \rightarrow HOI + IO_3^- + H^+. \tag{BL14}$$

As $[HIO_2]$ increases, $[I^-]$ decreases (BL2). Thus Process I, which is catalyzed by I^-, is progressively suppressed as Process II gains control.

In the FKN mechanism the shift from Process A to Process B occurs when $[Br^-]$ falls to a point at which it can no longer prevent [by reaction (R2)] $HBrO_2$ from initiating the radical reactions of Process B by reaction (R5). The situation in the SN mechanism is more complex. During the SN Process I, $[I^-]$ is often maintained at a relatively constant value. Indeed, Fig. 12 shows $[I^-]$ actually increasing before the cataclysmic drop that signals the onset of Process II. In the SN mechanism it is the *increasing concentrations* of I_2 and dissolved O_2 that triggers Process II. However, the ratio $[I^-]/[IO_3^-]$ affects the exact values of $[O_2]$ and $[I_2]$ that must accumulate before Process II becomes dominant.

Because the HIO_2 and HOI products of (BLv) become involved again with principal reactants in (BL5) and (BL3), the stoichiometry of Process II is independent of $HOO\cdot$ chain length and is given by (BLiv) + (BL8) + 3*(BLv) + (BL14) + 2*(BL4) + 2*(BL3). This sum yields the stoichiometry

$$I_2 + 11HOOH \rightarrow 2IO_3^- + 2H^+ + 10H_2O + 3O_2. \tag{BLvi}$$

(BLvi) differs from (BLii) because (BLvi) accomplishes the catalytic decomposition of 6 moles of HOOH. This rationalizes the experimental observation that the pulses of O_2 appear during the rapid oxidation of I_2. The presence of I· in Process II rationalizes the effects of visible light on the oscillations (Sharma and Noyes, 1975). I_2 is the only species present in the system that absorbs visible light. Noyes (1976b) has pointed out the pitfalls of calculating net stoichiometries in complex mechanisms, but the calculations in this section are carefully done.

When $[O_2]$ and $[I_2]$ reach values such that Process II becomes dominant, I_2 is removed thus lessening the efficiency of the radical chain. However, because the release of dissolved O_2 from a supersaturated solution, (BL15) is relatively slow with a time constant of the order of min^{-1}, the concentration of dissolved O_2 increases rapidly as Process II gains control.

$$O_2(aq) \rightleftharpoons O_2(g). \qquad \text{(BL15)}$$

Thus Process II continues even as $[I_2]$ falls. If (BL15) had a time constant of the order of Process II, then a steady state would be reached. As it is, this overall steady state is grossly overshot as $[I_2]$ is driven to a very low value. When Process II ceases and Process I regains control the supersaturation of dissolved O_2 is released, and the cycle is complete. Now Process I again starts O_2 and I_2 accumulation for the next cycle. The importance of dissolved O_2 to the SN mechanism rationalizes the sensitivity of the BL reaction to applied oxygen pressure (Sharma and Noyes, 1975).

Noyes and Edelson (1978) are attempting to simulate the BL reaction by numerical integration of the SN dynamic equations. The problem is difficult because the rate constants for many of the reactions have not been measured. Furthermore, the observed sensitivity of the BL reaction itself to experimental parameters suggests the model will be sensitive to the values of the rate constants used. Noyes and Edelson (1977) have not yet found the correct set of rate constants to reproduce all experimental features of the BL reaction. However, they have shown that the SN mechanism indeed does lead to oscillation.

Both the FKN mechanism and the SN mechanism involve switching between radical and nonradical processes. In both cases the occurrence of the nonradical process leads inevitably to the radical process autocatalytically gaining control of the system. There is assumed to be very little interaction between the processes, and normally only one process or the other is occurring at a measurable rate at a given time. It has been suggested (Field and Noyes, 1977) that this separation of radical and nonradical processes will be found to be common to a large class of chemical oscillators. Indeed all systems presently known to oscillate have the potential of such a mechanism. However, the exact method of switching between the nonradical and radical processes is quite different in the FKN and SN mechanisms, as is the mechanism for return of control to nonradical process. Both mechanisms rely upon delayed feedback to introduce steady state overshoots and oscillation.

3. Modifications of the BL Reaction

Modifications of the BL reaction are possible. These modifications are sufficiently severe, and the behaviors of the oscillators obtained so different from the BL reaction, that they should be treated separately. Briggs and Rauscher (1973) discovered that addition of malonic acid and either cerium (III) or manganese (II) to a BL reagent leads to vastly enhanced oscillatory behavior. If starch is used as an indicator, the color of the reagent oscillates through a colorless to gold to blue cycle. Unlike the original BL oscillations, these are very readily apparent with a convenient frequency of a few min^{-1} at room temperature. Cooke (1976) has discovered that malonic acid can be replaced by acetone. Briggs and Rauscher (1973) also found that invisible oscillations in $[I^-]$ can be induced in a nonoscillatory BL reagent at room temperature by the addition of only malonic acid. The range of concentrations over which these oscillations occur is quite small.

It is anticipated that much more work will be done on these modified BL oscillators because they are conveniently studied in the laboratory and present considerable mechanistic difficulties. No mechanisms have yet been proposed for them. The mechanism of the cerium or manganese catalyzed reduction of iodate by HOOH that is central to this class of oscillators has not been investigated. An investigation of the kinetics, mechanism, and equilibrium of the iodination of malonic acid has recently been reported (Leopold and Hairn, 1977).

C. Other Temporal Oscillators

There are a number of temporal oscillators that are not as well characterized as the preceding group.

1. The Morgan Reaction

Morgan (1916) discovered oscillations in the rate of carbon monoxide evaluation during the dehydration of formic acid in concentrated sulfuric acid, reaction (M1).

$$HCOOH \xrightarrow[H_2SO_4]{conc.} H_2O + CO. \tag{M1}$$

Reaction (M2) is also favored thermodynamically, but it does not occur to a measurable extent in concentrated sulfuric acid.

$$HCOOH \xrightarrow[H_2SO_4]{conc.} CO_2 + H_2 \tag{M2}$$

Showalter and Noyes (1978) are reinvestigating the Morgan reaction. They have concluded that the basis of the oscillations is indeed chemical in nature; the system is strongly affected by additions of species like formaldehyde, nitrate ion, or ferrous sulfate. A discussion of preliminary mechanistic interpretations of Showalter and Noyes (1977) has appeared (Noyes and Field, 1977), and it is expected that the Morgan reaction will shortly be as well understood as the BZ and BL reactions. Showalter and Noyes invoke a radical chain mechanism involving iron salts that are always present as PPM impurities in sulfuric acid. This mechanism features the catalytic decarbonylation of protonated formic acid, $H(CO)^+OH_2$, by $HO \cdot$ and chain branching induced by the enhanced ability of $Fe(CO)^{+3}$ over Fe^{+3} to produce $HO \cdot$ by the oxidation of water. The proposed mechanism resembles the SN mechanism more closely than the FKN mechanism, but mechanistic trends discerned in the FKN and SN mechanisms are mostly preserved. An interesting feature of the Morgan reaction is that at elevated CO pressure the system reaches equilibrium with substantial amounts of HCOOH remaining. Showalter and Noyes (1977) have estimated on the basis of experiments how far this system must differ from equilibrium for oscillation to occur. The Morgan reaction is the only known real chemical oscillator in which such an estimation can be made. Bowers and Rawji (1977) have also investigated the Morgan reaction. They concluded that the fundamental cause of the oscillations is physical, not chemical. They explain the oscillations in terms of supersaturation of the reaction mixture with CO and an enhanced reaction rate in the high surface area foam that appears during bursts of CO release. It should be possible to distinguish between these two mechanisms by careful experimentation.

2. Ammonium Nitrite Decomposition

Ammonium nitrite (NH_4NO_2) decomposes in aqueous solution to produce N_2 as a product. The production of N_2 may be oscillatory (Noyes and Field, 1977) with a frequency of a few min^{-1}. Radicals are implicated by the appearance of NO gas. The same phenomenon is observed during the decomposition of methyl ammonium nitrite $(CH_3NH_3NO_2)$. Little is presently known of the mechanisms of these decompositions, but investigations are under way (Smith and Noyes, 1978).

3. Dithionite Ion Decomposition

In aqueous solution, dithionite ion, $S_2O_4^{-2}$ decomposes according to (D1).

$$2S_2O_4^{-2} + H_2O \rightarrow 2HSO_3^- + S_2O_3^{-2}. \tag{D1}$$

The decomposition is sometimes oscillatory (Rinker *et al.*, 1965). In unbuffered solutions the pH oscillates, but the really unique feature is that the concentration of dithionite ion itself appears to oscillate. This can only be allowed thermodynamically if large concentrations of some intermediate develop and then disappear regenerating a *portion* of the initial dithionite ion. The concentration of dithionite at the peak of one cycle is always less than that at the peak of its predecessor. About 5 cycles are normally observed. The difficulty of studying this reaction experimentally (ESR techniques are used), as well as the very complex oxidation state and polymerization chemistry or sulfur in aqueous solution, has inhibited progress on this oscillator. However, Depoy and Mason (1974) have proposed a mechanism that stimulates the oscillations quite nicely.

4. Photochemical Oscillators

Nitzan and Ross (1973) have shown using model systems in which there is no net chemical change that oscillations, multiple steady states, and instabilities may be photochemically driven in chemical systems. Oscillations in photochemical systems have very recently been observed, but it is not yet certain that these systems are mechanistically analogous to the Ross models.

Nemzek and Guillet (1976) observed oscillations in light emission intensities from both acetone and biacetyl during the photolysis of acetone in acetonitrile solvent. Because the oscillatory emissions from acetone and biacetyl are 180° out of phase, they concluded that only the biacetyl concentration oscillates; oscillation in acetone emission intensity results from quenching of triplet acetone by biacetyl. Epstein *et al.* (1977) are continuing work on this system and have succeeded in better characterizing the oscillations experimentally.

Yamazaki *et al.* (1976) have observed oscillation in fluorescence intensity from a photoproduct during the photolysis of 1,5-naphthyridine in cyclohexane. The oscillations are very nicely formed for some experimental conditions and have a period of several minutes. No mechanism has been proposed for this system.

Turro (1977) has discovered oscillations in the photostationary state sometimes attained when the peroxide of 9,10-diphenyl anthracene is simultaneously being formed photochemically and returning to 9,10-diphenyl anthracene thermally. This system appears to have the best chance of being analogous to the Ross (Nitzan and Ross, 1973) models since no chemical change occurs. Noyes and Field (1977) have suggested that the mechanism of the oscillations may be similar to the FKN and SN mechanisms and involve singlet O_2. Bose *et al.*, (1977) have observed a similar phenomena upon the irradiation of anthracenes in chlorocarbons.

5. Oscillatory Gas Phase Reactions

Spontaneous oxidations of hydrocarbons and other materials in the gas phase not only show slow flames and explosions but also may proceed in a periodic manner. Oscillations are accompanied by weak pulses of light emission and temperature rises of less than 200°C. In closed systems up to 11 pulses have been observed. In flow systems pulses may continue indefinitely. Gray *et al.*, (1974) supply an entry into the literature of oscillatory flames and discuss in detail the oxidation of propane, a typical member of this class of reactions. One can imagine branched chain type mechanisms for these reactions that could, in principle, lead to unstable steady states and oscillations. However, in gas systems, another mechanism exists that may lead to unstable steady states. Because of the low densities of gases, relatively small exothermicities can lead to large temperature changes and large feedback effects resulting from the temperature dependence of rate constants. These thermal effects may lead to so-called thermokinetic oscillations. Gray *et al.*, (1974) point out that there has not been a chemically plausible, isothermal mechanism proposed for a gas phase oscillator. Successful models seem to involve the interaction of branched chain, autocatalytic, kinetics with temperature effects. Pilling and Noyes (1978) have suggested that the situation may be even more complex in some systems. They argue that the mechanism proposed by Yang and Berlad (1974) for the oscillatory gas phase oxidation of CO by O_2 is chemically implausible. They further concluded that even any nonisothermal mechanism involving only ground state species does not become unstable unless convection driven by thermal gradients is included.

IV. Flow Systems: A Diversity of Nonmonotonic Behavior

A. EXPERIMENTAL CHARACTERIZATION AND TYPES OF BEHAVIOR OBSERVED

The open system requirement for sustained nonmonotonic behavior in a chemical reaction may apparently be met in a continuously stirred flow reactor. However, chemical engineers have long known that reactions that show only stable steady states and monotonic behavior in a stirred batch reactor may exhibit nonmonotonic behavior in a continuously stirred flow reactor (Perlmutter, 1972). Reactions showing oscillations in flow systems are normally strongly exothermic. Feedback loops can thus result from the effect of temperature change on rate constants. Nonmonotonic behavior may result (Uppal *et al.*, 1974) from the linking of such feedback to the flow properties of the system. On the basis of the preceding discussion, it is not surprising that running the BZ and BR reactions in continuously stirred flow systems leads to very complex behavior. The BR flow system seems to be better characterized at this time than the equivalent BZ system.

In flow experiments, the reactants are normally led into the well-stirred reactor in two pressurized streams and a volume equal to the input volume is continuously drained off. The principal experimental variables are: the concentration of reactants in input streams, the flow rates, the residence time in the reactor (flow rate/volume of reactor), temperature, and ambient light intensity. The last variable is important in the BR reaction. Depending upon these experimental constraints, steady states can be either unique or multiple, and either stable or unstable. [Composite double oscillations and excitability (Section II,C,3,e) may also appear.] An experimental state diagram may be determined showing which behavior can be expected for a particular set of experimental conditions. Pacault *et al.*, (1976) have determined such a diagram for the BR reaction. Table II indicates the types of nonmonotonic behavior observed to this date in the BZ and BR reactions. Degn (1968) has found bistability in the horseradish peroxidase catalyzed oxidation of NADH by O_2. This reaction was carried out in an open system in which O_2 was continuously bubbled in.

Figures 13–16 serve to illustrate the behaviors of the BZ and BR reactions in flow systems.

B. MULTIPLE STEADY STATES

Figures 13 and 14 were obtained in experiments with the BZ reaction. Figure 13 shows a transition from a limit cycle, oscillatory trajectory around

Table II[a]

NONMONOTONIC BEHAVIOR OBSERVED IN FLOW SYSTEMS

Behavior	Belousov–Zhabotinskii	Briggs–Raucher
Sustained oscillation	1, 2, 3	4
Multiple steady states	2	5, 6
Excitability	—	7
Composite oscillations	1, 2, 3, 8	6
Phase synchronization	9	—
Chaos	10	—

[a] Table references:

(1) V. A. Vavilin, A. M. Zhabotinskii, and A. N. Zaikin (1968). *Russ. J. Phys. Chem.* **42**, 1649.

(2) M. Marek and E. Svobodova (1975). *Biophys. Chem.* **3**, 263.

(3) K. R. Graziani, J. L. Hudson, and R. A. Schmitz (1977). *Chem. Eng. J.* **12**, 9.

(4) A. Pacault, P. DeKepper, and P. Hanusse (1975). *C. R. Hebd. Seances Acad. Sci. Ser. B* **280**, 157.

(5) A. Pacault, P. DeKepper, P. Hanusse, and A. Rossi (1975). *C. R. Hebd. Seances Acad. Sci. Ser. C* **281**, 215.

(6) P. DeKepper, A. Pacault, and A. Rossi (1975). *C. R. Hebd. Seances Acad. Sci. Ser. C* **282**, 199.

(7) P. DeKepper (1976). *C. R. Hebd. Seances Acad. Sci. Ser. C* **283**, 25.

(8) P. G. Sorensen (1974). *Faraday Symp. Chem. Soc.* **9**, 88.

(9) M. Marek and I. Stucil (1975). *Biophys. Chem.* **3**, 274.

(10) R. A. Schmitz, K. R. Graziani, and J. L. Hudson (1977). *J. Chem. Phys.* **67**, 3040.

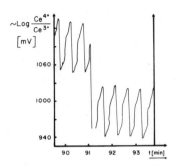

Fig. 13. Change from one oscillatory regime to another in the BZ reaction. Initial and inlet concentrations (reactor concentrations are different during reaction): 0.063 M KBrO$_3$; 0.032 M malonic acid; 10^{-3} M Ce(IV); reactor volume = 200 ml, flow rate = 1100 ml/hr. [Reproduced by permission from M. Marek and E. Svobodova, *Biophys. Chem.* **3**, 263 (1975).]

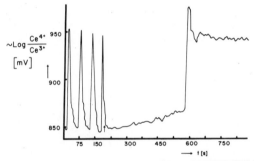

Fig. 14. Change from an oscillatory to a nonoscillatory regime followed by a transition between stable steady states in the BZ reaction. Initial and inlet concentrations (reaction concentrations are different during reaction): 0.01 M KBrO$_3$; 0.032 M malonic acid; 0.0003 M Ce(IV), reactor volume = 200 ml; flow rate = 600 ml/hr. [Reproduced by permission from M. Marek and E. Svobodova, *Biophys. Chem.* **3**, 263 (1975).]

one unstable steady state to a second limit cycle around another unstable steady state. Figure 14 illustrates a transition from one stable steady state to another. Both of these transitions occur as the result of deliberate perturbations in the concentrations of reactants. Graziani *et al.*, (1977) have demonstrated that addition of flow terms to the FKN dynamic equations introduces multiple steady state solutions for some values of the experimental parameters.

In a flow system, the BR reaction exhibits an unambiguous hysteresis effect (Pacault *et al.*, 1976) (Section II,C,3,d) as the flow rate is varied at constant input stream concentrations of the principal reactants. For a certain range of flow rates, both oscillatory and nonoscillatory states are observed. Transition between these two states is observed to occur at only two values of the flow rate, C_s and C. These values are at the extremes of the range of bistability and differ by over a factor of 10. Which point the transition occurs at as the flow rate is varied is determined by whether the flow rate is entering the bistable region from above or below. Within the bistable region, transitions can be triggered by appropriate addition or removal of I_2. Furthermore, the magnitude of the perturbation in $[I_2]$ required to cause transition is measurable and reproducible. These observations are those expected from a system closely related to the theoretical model of Fig. 4.

C. COMPOSITE DOUBLE OSCILLATIONS

Composite double oscillations appear as regions of oscillation separated by periods of steady reaction. Switching between the two states is as regular as are the oscillations. The phenomena appears over a range of flow rates near a point of transition between oscillatory and nonoscillatory behavior, and it is common to all flow systems studied (see Table II). Figure 15

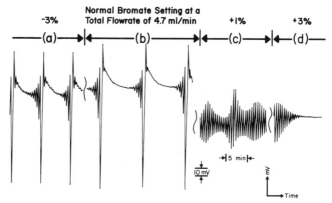

Fig. 15. Effect of $NaBrO_3$ feed rate on composite double oscillation in the BZ reaction: (a) 1.28 ml min^{-1}; (b) 1.32 ml min^{-1} (total flow = 4.7 ml min^{-1}); (c) 1.33 ml min^{-1}; (d) 1.36 ml min^{-1}. [Reproduced by permission from K. Graziani, J. L. Hudson, and R. A. Schmitz, *Chem. Eng. J.* **12**, 9 (1977).]

illustrates composite double oscillation and its sensitivity to flow rate. Boissonade (1976) has analyzed a flow system model that shows composite double oscillation. The fundamental mechanism of their appearance is related to a situation in which transition between a stable and an oscillatory state depends upon a bifurcation intermediate that is produced at different rates during stable and oscillatory phases. Switching between the two states may then occur for certain flow rates because the stationary state concentration of this intermediate may be on a different side of the bifurcation border in one state than in the other. Noyes (1974b) has suggested that in the BZ reaction this critical species may be bromomalonic acid. It is certainly a species involved in Process C.

D. EXCITABILITY

Excitability (Section II,C,3,e) is a property of a stable steady state (Troy and Field, 1977). It has been most clearly observed in the flowing BR reaction (Pacault *et al.*, 1976). Figure 16 illustrates excitability in the BR reac-

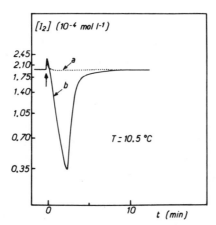

Fig. 16. Excitability in the BR reaction. (a) $[I_2]$ perturbation $= 0.08*10^{-4}$ M; (b) $[I_2]$ perturbation $= 0.10*10^{-4}$ M. Inlet concentrations (different than in the reactor during reaction): $H_2O_2 = 0.033$ M; $KIO_3 = 0.047$ M; $CH_2(COOH)_2 = 5*10^{-3}$ M. Metal ion and $HClO_4$ concentrations were not reported. [Reproduced by permission from A. Pacault, P. Hanusse, P. DeKepper, C. Vidal, and J. Boissonade, *Acc. Chem. Res.* **9**, 438 (1976).]

tion. When the steady state is perturbed by addition of I_2 under a threshold amount, the system rapidly decays back to the steady state. Perturbations that exceed the threshold lead to an excursion of about the same magnitude and shape as the limit-cycle oscillation in the system. The system returns to its steady state after a single excursion. Interestingly, a positive perturbation of $[I_2]$ leads to a negative excursion of $[I_2]$. This illustrates the ability of a negative feedback to destabilize a steady state by producing a counteracting influence to a perturbation much larger than the original perturbation.

E. PHASE SYNCHRONIZATION

Synchronization of phase between two connected, continuously stirred flow reactors has been experimentally observed in the BZ reaction by Marek and Stucil (1975). Phase synchronization among coupled oscillators has often been used to model biological systems (Nicolis and Portnow, 1973). In Marek and Stucil's experiments, the initial oscillatory frequency in each reactor was controlled by the temperature (different) of each reactor. The reactors shared a common perforated wall, and the degree of interaction could be controlled by varying the level of perforation of this common wall. Depending upon the degree of interaction and the initial difference in frequency, phenomena observed included: synchronization of oscillations in both reactors at either a common frequency or at multiples of a common frequency, irregular rhythm splitting (suppression of one cycle) of the lower frequency oscillator, irregular synchronization, and amplitude amplification. Synchronization at multiples of a common frequency has been discussed theoretically by Ruelle (1973). Attempts (Marek and Stucil, 1975) to model this system using the model of Zhabotinskii *et al.* (1971) for the chemical portion failed.

F. Chaos

The unambiguous experimental observation of intrinsic chaos (Section II,C,3,f) is essentially impossible. Systems may approach periodic limit-cycle oscillation asymptotically. Thus one cannot be certain that a system apparently in a chaotic state is not simply on such an approach to periodic oscillation. Furthermore, even if a globally attractive limit cycle exists, finite perturbations can keep a system from actually approaching it. Neither of these situations is truly chaotic in the sense discussed by Rössler (1976a,b). It has also been pointed out (Rössler, 1976b) that these same factors enter into simulation of chaos by numerical integration of appropriate dynamic laws. Finally, this author has been associated with several experimental ventures whose results were chaotic. The cause of this chaos was unknown but certainly unrelated to the phenomenon under discussion here.

However, Schmitz *et al.* (1977) have presented strong evidence for the existence of chaotic states in the BZ reaction. They used ferroin as the metal ion catalyst and ran the reaction in a continuously stirred flow reactor. These authors present three reasons to believe that the aperiodic oscillations that they observed are indeed the result of the existence of intrinsically chaotic states. These reasons are: (1) the flow rates at which the apparently chaotic behavior occurs are bracketed by flow rates at which normal periodic oscillations occur, (2) there is a theoretical basis (Rössler, 1976a) to expect chaotic states in the BZ reaction, and (3) the strong resemblance of the observed aperiodic oscillations to the simulations of Rössler (1976b).

G. Unstirred Flow Systems

When an oscillatory reaction is run in an unstirred, plug flow reactor, the time variable becomes a distance variable. Thus stationary nonmonotonic concentration profiles appear along the length of the reactor. Experiments of this type have been carried out by several groups (Marek and Svobodova, 1975; Bowers *et al.*, 1972). Such a situation can, in principle, supply a leisurely way to study an oscillating reaction at any point in its cycle. However, little advantage has been taken of this technique to date.

V. Unstirred Solutions: Traveling Waves of Chemical Activity

A. PHASE WAVES AND TRIGGER WAVES

Chemical activity need not occur uniformly throughout the volume of an unstirred solution. In a reagent showing temporal oscillation, this fact may manifest itself in as simple a manner as different parts of the reagent oscillating with different frequencies or phase. However, much more complex spatial behavior can occur in such systems. The most striking example of such a phenomenon is traveling waves of chemical activity. Such waves have only been observed in modifications of the BZ reaction, and this section will be entirely devoted to discussion of these systems.

There are two mechanisms by which traveling waves may develop in an unstirred, spatially distributed system. The first leads to so-called phase waves and is not essentially dependent upon diffusion. Phase waves result simply from the existence of a monotonic phase gradient in an oscillatory medium. A mechanism for the establishment of such a gradient will be discussed in later paragraphs. If an oscillator subject to such a phase gradient possesses an identifying feature, e.g., a color change at some point in its cycle, then this feature will appear to move through the reagent as a wave. In fact nothing actually moves, and phase waves will appear to move directly through an impenetrable barrier. These waves are analogous to the spot of light that gives the appearance of motion around a theatre marquee or down the approach to an airport runway. The velocity of a phase wave is inversely dependent upon the phase gradient in the medium. Because there is no limit to the shallowness of the phase gradient, and because nothing is required to physically move, there is no upper limit to the velocity of a phase wave. As we shall see in the next section, however, there is a lower limit in BZ-like reagents.

The second mechanism of traveling wave propagation depends upon the interaction of chemical reaction and diffusion and leads to so-called trigger waves. These waves may appear in either oscillatory or nonoscillatory reagents, although a nonoscillatory reagent must retain excitability (Section II,C,3,e) if trigger waves are to appear. Propagation of trigger waves can occur only if the chemical system undergoes very sudden and very large changes in the concentrations of intermediates and if these changes can be triggered in a volume of reagent by addition or removal of the intermediates involved. The sharp changes may either occur at some point in the cycle of an oscillatory system or they may occur as the result of a perturbation in an excitable system. In a spatially distributed reagent, sharp changes in intermediate concentrations in one volume can lead to steep diffusion gradients. It is movement of intermediates across these gradients that causes the triggerable concentration changes (and their accompanying diffusion gradients) to move through the reagent. Trigger waves propagate through a reagent at a rate determined by the rates of chemical reactions and the diffusive properties of intermediates. Thus, unlike phase waves, they exhibit a unique velocity in a reagent of given composition. A trigger wave can travel microscopic distances no faster than the mean velocity of diffusing intermediates *between collisions*. However, because of the interaction of reaction with diffusion, they can travel macroscopic distances much more rapidly than a *particular* intermediate molecule could diffuse that distance.

In an oscillatory reagent, trigger wave propagation occurs by shifting of the phase of adjacent areas by diffusion of intermediates. When a trigger wave passes through an oscillatory reagent, it sets up a continuous phase gradient. Thus the trigger wave is followed by phase waves traveling at the same velocity. However, the phase waves appear at the frequency of oscillation of the medium. A phase wave cannot move slower than a trigger wave does in a certain oscillatory medium because the phase wave will initiate a trigger wave.

In a both oscillatory and excitable media, a trigger wave must be initiated by some process that causes steep concentration gradients to form. Once these gradients are formed, diffusion perturbs adjacent excitable areas to take the same excursion, and a trigger wave moves through the reagent at the velocity characteristic of reagent composition. Each trigger wave must be initiated individually, and trigger waves propagate through a point in space at a frequency depending upon the rate of initiation. Trigger waves are analogous to the spread of an infectious disease or, more importantly, to the conduction of an electrical impulse through a nerve or muscle. The latter analogy is discussed in depth in this volume by Troy and by DeHaan and DeFelice.

B. TRAVELING WAVES IN THE BELOUSOV–ZHABOTINSKII REACTION

Busse (1969) first observed traveling waves in the cerium catalyzed BZ reaction with malonic acid. In his experiment, an oscillatory reagent was contained in a vertical cylinder and horizontal traveling waves were induced by the formation of a vertical concentration gradient of sulfuric acid. These waves are indubitably phase waves. They are not blocked by an impenetrable barrier (Kopell and Howard, 1973a). The sulfuric acid concentration gradient introduces a frequency gradient as well as a phase gradient, and this leads to an apparent piling up of phase waves at the higher frequency end of the gradient. This system has been studied experimentally by several groups (Herschkowitz–Kaufman, 1970; Rastogi and Yadava, 1972) and Kopell and Howard (1973b, 1974) have treated the mathematics of these phase waves.

A much more interesting system is one first investigated by Zaikin and Zhabotinskii (1970) and quantitatively modified by Winfree (1972). In this reagent Ferroin[4] is used as catalyst with a mixture of malonic acid and bromomalonic acids. As the experiment is generally run, the normally red (reduced form of the catalyst) reagent is spread in a thin layer (~ 2 mm) over a flat surface. Blue waves (oxidized form of the catalyst) with sharp leading edges and diffuse trailing edges may then propagate through the reagent with characteristic velocities of 3–8 mm/min. If the width of the traveling waves is greater than the depth of the reagent, the system is essentially two-dimensional.

Field and Noyes (1972) rationalized the appearance of these waves in terms of the FKN mechanism. Based upon the profiles of the temporal BZ oscillations, they suggested the concentration cross-sections shown in Fig. 17 for the traveling waves in the cerium catalyzed system. Note the very steep concentration gradients in space of Br^- and $HBrO_2$. The wave moves from left to right by consuming Br^-, but leaves behind itself a very much

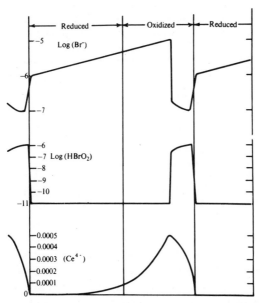

Fig. 17. Approximate concentration profiles for Br^- and $HBrO_2$ (logarithmic scales) and for Ce^{4+} (linear scale) in a typical train of waves moving from left to right. [Reproduced by permission from R. J. Field and R. M. Noyes, *Nature (London)* **237**, 390 (1972).]

higher [Br$^-$] than it is moving into. This area of high [Br$^-$] behind a wave is temporarily refractory to the passage of another wave. Similar waves appear in the cerium (Wolfe and Field, 1976) and manganese (Rastogi *et al.*, 1975) ion catalyzed systems, but they are not as well defined as in the Ferroin catalyzed system.

When spread in a thin layer in contact with the atmosphere, the ferroin catalyzed system can be made to be either oscillatory or excitable by varying the composition of the reagent (Winfree, 1972). Winfree (1974a) has shown that in an oscillatory reagent both phase and trigger waves can exist. Field and Noyes (1974b) and Showalter and Noyes (1976) have investigated the trigger waves in excitable reagents. Waves are normally initiated at so-called pacemaker centers, and they propagate outwards leading to the formation of circular patterns around pacemakers. Pacemakers are apparently surface imperfections such as scratches on the surface, gas bubbles, or dust motes. Repetitive initiation of waves at a particular pacemaker center leads to the formation of a set of concentric rings around it as is shown in Fig. 18. In an oscillatory reagent, the first wave initiated by a pacemaker is a trigger wave that travels at the characteristic frequency of the reagent. If no further trigger waves are generated, then the initial trigger wave will be followed by phase waves traveling at the characteristic trigger wave velocity, but appearing at the frequency of oscillation of the reagent. In a merely excitable reagent, each wave must be initiated individually; and their rate of appearance is controlled by the rate of initiation at the pacemaker.

Showalter and Noyes (1976) have invented a method by which trigger waves can be initiated at will in an excitable reagent from which all adventitious pacemakers have been excluded. They found that pulsing a silver electrode with a positive bias of 2 V for several milliseconds initiates a trigger wave. A continuous negative bias of 0.6 V suppresses trigger wave initiation. The mechanism of initiation is apparently removal of Br$^-$ by reaction (TWI).

$$Br^- + Ag \rightarrow AgBr + e^-. \tag{TWI}$$

These waves travel at the characteristic velocity of the reagent unless [Br$^-$] has not relaxed back down to the steady-state value following the passage of an earlier wave. Elevated [Br$^-$] leads to smaller velocities.

The refractory zone of high [Br$^-$] behind a wave leads to several phenomena besides the simple slowing down of following waves. One of these is the observation that these waves are annihilated not reflected at physical barriers. However, a much more interesting result occurs if a wave is broken by physical shearing (perhaps by momentarily tilting the container). Then the area of oxidation at the broken end curls around the refractory zone to produce spirals which rotate at a velocity characteristic of the reagent. If the

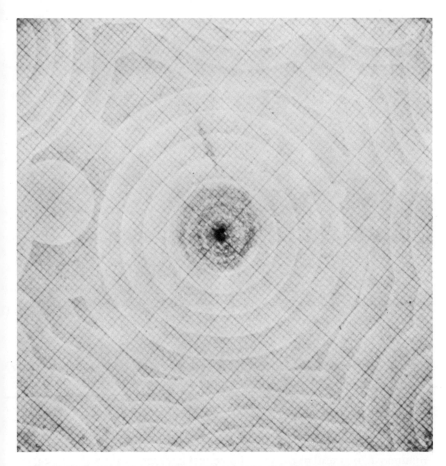

Fig. 18. Typical set of concentric traveling bands of oxidation developing in the thermostated Petri dish. A Nichrome wire at the center is serving as pacemaker: $[H_2SO_4]_0 = 0.26$ M, $[NaBrO_3]_0 = 0.23$ M, [malonic acid]$_0 = 0.024$ M, [bromomalonic acid]$_0 = 0.075$ M, [ferroin]$_0 = 5.45 \times 10^{-3}$ M, temperature $= 25.0 \pm 0.1°C$. [Reproduced by permission from R. J. Field and R. M. Noyes, *J. Am. Chem. Soc.* **96**, 2001 (1974).]

reagent is deep enough to be three-dimensional, rotating scrolls are produced. Winfree (1974b,c) has carefully characterized these rotating structures and modeled (1974d) them in terms of simple excitable kinetics, and he discusses these structures in more depth and relates them to biological systems in the article by Winfree in this volume.

Field and Noyes (1974b) have measured the velocity of trigger waves propagating into virgin territory and determined the relationship between the characteristic velocity of trigger waves in a reagent and the chemical composition of the reagent. They found that trigger wave velocity increases with $([H^+][BrO_3^-])^{1/2}$ and is essentially independent of the concentrations of other components of the reagent. A great deal of effort has gone into rationalizing their results on the basis of the Oregonator reaction–diffusion equations (TW2) and (TW3).

$$\left(\frac{\partial X}{\partial t}\right)_l = D_x \left(\frac{\partial^2 X}{\partial l^2}\right)_t + k_{(O1)} Y - k_{(O2)} XY + k_{(O3)} BX - 2k_{(O4)} X^2 \qquad \text{(TW2)}$$

$$\left(\frac{\partial Y}{\partial t}\right)_l = D_Y \left(\frac{\partial^2 Y}{\partial l^2}\right)_t - k_{(O1)} Y - k_{(O2)} XY + fk_{(O5)} Z \qquad \text{(TW3)}$$

$$\left(\frac{\partial Z}{\partial t}\right)_l = D_Z \left(\frac{\partial^2 Z}{\partial l^2}\right)_t + k_{(O3)} BX - k_{(O5)} Z. \qquad \text{(TW4)}$$

Solution of these equations either analytically or numerically is quite difficult. However, solutions of various simplifications have appeared and are due to Field and Noyes (1974b), Tilden (1974), Murray (1976), Tyson (1976), and Field and Troy (1978). Simplifications normally involve suppression of diffusion terms for one or more variables and suppression of some kinetic terms. This work is reviewed in detail by Troy in this volume. However, in summary, it can be said that most treatments yield the correct proportionality of trigger wave velocity to $([H^+][BrO_3^-])^{1/2}$, but with proportionality constants 10–20 times higher than the experimental.

We (Reusser and Field, 1978) are now able to solve Eqs. (TW2)–(TW4) in their complete form by a numerical integration technique. The procedure uses a discretization of the spatial variable technique based upon the method of lines (Liskovets, 1965) to remove the partial differentials. The resulting large set of ordinary differential equations in time is solved by a version (Hindmarsh, no date) of the Gear numerical integration algorithm designed for use with systems having a banded Jacobian matrix.

We have investigated the behavior of Eqs. (TW2)–(TW4) for the values of $k_{(O1)}$–$k_{(O4)}$ suggested by Field and Noyes (1974a), $D_X = D_Y = D_Z = 1.0 \times 10^{-5}$ cm^2/sec and $f = k_{(O5)} = 1$. As indicated in Fig. 7, these values lead to an unstable steady state and oscillations in the space independent Oregonator. The appearance of both phase and trigger waves is expected under these conditions. We define the spatial phase gradient, ϕ, of a distributed system by Eq. (TW5).

$$\phi = \frac{\text{seconds of Oregonator period}}{\text{linear millimeters}} \qquad \text{(TW5)}$$

For the parameters used, the period of the Oregonator temporal oscillations is about 49 seconds. Thus if a layer of reagent has a spatial phase gradient such that the Oregonator cycle is complete over 49 mm, ϕ is equal to one. A pure phase wave traveling along this gradient would have a velocity of 1 mm/sec.

The initial $(t = 0)$ values of X, Y, and Z assigned to each space point are simply the values of these variables at the appropriate phase of the Oregonator temporal oscillation (Fig. 8). This procedure yields initial waveforms very similar to those exhibited in Fig. 17. Upon the start of integration, waves may initially undergo modest changes if the effects of diffusion are important at the phase gradient chosen. The principal changes are a small general lowering of intermediate concentrations and a lessening of the steepness of the spatial concentration gradients. The stabilized wave then moves across the grid of space points. At sufficiently low values of ϕ (<0.1 sec/mm), changes at adjacent points due to temporal oscillation grossly outweigh changes due to diffusion. Thus waves move at a constant velocity that is exactly the reciprocal of ϕ. Furthermore, in these cases diffusion does not perturb the initial waveform. These are phase waves; their velocity does not change if D_X, D_Y, and D_Z are set equal to zero. As ϕ increases phase wave velocity decreases linearly until the effects of diffusion start to become important at $\phi \cong 0.3$ sec/mm. When this occurs the velocity of the traveling wave is not constant but instead approaches a limiting value as it crosses the grid. This limiting velocity is the same regardless of the initial ϕ greater than 0.3 sec/mm chosen. We presume this velocity to be the trigger wave velocity characteristic of the set of rate constants, concentrations, and diffusion coefficients used. For the values of these parameters used here, the limiting velocity is approximately 52 mm/min. This result is about 9.5 times higher than the velocity actually observed in a reagent of the corresponding composition. The difference probably arises from the neglect of reversibility in the Oregonator and errors in FKN's estimated rate constants.

We are currently expanding our use of this method to further explore the properties of solutions of Eqs. (TW2)–(TW4) and to relate these to the experimental behavior of the BZ traveling waves.

VI. Conclusion

Beyond a critical distance from equilibrium, it is possible for nonmonotonic behavior to appear in systems with nonlinear dynamic laws. In the case of chemical reactions, such behavior appears in the generally small concentrations of intermediate species. In Section II of this chapter various models are used to illustrate the types of nonmonotonic behavior likely to appear in chemical systems. These include: limit-cycle temporal oscillations of intermediate concentrations around either unstable or, occasionally, around stable steady states, multiple steady states and associated hysteresis phenomena, excitability, and, in unstirred reagents, traveling spatial inhomogeneities in intermediate concentrations. Reactions run in a flow reactor often exhibit a wider variety of these nonmonotonic behaviors than does the same reaction run in a batch reactor.

The mechanistically best understood of the chemical reactions that exhibit nonmonotonic behavior are the Belousov–Zhabotinskii (BZ) reaction (the metal ion catalyzed oxidation of organic materials by bromate ion) and the Bray–Liebhafsky (BL) reaction (the iodate catalyzed decomposition of hydrogen peroxide). In a stirred aqueous medium, both of these reactions exhibit temporal oscillations in intermediate concentrations. In Section III the currently accepted mechanisms of these reactions are discussed in detail. These mechanisms are similar in that both involve switching between dominance by one of two nearly independent sets of reactions. Which process is dominant at a particular time depends upon the concentration of one or more "control" intermediates. The independence of these overall processes is guaranteed because one involves only species with even numbers of electrons while the other involves radical species with odd numbers of electrons. Switching between processes, and thus oscillation, occurs because the feedback that would be expected to drive the system to an overall steady state is delayed such that this state may be grossly overshot before feedback becomes effective. Rapid shifts between processes resulting from autocatalysis in the radical process expedite overshooting of the steady state even if feedback is not greatly delayed.

Section III also contains a discussion of several other chemical oscillators that are mechanistically less well understood than the BZ and BL reactions. However, nothing presently known concerning these latter oscillators appears to contradict the above generalizations based upon the two well-understood chemical oscillators. Gas phase chemical oscillators apparently differ from solution phase chemical oscillators in that feedback can arise through the temperature dependence of rate constants or through heat caused turbulence in these low heat capacity systems.

A modification of the Bray–Liebharsky reaction due to Briggs and Rauscher has been shown (Section IV) to exhibit in a flow reactor essentially the entire gamut of possible nonmonotonic behavior discussed in Section II.

In Section V we discuss the traveling waves of chemical activity that appear in the Belousov–Zhabotinskii reaction. We distinguish between the two types of waves, phase and trigger, that may appear in an unstirred reagent in which oscillation is possible. Phase waves (diffusion independent) only appear in a reagent that is oscillatory, but trigger waves (diffusion dependent) may appear in either an oscillatory reagent or in a reagent that is merely excitable. We show that these traveling waves can be quantitatively rationalized by a model based upon the mechanism of the Belousov–Zhabotinskii reaction discussed in Section III.

References

Aldridge, J. (1976). *In* " Handbook of Engineering in Medicine and Biology " (D. Fleming, ed.). Chem. Rubber Co. Press, Cleveland, Ohio.

Barkin, S., Bixon, M., Bar-Eli, K., and Noyes, R. M. (1977). *Int. J. Chem. Kinet.* **9**, 841.

Beck, M., and Varadi, Z. (1976). *Magy. Kem. Foly.* **81**, 519 [*Chem. Abstr.* **84**, 120716μ (1975)].

Belousov, B. P. (1959). *In* " Ref. Radiats. Med. 1958," p. 145. Medgiz, Moscow.

Benson, S. W. (1960). *In* " Foundations of Chemical Kinetics," p. 11. McGraw-Hill, New York.

Betts, R. H., and MacKenzie, A. N. (1951). *Can. J. Chem.* **29**, 655.

Boissonade, J. (1976). *J. Chim. Phys.* **73**, 540.

Bornmann, L., Busse, H., and Hess, B. (1973). *Z. Naturforsch.* **C28**, 514.

Bose, R. L., Ross, J., and Wrighton, M. S. (1977). *J. Am. Chem. Soc.* **99**, 6119.

Bowers, P. G., Caldwell, K. E., and Prendergast, D. F. (1972). *J. Phys. Chem.* **76**, 2185.

Bowers, P. G., and Rawji, G. (1977). *J. Phys. Chem.* **81**, 1549.

Bray, W. C. (1921). *J. Am. Chem. Soc.* **43**, 1262.

Briggs, T. S., and Rauscher, W. C. (1973). *J. Chem. Educ.* **50**, 496.

Busse, H. G. (1969). *J. Phys. Chem.* **73**, 750.

Busse, H. G. (1971). *Nature (London), Phys. Sci.* **233**, 137.

Buxton, G. V., and Dainton, F. S. (1968). *Proc. R. Soc. London, Ser. A.* **304**, 427.

Cesari, L. (1962). " Asymptotic Behavior and Stability Properties in Ordinary Differential Equations," Ch. 1, Erg. Mathem. New Series, Vol. 16. Springer-Verlag, Berlin and New York.

Chance, B., Pye, E. K., Ghosh, A. K., and Hess, B. (eds). (1973). " Biological and Chemical Oscillators." Academic Press, New York.

Clarke, B. L. (1974a). *J. Chem. Phys.* **60**, 1481.

Clarke, B. L. (1974b). *J. Chem. Phys.* **60**, 1493.

Clarke, B. L. (1976a). *J. Chem. Phys.* **64**, 4165.

Clarke, B. L. (1976b). *J. Chem. Phys.* **64**, 4179.

Clarke, B. L. (1975a). *J. Chem. Phys.* **62**, 773.

Clarke, B. L. (1975b). *J. Chem. Phys.* **62**, 3726.

Cooke, D. O. (1976). *J. Chem. Soc. D.*, 27.

Cronin, J. (1977). *SIAM Rev.* **19**, 100.

Dayantis, J., and Sturm, J. (1975). *C. R. Hebd. Seances Acad. Sci., Paris* **280C**, 1447.

Degn, H. (1967). *Nature (London)* **213**, 589.

Degn, H. (1968). *Nature (London)* **217**, 1047.

Demas, J. N., and Diemente, D. (1973). *J. Chem. Educ.* **50**, 357.

DePoy, P. E., and Mason, D. M. (1974). *Faraday Symp. Chem. Soc.* **9**, 47.

Edelson, D., and Noyes, R. M. (1978). *Int. J. Chem. Kinet.* (submitted).

Edelson, D., Field, R. J., and Noyes, R. M. (1975). *Int. J. Chem. Kinet.* **7**, 417.

Edelstein, B. B. (1970). *J. Theor. Biol.* **29**, 57.

Epstein, I., Steel, C., and Popok, S. (1977). Brandeis Univ., Waltham, Mass. (work in progress).

Field, R. J. (1975). *J. Chem. Phys.* **63**, 2289.

Field, R. J., and Noyes, R. M. (1972). *Nature (London)* **237**, 390.

Field, R. J., and Noyes, R. M. (1974a). *J. Chem. Phys.* **60**, 1877.

Field, R. J., and Noyes, R. M. (1974b). *J. Am. Chem. Soc.* **96**, 2001.

Field, R. J., and Noyes, R. M. (1977). *Acc. Chem. Res.* **10**, 214.

108   Richard J. Field

Field, R. J., and Troy, W. C. (1978). *SIAM J. Appl. Math.* (in press).
Field, R. J., Körös, E., and Noyes, R. M. (1972). *J. Am. Chem. Soc.* **94**, 8649.
Field, R. J., Brummer, J. G., and Cerino, M. (1977). (Work in progress).
Franck, U., and Geiseler, W. (1970). *Naturwissenschaften* **58**, 52.
Gear, C. W. (1971). "Numerical Initial Value Problems in Ordinary Differential Equations," Chap. 11. Prentice-Hall, Englewood Cliffs, New Jersey.
Glansdorff, P., and Prigogine, I. (1971). "Thermodynamic Theory of Structure, Stability and Fluctuations." Wiley (Interscience), New York.
Gray, B. F. (1970). *Trans. Faraday Soc.* **66**, 363.
Gray, P., Griffiths, J. F., and Moule, R. J. (1974). *Faraday Symp. Chem. Soc.* **9**, 103.
Graziani, K. R., Hudson, J. L., and Schmitz, R. A. (1977). *Chem. Eng. J.* **12**, 9.
Hanusse, M. P. (1972). *C. R. Hebd. Seances Acad. Sci., Ser. C.* **274**, 1245.
Hastings, S. P., and Murray, J. D. (1975). *SIAM J. Appl. Math.* **28**, 678.
Herbo, C., Schmitz, G., and van Glabbeke, M. (1976). *Can. J. Chem.* **54**, 2628.
Herschkowitz-Kaufman, M. (1970). *C. R. Hebd. Seances Acad. Sci., Ser. C.* **270**, 1049.
Hindmarsh, A. C. (no date). "Solution of Ordinary Differential Equations Having Banded Jacobian." Lawrence Livermore Laboratory. Rep. UCID-30059, Rev. 1. Program available as ACC No. 661 from: Argonne Code Center, 9700 South Cass Ave., Argonne, Illinois 60439.
Hirschfelder, J. O. (1952). *Proc. Natl. Acad. Sci. U.S.A.* **38**, 235.
Hodgkin, A. L. (1967). "The Conduction of the Nervous Impulse." Liverpool Univ. Press, Liverpool, England.
Hopf, E. (1942). *Ber. Verh. Saechs. Akad. Wiss. Liepzig., Math. Naturviss. Kl.* **94**, 3.
Hsü, I-D., and Kazarinoff, N. D. (1976). *J. Math. Anal. Appl.* **55**, 61.
Jacobs, S. S., and Epstein, I. R. (1976). *J. Am. Chem. Soc.* **98**, 1721.
Jwo, J-J., and Noyes, R. M. (1975). *J. Am. Chem. Soc.* **97**, 5422.
Kaner, R. J., and Epstein, I. R. (1978). *J. Am. Chem. Soc.* (submitted).
Kasperek, G. J., and Bruice, T. C. (1971). *Inorg. Chem.* **10**, 382.
Knight, G. C., and Thompson, R. C. (1973). *Inorg. Chem.* **12**, 63.
Kopell, N., and Howard, L. N. (1973a). *Science* **180**, 1171.
Kopell, N., and Howard, L. N. (1973b). *Stud. Appl. Math.* **52**, 291.
Kopell, N., and Howard, L. N. (1974). *Proc. SIAM-AMS* **8**, 1.
Körös, E. (1977). Institute of Inorganic and Analytical Chemistry, L. Eötvös Univ., Budapest, Hungary, private communication.
Körös, E., and Burger, M. (1973). In "Ion Selective Electrodes" (E. Pungor, ed.), p. 191. Hungarian Acad. Sci., Budapest.
Körös, E., Orban, M., and Nagy, Zs. (1973). *J. Phys. Chem.* **77**, 3122.
Lefever, R., and Deneubourg, J. L. (1975). *Adv. Chem. Phys.* **29**, 349.
Leopold, H., and Haim, A. (1977). *Int. J. Chem. Kinet.* **9**, 83.
Liskovets, O. A. (1965). *Diff. Eqs.* **1**, 1308.
Marek, M., and Stucil, I. (1975). *Biophys. Chem.* **3**, 241.
Marek, M., and Svobodova, E. (1975). *Biophys. Chem.* **3**, 263.
Massagli, A., Indelli, A., and Pergola, F. (1970). *Inorg. Chim. Acta.* **4**, 593.
Minorsky, N. (1962). "Nonlinear Oscillations." Van Nostrand, Princeton, New Jersey.
Morgan, J. S. (1916). *J. Chem. Soc., Trans.* **109**, 274.
Murray, J. D. (1976). *J. Theor. Biol.* **56**, 329.
Nemzek, T. L., and Guillet, J. E. (1976). *J. Am. Chem. Soc.* **98**, 1032.
Nicolis, G. (1971). *Adv. Chem. Phys.* **19**, 209.
Nicolis, G. (1975). *Adv. Chem. Phys.* **29**, 29.
Nicolis, G., and Portnow, J. (1973). *Chem. Rev.* **73**, 365.

Nicolis, G., and Auchmuty, G. (1974). *Proc. Natl. Acad. Sci. U.S.A.* **71**, 2748.
Nicolis, G., Malek-Mansour, M., Kitahara, K., and Van Nypelseer, A. (1974). *Phys. Lett. A* **48**, 217.
Nicolis, G., Prigogine, I., and Glansdorff, P. (1975). *Adv. Chem. Phys.* **32**, 1.
Nitzan, A., and Ross, J. (1973). *J. Chem. Phys.* **59**, 241.
Noszticzius, Z. (1977). *J. Phys. Chem.* **81**, 185.
Noyes, R. M. (1974a). *In* "Techniques of Chemistry" (E. S. Lewis, ed.), Vol. VI, Pt. II, Ch. IX. Wiley (Interscience), New York.
Noyes, R. M. (1974b). *Faraday Symp. Chem. Soc.* **9**, 89.
Noyes, R. M. (1976a). *J. Chem. Phys.* **64**, 1266.
Noyes, R. M. (1976b). *J. Chem. Phys.* **65**, 848.
Noyes, R. M., and Jwo, J-J. (1975). *J. Am. Chem. Soc.* **97**, 5431.
Noyes, R. M., and Edelson, D. (1978). (Work in progress).
Noyes, R. M., and Bar-Eli, K. (1977). *Can. J. Chem.* **55**, 3156.
Noyes, R. M., and Field, R. J. (1977). *Acc. Chem. Res.* **10**, 273.
Noyes, R. M., Field, R. J., and Thompson, R. C. (1971). *J. Am. Chem. Soc.* **93**, 7315.
Othmer, H. (1975). *Math. Biosci.* **24**, 205.
Pacault, A., Hanusse, P., DeKepper, P., Vidal, C., and Boissonade, J. (1976). *Acc. Chem. Res.* **9**, 438.
Pauling, L. (1970). "General Chemistry," 3rd ed. p. 501. Freeman, San Francisco.
Perlmutter, D. D. (1972). "Stability of Chemical Reactors." Prentice-Hall, Englewood Cliffs, New Jersey.
Pilling, M. J., and Noyes, R. M. (1978). *J. Chem. Soc., Faraday I* **74** (in press).
Prigogine, I. (1967). "Thermodynamics of Irreversible Processes," Ch. VI. Wiley (Interscience), New York.
Prigogine, I., and Lefever, R. (1968). *J. Chem. Phys.* **48**, 1695.
Rastogi, R. P., and Yadava, K. D. S. (1972). *Nature (London), Phys. Sci.* **240**, 19.
Rastogi, R. P., and Yadava, K. D. S. (1974). *Indian J. Chem.* **12**, 687.
Rastogi, R. P., and Kumar, A. (1976). *J. Phys. Chem.* **80**, 2548.
Rastogi, R. P., Yadava, K. D. S., and Prasad, K. (1974). *Indian J. Chem.* **12**, 974.
Rastogi, R. P., Yadava, K., and Prasad, K. (1975). *Indian J. Chem.* **13**, 352.
Reusser, E., and Field, R. J. (1978). *J. Am. Chem. Soc.* (in press).
Rinker, R. G., Lynn, S., Mason, D. M., and Corcoran, W. C. (1965). *Ind. Eng. Chem. Fund.* **4**, 282.
Rössler, O. E. (1976a). *Z. Naturforsch.* **31a**, 259.
Rössler, O. E. (1976b). *Z. Naturforsch.* **31a**, 1168.
Ruelle, D. (1973). *Trans. N.Y. Acad. Sci.* **70**, 66.
Sattinger, D. (1973). "Topics in Stability and Bifurcation Theory." Springer-Verlag, Berlin and New York.
Schmitz, G. (1973). *J. Chim. Phys. Phys.—Chim. Biol.* **70**, 997.
Schmitz, R. A., Graziani, K. R., and Hudson, J. L. (1977). *J. Chem. Phys.* **67**, 3040.
Sharma, K. R., and Noyes, R. M. (1975). *J. Am. Chem. Soc.* **97**, 202.
Sharma, K. R., and Noyes, R. M. (1976). *J. Am. Chem. Soc.* **98**, 4345.
Shear, D. (1967). *J. Theo. Biol.* **59**, 19.
Showalter, K., and Noyes, R. M. (1976). *J. Am. Chem. Soc.* **98**, 3730.
Showalter, K., and Noyes, R. M. (1978). *J. Am. Chem. Soc.* (in press).
Smith, K., and Noyes, R. M. (1978). Univ. Oregon (in press).
Stroot, P., and Janjic, D. (1976). *Helv. Chim. Acta.* **58**, 116.
Thom, R. (1969). *Bull. Am. Math. Soc.* **75**, 2.
Thompson, R. C. (1971). *J. Am. Chem. Soc.* **93**, 7315.

Thompson, R. C. (1973). *Inorg. Chem.* **12**, 1905.

Tilden, J. (1974). *J. Chem. Phys.* **60**, 3349.

Turro, N. J. (1977). Columbia University, New York (work in progress).

Troy, W. C. (1977). Department of Mathematics, University of Pittsburgh, Pittsburgh, Pa.

Troy, W. C., and Field, R. J. (1977). *SIAM J. Appl. Math.* **32**, 306.

Turner, J. S. (1975). *Adv. Chem. Phys.* **29**, 63.

Turner, J. S. (1976). *Phys. Lett. A* **56**, 155.

Turner, J. S., Mielczarek, E. V., and Mushrush, G. W. (1977). *J. Chem. Phys.* **66**, 2217.

Tyson, J. J. (1975). *J. Chem. Phys.* **62**, 1010.

Tyson, J. J. (1976). "Lecture Notes on Biomathematics" (S. Levin, ed.), Vol. 10. Springer-Verlag, Berlin and New York.

Tyson, J. J. (1977). *J. Chem. Phys.* **66**, 905.

Tyson, J. J. (1978). Preprint.

Tyson, J. J., and Light, J. C. (1973). *J. Chem. Phys.* **59**, 4164.

Uppal, A., Ray, W. H., and Poore, A. B. (1974). *Chem. Eng. Sci.* **29**, 967.

Weisbuch, G., Salomon, J., and Atlan, H. (1975). *J. Chim. Phys. Phys.—Chim. Biol.* **72**, 71.

Weston, R. E., and Schwarz, H. A. (1972). "Chemical Kinetics." Prentice-Hall, Englewood Cliffs, New Jersey.

Winfree, A. T. (1972). *Science* **175**, 634.

Winfree, A. T. (1974a). *Faraday Symp. Chem. Soc.* **9**, 38.

Winfree, A. T. (1974b). *In* "Mathematical Problems in Biology, Lecture Notes in Biomathematics" (P. van den Driessche, ed.), Vol. 2, p. 241. Springer-Verlag, Berlin and New York.

Winfree, A. T. (1974c). *In* "Science and Humanism: Partners in Human Progress" (H. Mel, ed.). Univ. California Press, Berkeley, California.

Winfree, A. T. (1974d). *Proc. SIAM-AMS* **8**, 13.

Wolfe, H., and Field, R. J. (1976). (Unpublished results).

Yamazaki, I., Fujita, M., and Baba, H. (1976). *Photochem. Photobiol.* **23**, 69.

Yang, C. H., and Berlad, A. L. (1974). *J. Chem. Soc., Faraday Trans. I* **70**, 1661.

Yatsimirskii, K. B., Grageror, L. P., Tikhonova, L. P., Kiprianova, L. N., Zakrevskaya, L. N., and Levit, A. F. (1976). *Dokl. Akad. Nauk. SSSR* **226**, 1133.

Zaikin, A. N., and Zhabotinskii, A. M. (1970). *Nature (London)* **225**, 535.

Zhabotinskii, A. M. (1964a). *Dokl. Akad. Nauk. SSSR* **157**, 392.

Zhabotinskii, A. M. (1964b). *Biofizika* **9**, 306.

Zhabotinskii, A. M., Zaikin, A. N., Korzukhin, M. D., and Kreitser, G. P. (1971). *Kinet. Catal.* **12**, 516.

Population Cycles

G. Oster
Division of Entomology and Parasitology, College of Natural Resources
University of California, Berkeley, California

A. Ipaktchi
Department of Mechanical Engineering
University of California, Berkeley, California

I. Introduction

Populations whose abundance is apparently periodic have been the inspiration for a large part of ecological theory. Volterra's well known models for cyclical predator–prey patterns provided an explanation that seemed satisfactory for awhile; however, as more and better data became available, and as ecologists became more sophisticated mathematically, it became clear that the phenomenon of population oscillations was considerably more complicated than it had appeared to Volterra. In this chapter we will review some of these complications by focussing on a very detailed set of experiments and a simple model that has been proposed several times in the literature to explain the data.

II. The Population Balance Equation

For the purposes of population modeling we shall define the "state" of an individual as that collection of characteristics that affect its contribution to the population growth. These characteristics include the individual's age, physiological state, and genetic constitution. Therefore, we can represent an individual's state as a point $z(a, x, y)$ in a phase space whose coordinates are age, a, physiological condition, x (nutritional state, etc.), and genotype, y (e.g., frequencies of genes affecting reproductive potential). Then the state of the population can be summarized in a density function $n(t, a, x, y)$ specifying the number of individuals in each volume element dz, and where

$$N(t) = \int_V n \, dz \tag{2.1}$$

is the total population under consideration at time t. The density $n(\cdot)$ must obey a conservation law of the form

$$\partial n/\partial t = -\nabla \cdot (gn) - \text{death rate} + \text{birth rate} \tag{2.2}$$

where

$$\frac{dz}{dt} = g = \left(\frac{da}{dt} = 1, \frac{dx}{dt}, \frac{dy}{dt}\right)$$

is the growth rate (or phase velocity) in a, x, and y (Oster, 1976).

The subject of theoretical population dynamics is largely the study of Eq. (2.2). For our purposes here we shall make the dramatic simplification of neglecting the dependence of the population dynamics on its physiological (x) and genetic (y) composition, and focus on the effects of age structure alone. Even so, we shall find that Eq. (2.2) is capable of a surprisingly complicated repertoire of dynamical behaviors.

Considering only the age density function $n(t, a)$, the balance equation takes the form

$$(\partial n/\partial t) + (\partial n/\partial a) = -\mu n + \delta(a) \int_0^\infty da' b(t, a') n(t, a') \tag{2.3}$$

where $\mu(\cdot)$ is the per capita death rate, $b(t, a')$ is the per capita birth rate, and $\delta(a)$ is the delta function at $a = 0$. In addition to their dependence on age, $\mu(\cdot)$ and $b(\cdot)$ are influenced by the population density $N(t)$; it is this dependence that is responsible for the dynamic complexity we shall encounter below. With appropriate assumptions of smoothness and boundedness on the constitive relations $\mu(\cdot)$ and $b(\cdot)$, Eq. (2.3) can be shown to possess a unique solution (Gurtin and MacCamy, 1974).

Several reduced descriptions of Eq. (2.3) are commonly employed in population modeling. The most drastic simplification is to integrate over all ages, thus reducing (2.3) to an ordinary differential equation for the total population $N_T(t)$:

$$dN_T/dt = \int_0^\infty bn \, da' - \int_0^\infty \mu n \, da'. \tag{2.4}$$

With appropriate assumptions on $b(\cdot)$ and $\mu(\cdot)$, one can recover most of the commonly employed lumped parameter population models in the literature. However, this is too drastic a simplification for most purposes, since it averages out the maturation delay between birth and sex reproduction. This delay, coupled with the nonlinearity in $b(\cdot)$ and/or $\mu(\cdot)$, turns out to be crucial for understanding many features of population behavior.

III. Oscillations in a Simple Ecosystem

A. THE EXPERIMENTS

The Australian entomologist A. J. Nicholson maintained in his laboratory a population of sheep blowflies for several years under carefully controlled conditions and obtained an accurate census of the various life stages of the flies. Two selections from his data are shown in Fig. 1a,b (Nicholson, 1954 and 1957). Although the flies lived in a constant environment with a steady food supply, nevertheless, the population waxed and waned in a more or less regular fashion with a period of between 35 to 40 days. In the absence of any external stimulus to entrain the population, he concluded that the capacity to oscillate can be endogenous to a population, a reflection of its own intrinsic regulatory processes.

In addition to this basic cycle, one can discern several other features of the dynamics that beg explanation: (a) the population peaks appear to have a regular "fine structure" consisting of a double, or occasionally triple, peak; (b) there appears to be a "subharmonic" superimposed on and spanning three periods of the basic limit cycle; (c) the peaks exhibit a degree of irregularity suggesting stochastic effects are present despite the carefully controlled experimental conditions. In Fig. 1c,d we have plotted Fourier spectra of the data; these will prove useful later in comparing the data with model simulations.

B. A "SIMPLE" MODEL

A detailed study of this data has been performed, and an elaborate model of the form (2.2) constructed. This will be reported elsewhere (Auslander *et al.*, 1977); here we shall sketch how the major features of the data can be explained by a much simpler model. One major purpose of our study will be to demonstrate that an accurate fit to the data may well be impossible *in principle*, due to the nature of the population balance equations. The consequences of this fact may have implications for modeling of other physical and biological phenomena as well.

The model we shall study here is obtained from (2.3) by first averaging the birth and death rates over all ages; and then integrating over the adult age

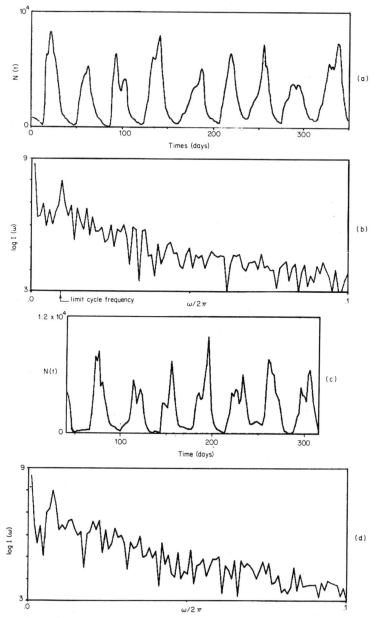

Fig. 1. (a)–(d) Nicholson's experimental data on sheep blowflies and corresponding frequency spectra. The oscillation in the adult population was generated by a daily supply of 0.4 g and 0.5 g of ground liver in (a) and (c), respectively. [Due to the linear relation of food supply and the total population (Nicholson, 1957). Fig. (c) is rescaled to match the 0.4 g experiment.]

classes to obtain†

$$dN(t)/dt = l(\alpha)\bar{b}[N(t - \alpha)]N(t - \alpha) - \bar{\mu}N(t) \tag{3.1}$$

where $N(t) = \int_\alpha^\infty n(t, a)\,da$ is the total adult population and α is the age at which adult flies emerge from their pupal cases. $l(\alpha)$ is the fraction of eggs surviving to adulthood, and $\bar{b}(\cdot)$ and $\bar{\mu}(\cdot)$ are the age-averaged birth and death rates. Thus we have simplified the age structure in Eq. (2.3) to a maturation delay. In the following development we shall assume $\bar{\mu}$ is nearly constant, but we shall retain the crucial dependence of the birthrate on the adult population density. In particular, independent experiments show that the form of $\bar{b}(\cdot)$ can be approximated by a function of the form (Auslander *et al.*, 1977; Wu, Oster, and Guckenheimer, 1977)

$$\bar{b}[N(t)]N(t) = b_m e^{-N(t)/K} \cdot N(t). \tag{3.2}$$

The shape of Eq. (3.2) is sketched in Fig. 2. The crucial feature is the critical point at N_c; when the adult population exceeds this density, the competition

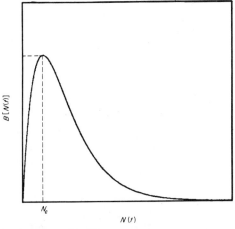

Fig. 2. A typical birth curve, $B[N(t)]$. The presence of a critical point $dB[N_c]/dt = 0$ (a "hump") is the critical features controlling the dynamics.

for food so reduces the average adult fecundity that the birth rate of the entire population begins to fall off. The shape of this curve must summarize all of the complexities associated with birth and food competition, and so we shall have to sacrifice some degree of realism in this approximation. The equation we will study here is

$$dN(t)/dt = \bar{b}e^{-N(t-\alpha)/K} \cdot N(t - \alpha) - \bar{\mu}N(t) \tag{3.3}$$

† The averaged birth and death may be functions of the moments of the age distribution, or other variables of the population. A set of differential equations can be derived to describe the variation of these moments in time. A detailed discussion of this can be found in Auslander *et al.* (1978).

where $\hat{b} = b_m l(\alpha)$. This may be conveniently rescaled by letting $y = N/K$ and measuring time in units of α, $\tau = t/\alpha$, obtaining

$$dy(\tau)/d\tau = by(\tau - 1)e^{-y(\tau - 1)} - \mu y(\tau) \tag{3.4}$$

where $b = \alpha l(\alpha)b_m$ and $\mu = \alpha\bar{\mu}$. The structure of the model can be represented schematically as shown in Fig. 3.

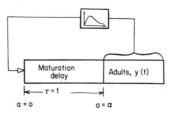

Fig. 3. A population may be viewed as a delay line with birth represented as a positive feedback.

C. Dynamic Behavior of the Model

In order to study the dynamics of Eq. (3.4) we shall compare its behavior with an even simpler model. Referring back to Eq. (2.3), let us make the following approximation. Assume that the adults lay their eggs in one burst when they reach age $a = a^*$. Then the equation for the birthrate $n(t, o) \equiv B(t)$ reduces to

$$B(t) = \bar{b}(N)l(a^*)B(t - a^*). \tag{3.5}$$

Choosing a^* as our time, step (3.5) is a difference equation of the form

$$x_{t+1} = F(x_t) \tag{3.6}$$

where $F(\cdot)$ has the shape sketched in Fig. 4. x_t is the number of newborn (eggs) in generation t [which is proportional to the adult population by the mortality factor $l(a^*)$].

While the model (3.6) is far removed from the original model, it retains the two essential features which control the dynamical behavior: the maturation delay and the "1-hump" nonlinearity in the birthrate feedback. Therefore, we can expect that the dynamical behavior of the difference equation

$$x_{t+1} = x_t e^{-rx_t} \tag{3.7}$$

will capture some of the qualitative features of the delay equation (3.4).

A complete discussion of Eq. (3.6) can be found in May and Oster (1976) and Guckenheimer, Oster, and Ipaktchi (1977). Here we shall only cite the relevant features.

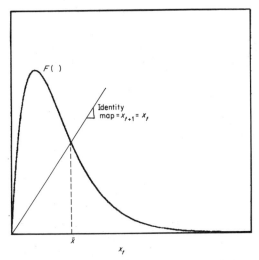

Fig. 4. The discrete generation difference equation $x_{t+1} = F(x_t) = x_t e^{r(1-x_t)}$.

(i) For values of the birthrate parameter r below a critical value r_1, the fixed point $0 = F(0)$ at $\bar{x}_0 = 0$ is the only stable equilibrium.

(ii) As r increases past r_1, the origin becomes unstable and is replaced by the fixed point at $\bar{x} = F(\bar{x})$ as the only stable equilibrium (cf. Fig. 4). $\bar{x}$ remains stable until a second critical value, r_2, is reached, whereupon it too becomes unstable.

(iii) For values $r_1 < r < r_2$ the orbit X_t exhibits a 2-point cycle between the two fixed points of the "period-2 map" $X_2 = F(F(X_2))$. As r is increased past r_2 the period-2 orbit bifurcates to a period-4 cycle. This cycle is stable for values of r between $r_2 < r < r_4$.

(iv) This sequence of bifurcations continues as r is increased, generating a progression of cycles whose period is 2^k ($k = 0, 1, 2, \cdots$) until a critical value, $\hat{r}$, is reached, whereupon cycles of odd period commence to occur. By the time the parameter r reaches the value r_3 where a period-3 cycle has made its appearance, cycles of *all* periods are present, and (even more surprising) there are *aperiodic* orbits, i.e., trajectories which wander "chaotically," never settling into any repeating sequence of points. This sequence of dynamic behaviors can be represented graphically on a bifurcation diagram, as shown in Fig. 5.

(v) For the one-dimensional difference equation (3.7) only one stable periodic orbit exists for each value of r (although many unstable orbits may coexist). Moreover, the chaotic orbits have measure zero, i.e., they exist only for isolated initial conditions. Thus chaotic dynamics may not be a structurally stable feature of such equations. However, if instead of considering

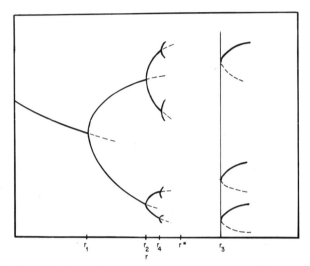

Fig. 5. The hierarchy of stable fixed points which are produced as the parameter of Eq. (3.7) increases. Each pair arises by a bifurcation as a previous point becomes unstable.

only a single age class, we discretize equations (2.3) into two age classes, then the dynamics are much richer. That is, the two-dimensional generalization of Eq. (3.7) is

$$\mathbf{F} : \mathbb{R}^2 \to \mathbb{R}^2$$

$$(x_t, y_t) \mapsto (b_1 e^{-b_2(x_t + y_t)} \cdot y_t, lx_t) = (x_{t+1}, y_{t+1}) \qquad (3.8)$$

where y_t is the adult population, x_t is the preadults, and l is the preadult mortality. This equation is also analyzed in Guckenheimer, Oster, and Ipaktchi (op. cit.); its bifurcation diagram exhibits at least four types of dynamical behavior: (1) stable equilibria, (2) one or more stable periodic orbits, whose periods depend on the parameter values, and which arise by a bifurcation mechanism somewhat different from the one-dimensional case, (3) "almost-periodic" orbits: trajectories which appear nearly periodic, but which have superimposed on them a low-frequency "beatlike" cycle, and (4) regions of stable "chaos": i.e., bounded regions which are attracting, but within which the orbit wanders in an apparent random fashion.

An important feature of the chaotic dynamics is its "graded" character. As the bifurcation parameter is varied, an existing periodic orbit gradually begins to dissolve into randomness, as if noise of increasing strength were being imposed on the system. Thus, the "degree of chaos" varies more or less continuously as the birthrate parameter changes. For this reason the Fourier analysis of trajectories is a useful tool in understanding the dynamics. Periodic and almost periodic orbits show up as definite peaks in the frequency domain, while aperiodic behavior is reflected by the broadening of the peaks and the appearance of higher frequency "noise" components, and low-frequency beatlike components.

For the difference equations discussed above the mechanism by which the various dynamic regimes appear and disappear is covered by the Hopf Bifurcation Theorem (Marsden and McCracken, 1976). A precise statement of this crucial result is given in the Appendix; heuristically the theorem says the following. Given a discrete dynamical system $\mathbf{F}_r : \mathbb{R}^n \to \mathbb{R}^n$ which contains an adjustable parameter, r, we investigate how the qualitative character of the orbits generated by $\mathbf{F}$ changes as we smoothly change the parameter. Assume that $\bar{x}$ is a stable periodic point of period p $(p = 1, 2, \cdots)$, i.e., a fixed point of $\mathbf{F}^p(\bar{x}) = \bar{x}$, where $\mathbf{F}^p$ denotes the pth iterate of $\mathbf{F}$. The local stability of the orbit (or equilibrium, if $p = 1$) is governed by the eigenvalues $\lambda_k(r)$, $(k = 1, 2, \cdots, n)$ of the derivative (Jacobian) of $\mathbf{F}^r(\cdot)$ at $\bar{x}$, $DF^p(\bar{x})$. So long as the spectrum of $\mathbf{F}^p$ lies wholly within the unit circle, $|\lambda_k(r)| < 1$ $(k = 1, 2, \cdots, n)$, $\bar{x}$ is locally attracting. As r is changed, the eigenvalues move in the complex plane; when a single pair of roots passes through the unit circle the orbit becomes locally unstable in that eigendirection. The content of the Hopf theorem is that if the nonlinear terms omitted are "strong enough" so that the equilibrium $\bar{x}$ is still attracting when leading eigenvalue lies on the unit circle $|\lambda_1(r^*)| = 1$, then in the neighborhood of the critical parameter value, r^*, there is a stable periodic orbit. This "vague attractor" condition on $\mathbf{F}^p(\cdot)$ can be checked, in principle, by calculating a complicated formula involving the higher order expansion terms for $F^p(\cdot)$ (Marsden and McCracken, op. cit.). In practice, however, this calculation is generally not practical except in the simplest cases. Therefore, one is generally forced to employ numerical simulation in the vicinity of the parameter values of interest. This places a large premium on working closely with experimental data to avoid endless searches of parameter space.

The nature of the periodic orbit that is born when the critical parameter value, r^*, is passed depends on where the first eigenvalue passes through the unit circle. The period-doubling bifurcation discussed in connection with the difference equations occurs when $\lambda_1(r)$ passes -1. Bifurcations through $\lambda(r) = +1$ create new fixed points rather than splitting existing ones. This is the mechanism that created the period -3 orbit in the one-dimensional map (2.3). For maps on R^1 these are the only types of bifurcation. In higher dimensions, where complex eigenvalues are admissible, bifurcations through $|\lambda(r^*) = 1|$ at an angle $\theta = 2\pi/k$ will usually result in orbits of period k (especially if k is near a small integer). The bifurcation that produces the chaotic attractor is a more complicated phenomenon.

All of these results for difference equations (maps on $\mathbb{R}^n$) have a counterpart in continuous time systems (differential equations on $\mathbb{R}^n$) (cf. Marsden and McCracken, op. cit.). The bifurcation point occurs when an eigenvalue of the linearized system passes through the imaginary axis to the right half-plane. In our subsequent discussion we shall apply this principle to interpret our simulations of the model (3.4). Although this is an infinite dimensional system, it turns out that the theorem also applies in this situation; that is, as we increase the number of age classes no pathologies arise in the limit (Marsden, 1973).

Equation (3.4) has two equilibria at $\bar{y}_1 = 0$ and $\bar{y}_2 = \ln(b/\mu)$. The linearizations about these equilibria are

$$dy_1/d\tau = by_1(\tau - 1) - \mu y_1(\tau) \tag{3.9}$$

$$dy_2/d\tau = \mu[1 - \ln(b/\mu)]y_2(\tau - 1) - \mu y_2(\tau). \tag{3.10}$$

The characteristic equations governing the stability of $\bar{y} = 0$ in Eq. (3.9) and $\bar{y} = \ln(b/\mu)$ in Eq. (3.10) are obtained by assuming a solution of the form $y \simeq e^{s\tau}, s \in \mathbb{C}$:

$$s + \mu - be^{-s} = 0 \tag{3.11}$$

$$s + \mu - \mu[1 - \ln(b/\mu)]e^{-s} = 0. \tag{3.12}$$

From Eq. (3.11) we see that the origin becomes unstable when $b > \mu$. In Fig. 6 we have sketched the root locus of Eq. (3.12) as b is increased while $\mu = $ constant. There infinite sets of roots, S_1 and S_2; as b increased S_1 moves to the left and S_2 to the right as shown. The last left-going root in S_1 crosses the imaginary axis at $b = \mu$, at which point the fixed point $\bar{y}_2$ becomes

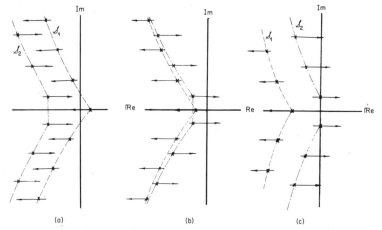

(a)   (b)   (c)

Fig. 6. Root locus of the characteristic equation (3.12) for fixed μ and positive increasing b. The equation has two sets of roots, S_1 and S_2; S_1 moves to the left and S_2 to the right as b increases. Case (a) y_{e_1} stable and y_{e_2} unstable. Case (b) y_{e_1} unstable and y_{e_2} stable. Case (c) y_{e_1} unstable, y_{e_2} is about to become unstable and give birth to a limit cycle.

locally stable. At the same time, Eq. (3.11) shows that $\bar{y}_1 = 0$ becomes unstable. In the parameter interval $\mu < b < b_1$, all the roots of Eq. (3.12) are in the left half-plane, and consequently $\bar{y}_2$ is locally attracting. As b is increased past b_1, two conjugate roots of Eq. (3.12) cross the imaginary axis. At this point a stable limit cycle is born by a mechanism analogous to that described for the difference equation. The bifurcation frequency as well as the critical parameter value b can be calculated by setting $s = i\omega$ in Eq. (3.12), which yields

$$\omega = -\mu[1 - \ln(b/\mu)] \sin \omega \qquad (3.13)$$

$$\mu = -\mu[1 - \ln(b/\mu)] \cos \omega. \qquad (3.14)$$

Dividing Eq. (3.13) by Eq. (3.14) yields the equation for ω:

$$\omega = -\mu \tan \omega. \qquad (3.15)$$

The smallest solution of Eq. (3.15) corresponds to the bifurcation frequency. The bifurcation value b_1 is obtained by solving Eqs. (3.13) and (3.14); yielding

$$b_1 = \mu e^{(\cos \omega - 1)/\cos \omega}. \qquad (3.16)$$

The vague attractor condition guaranteeing the stability of the incipient limit cycle cannot be easily computed explicitly (cf. Marsden and McCracken, op. cit.). However, since we shall be working directly with Nicholson's data, we can verify numerically that in the neighborhood of the bifurcation value the system is indeed stably cyclic, with the approximate period given by Eq. (3.15).

Equations (3.15) and (3.16) describe a family of curves in the (b, μ) plane. These curves divide the parameter space into regions, each region specifying the number of conjugate poles in the right half-plane. These regions are shown in Fig. 7. So far we have discussed only the situation that occurs when the first root crossing creates the periodic orbit. Later we shall see the significance of the subsequent root crossings.

The attracting periodic orbit created by the first bifurcation is really a limit cycle in an infinite-dimensional function space, since the system state is not a point in $\mathbb{R}^n$, but rather a function $y(t)$ on the unit interval, $t \in [0, 1]$. Nevertheless, it is still instructive for some purposes to plot the orbit in a "phase plane" $(y(t), \dot{y}(t))$ as if it were a second-order system. A typical trajectory subsequent to the first bifurcation is shown in Fig. 8a. Because this is a projection of the orbit, it is possible for the orbit to cross itself, or even overlap itself for some time interval. In Fig. 8b the crossings of the orbit on itself correspond to the "double peak" shown in the time plot in Fig. 8a. Recall that Nicholson's data also exhibited this type of "high-frequency" behavior (cf. Fig. 1). A consideration of the model (3.4) shows that we can

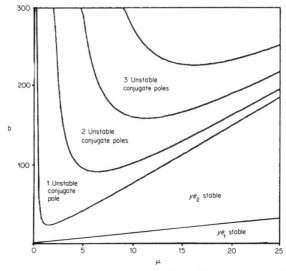

Fig. 7. Regions in the parameter space of Eq. (3.10) specifying the number of conjugate poles in the right half-plane.

link this phenomenon to the nature of the nonlinearity in the birth rate function shown in Fig. 2. Because the birth rate $B(N)$ has a critical point, $(dB/dN)(N_c) = 0$, if $N(t)$ is periodic, then B will have a double " period " (or rather a double hump) providing the death rate is sufficient to lower the population level below N_c at the cycle's minimum. This is sketched in Fig. 9. Since the amplitude of the limit cycles is quite large, we can safely presume that this explains the high-frequency double-peaks observed in the data. (Indeed, this effect can be used as a diagnostic for this type of density dependent birth rate.)

Next, let us increase the parameter, b, beyond the first bifurcation point, b_1. The limit cycle first increases in amplitude then, at $b = b_2$, a second bifurcation takes place. This bifurcation doubles the period; Fig. 10a, b, and c shows time plot, phase plane, and frequency spectrum. For finite-dimensional systems this second bifurcation occurs when the leading eigenvalue of the Poincare map (or period-1 map) crosses the unit circle (Marsden and McCracken, op. cit.). For infinite-dimensional system such as Eq. (3.4), it is not practical to compute the Poincare map; therefore, we need an approximate method for estimating b_2. In Fig. 11 we have superimposed on the parameter plane the locus of the bifurcation boundaries as determined numerically. For moderate death rates we see that the root locus of the second crossing pair provides an excellent approximation to the second bifurcation. The approximation is worse for the third root pair, and the

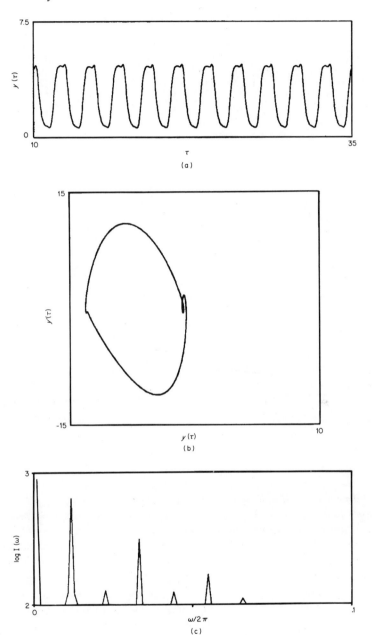

Fig. 8. A typical response of Eq. (3.4) after the first bifurcation ($b = 80$ and $\mu = 6$). (a) Time trajectory, (b) phase plane, and (c) Fourier spectrum.

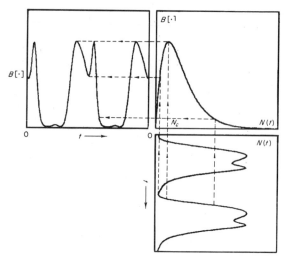

Fig. 9. A decrease in the population level beyond the critical point N_c, will cause a "double hump" in the birth, $B[N(t)]$, which will be reflected in the next generation as a "double peak" in the adult population.

fourth root locus is essentially useless. While this procedure is surely not a universally valid approximation scheme, for this system (and probably many others of similar type) it provides a simple method for estimating the first few bifurcation loci in the parameter plane.

In Fig. 12a and b we have plotted time plots and phase plane for the third bifurcation as the birth rate b is increased along the line shown in Fig. 11. Below the time plot is a Fourier spectrum showing the frequency components contained in the orbit (Fig. 12c). We see that the dynamic behavior of this system parallels that of the difference equations discussed above. The second bifurcation produces additional trajectory crossings in the phase plane corresponding to peaks with more high-frequency components. A low-frequency component also appears analogous to the "period-4" orbit generated by the second bifurcation of the difference equation (3.7). Further bifurcations complicate the dynamics further, and this is reflected in the Fourier spectrum by additional frequency components, both higher and lower.

It is clear that for sufficiently high birth rates b the trajectory appears so chaotic that is it difficult to distinguish the deterministic dynamics from stochastic effects. In Fig. 13a and b we have plotted such a chaotic orbit along with its spectrum. In Fig. 14a and b we have plotted the orbit, and its spectrum, corresponding to a value of b just after the first bifurcation, but where we have superimposed a small amount of Gaussian noise on the parameter b. A comparison of the orbits and the spectra of each illustrates

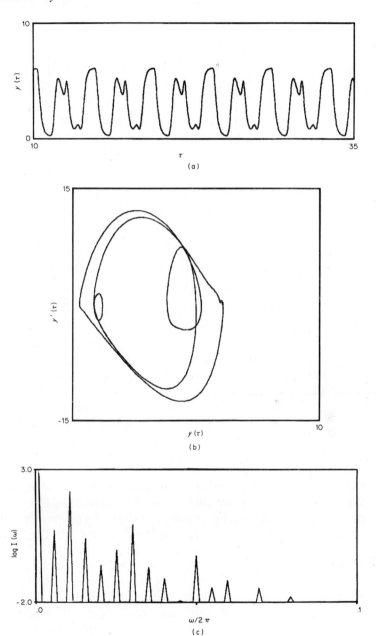

Fig. 10. Response of the system after the second bifurcation ($b = 100$, $\mu = 6$). (a) Time trajectory, (b) phase plane, and (c) frequency power spectrum.

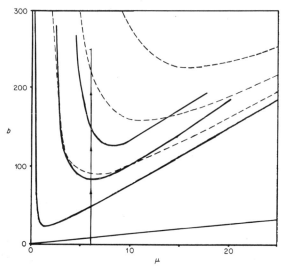

Fig. 11. Bifurcation boundaries (solid lines) superimposed on the results obtained by the linear analysis (dashed lines).

the difficulty of distinguishing deterministic from stochastic events in such circumstances. There may be differences on the power spectra between the two cases, or other statistical tests, which might serve as diagnostics for separating random from deterministic influences, but this is an area which is largely unexplored at present.

There are several important features of the dynamic behavior of the model in the chaotic regime. (1) The orbit is extraordinarily sensitive to small perturbations. That is, just like the difference equations, nearby orbits diverge rapidly. This is because the chaotic attractor is composed mostly, or entirely, of "hyperbolic points": points with at least one unstable eigendirection (that is, if we linearize about any point in the chaotic attractor there will always be at least one unstable eigenvalue). Because of this property even the slightest amount of noise, which must always be present (even in the computer as roundoff error), will be amplified in time. (2) The transient response is generally quite long, so that it takes many cycles before the system approaches its asymptotic, steady-state behavior. Because of these properties it is not possible to fit the model to data, *even in principle*. That is, even if the experiments were so controlled as to be virtually free of both experimental error and random influences, nevertheless, it would be quite impossible for the model to "track" the data indefinitely since small errors in setting the model's initial conditions will be amplified exponentially. Therefore, it is necessary to view the deterministic trajectory in the chaotic

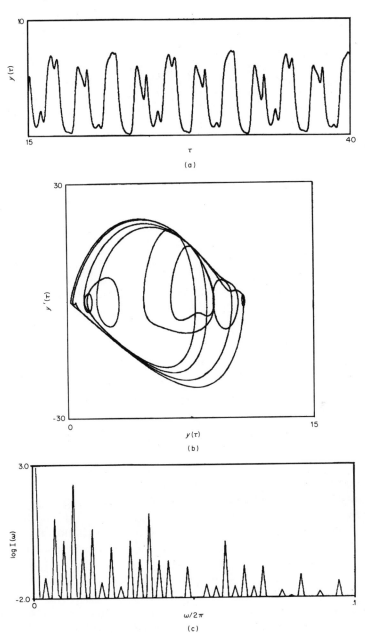

Fig. 12. Response of the system after the third bifurcation ($b = 120$, $\mu = 6$). (a) Time trajectory, (b) phase plane, and (c) frequency power spectrum.

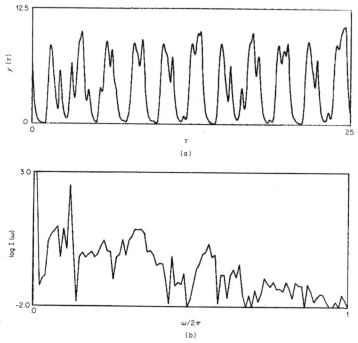

Fig. 13. A chaotic trajectory of Eq. (3.4) for $b = 180$ and $\mu = 6$. (a) Time trajectory and (b) frequency power spectrum.

regime as being analogous to a sample path in a random process. We cannot demand a pointwise fit of model and data, regardless of how small is the experimental error. Rather, we must be satisfied by qualitative agreement, coached in statistical terms. This is a radical shift in viewpoint for those accustomed to distinguishing between deterministic and stochastic dynamical processes. A deterministic system can look quite random.

Recalling our original discussion of Nicholson's experiments, let us now see how well our "simple" model can match the data. Fig. 15a shows Eq. (3.4) for parameter values corresponding to the second bifurcation (values well within the experimental range) along with the experimental data. The trajectory in Fig. 15a shows the qualitative features of (1) the major limit cycle, (2) the high-frequency multiple peaks, and (3) the low-frequency components. A better way to assess the match is to plot the difference between the model and data. This is shown in Fig. 15b, along with the frequency spectra in Fig. 15c. From this we see that there are systematic discrepancies that can only be ascribable to the oversimplified nature of the model. A more elaborate model does much better (Auslander *et al.*, op. cit.); this model includes the complete age-structure and nutritional state of the

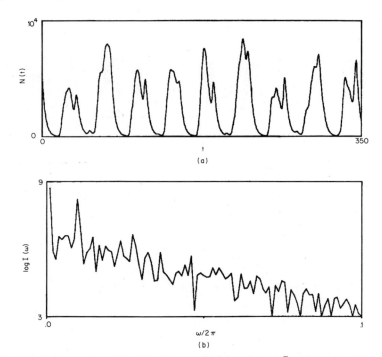

Fig. 14. (a) A trajectory of Eq. (3.3) with a stochastic parameter. $\bar{b} = 12 + n$, $\bar{\mu} = 0.3$, $\alpha = 16$, and $k = 500$ where n is filtered (autocorrelated) white noise. (b) Frequency power spectrum of the trajectory.

population, the age dependencies of the birth and death rates as well as a dynamical structure for the reproductive mechanism. These elaborations greatly complicate the model, and one might try replacing them by some randomness in the parameters as we did in Fig. 14. A comparison of the spectra with that of the data again reveals the difficulty of distinguishing cleanly between randomness, chaos and real experimental uncertainty.

The model (3.4) has the general form $dx(\tau)/d\tau = f[x(\tau), x(\tau - 1)]$. Simulations of other models of this type show dynamics similar to those we have described, providing $f(\cdot)$ is a continuous, differential function satisfying:

(i) $f(o, o) = 0$

(ii) $f(o, y) < -B$ for some $B > 0$

(iii) $(\partial f/\partial x)(x, y) \leq 0$ $\forall(x, y) \in \mathbb{R}_x \mathbb{R}$

(iv) $(\partial f/\partial y)(x, y) < 0$ $\forall x \in \mathbb{R}$

$\forall y$ such that $|y| > \delta$ for some $\delta > 0$.

A chemical analog of such a system is discussed in Oster and Perelson (in preparation).

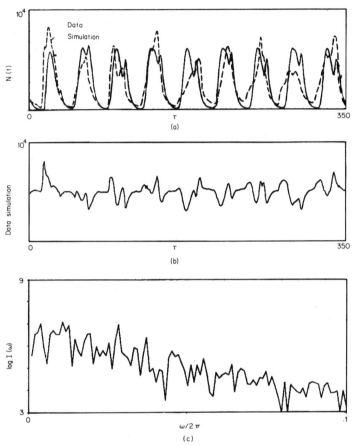

Fig. 15. (a) Comparison of a response of Eq. (3.3) for the parameter values of $\bar{b} = 12$, $\bar{\mu} = 0.28$, $K = 500$, and $x = 16$. (b) Difference between the data and the simulation. (c) Frequency power spectrum of the difference.

IV. Discussion

A dynamical system model can exhibit four classes of temporal behavior: (1) stable equilibria; (2) stable periodic orbits; (3) almost-periodic motion; (4) stable chaotic motion: i.e., apparently random motion within a bounded attracting region. Typically, by tuning one or more parameters, one can cause the system to pass through each of these regimes. The focus of our discussion has been on the difficulties that arise when using such a system as a model for a particular set of data. In practice, it is generally quite impossible to distinguish between the different types of asymptotic behavior. This is because ecological data are seldom extensive and detailed enough to distinguish between long cycles and chaotic dynamics, nor, when random influences are present, to distinguish between weakly stable equilibria, noisy limit cycles, and chaotic motion (cf. Poole, 1977). Fourier spectra of orbits can be of some assistance in this regard, but there is frequently a level of resolution in modeling ecological phenomena that cannot be overcome by refining empirical techniques. This calls for some revisions in our modeling philosophies when approaching data where time delays (or high dimensionality) and nonlinear effects are important. Such considerations are likely to apply to other natural phenomena as well.

Appendix

THE HOPF BIFURCATION

The Hopf Bifurcation Theorem was originally proved for bifurcations of singular points of vector fields (Hopf, 1942; Marsden and McCracken, 1976). The theorem was generalized by Ruelle and Takens (1971) to deal with bifurcations of diffeomorphisms about a fixed point or a periodic orbit.

Theorem: Let $\Phi_\mu : \mathbb{R}^2 \to \mathbb{R}^2$ be a one-parameter family of diffeomorphisms satisfying the following:

(i) The origin is a fixed point of Φ_μ.
(ii) For $\mu < 0$, the spectrum of Φ_μ at the origin is contained in $\{z \in \mathbb{C} \,|\, |z| < 1\}$.
(iii) For $\mu = 0$ the spectrum of Φ_μ at the origin has two isolated eigenvalues $\lambda(\mu)$ and $\bar{\lambda}(\mu)$ with multiplicity one and $|\lambda(\mu)| = 1$. The remaining part of the spectrum is contained in $\{z \in \mathbb{C} \,|\, |z| < 1\}$.
(iv) $(d/d\mu)(|\lambda(\mu)|)_{\mu=0} > 0$.
(v) $\arg \lambda(\mu) \neq (k/l) \cdot 2\pi$ for all $k, l \leq 5$.

Then there is a μ-dependent change of coordinates bringing Φ_μ to the following form in polar coordinates:

$$\Phi_\mu(r, \alpha) = ((1 + \mu)r - f_1(\mu)r^3, \; \alpha + f_2(\mu) + f_3(\mu)r^2) + \text{terms of order } r^5.$$

If $f_1(\mu) > 0$, then there is a continuous one parameter family of invariant attracting circles of Φ_μ, one for each $\mu \in (0, \varepsilon)$, for ε small enough.

References

Auslander, D., Guckenheimer, J., Ipaktchi, A., and Oster, G. (1977). To be published.
Auslander, D., Guckenheimer, J., Ipaktchi, A., and Oster, G. (1978). To be published.
Guckenheimer, J., Oster, G., and Ipaktchi, A. (1976). *J. Math. Biol.* **4**, 101–147 (1977).

Gurtin, M., and MacCamy, R. (1974). *Arch. Ratl. Mech. Anal.* **54**, 281–300.

Hopf, E. (1942). *Akad. Wiss. Leipzig* **94**, 1–2.

Marsden, J. (1973). *Bull. Amer. Math. Soc.* **79**, 537–541.

Marsden, J., and McCracken, M. (1976). *In* "Applied Mathematical Sciences," Vol. 19. Springer-Verlag, New York.

Nicholson, A. J. (1954). *Austr. J. Zool.* **2**, 9–65.

Nicholson, A. J. (1957). *Cold Spring Harbor Symp. Quant. Biol.* **22**, 153–173.

May, R., and Oster, G. (1976). *Am. Nat.* **110**, 573–599.

Oster, G. (1976). *In* "Modern Modelling of Continuum Phenomena" (R. DiPrima, ed.). Amer. Math. Soc., Providence, R.I.

Poole, R. (1977). *Ecology* **58**, 210–213.

Ruelle, D., and Takens, F. (1971). *Commun. Math. Phys.* **20**, 167–192.

Mathematical Modeling of
Excitable Media in Neurobiology and Chemistry

William C. Troy

Department of Mathematics,
University of Pittsburgh, Pittsburgh, Pennsylvania

I. Introduction

Since the early 1950s an increasing amount of attention has been focused on excitable media in which the interaction of chemical or electrical excitation with diffusion may cause the formation of complex spatial structures. Much of this interest is due to the early work of Turing (1952) who suggested that such a phenomenon may play an important role in morphogenisis.

In this article we discuss mathematical models of two excitable systems, namely nerve impulse transmission and the Belousov–Zhabotinskii chemical reaction.

In the next section we introduce the mathematical theory of nerve conduction developed by the British physiologists Hodgkin and Huxley (1952).

Their model consists of a system of nonlinear partial differential equations that describe the transmission of an electrical signal along the giant axon of the squid. Significant progress in the mathematical analysis of the Hodgkin–Huxley equations has been made during the past decade. We shall present several of these results with particular emphasis on the threshold of excitation, traveling wave solutions, and oscillatory phenomena. We shall also make a few remarks on a modification of the Hodgkin–Huxley system which has been developed to model the electrical activity in cardiac Purkinje fibers.

Section III begins with a brief description of the remarkable temporal and spatial phenomena observed in the Belousov–Zhabotinskii chemical reaction. Under a variety of experimental conditions this reaction exhibits self-sustained temporal oscillations, solitary and periodic plane waves, and rotating spiral waves of chemical activity. We shall discuss the mathematical model developed by Field and Noyes (1974a). Of particular interest is the reaction diffusion mechanism that causes the formation of spatial patterns. A mathematical analysis shows that this excitation phenomenon is similar to that observed in the Hodgkin–Huxley system and we point out these similarities as we proceed.

II. Nerve Impulse Transmission

A. The Hodgkin–Huxley Theory

The development of modern neurophysiology began in the early 1950s with the experiments performed by Hodgkin and Huxley on the giant axon of the squid "Loligo." Their investigations culminated with a complex system of nonlinear partial differential equations (Hodgkin and Huxley, 1952) that describes the propagation of an electrical signal along the nerve axon. The Hodgkin–Huxley (HH) model has the form

$$\partial^2 v/\partial x^2 = \{g_K(n)(v - v_K) + g_{Na}(m, h)(v - v_{Na}) + g_l(v - v_l)\} + c(\partial v/\partial t) \quad (2.1)$$

$$\partial m/\partial t = \gamma_m(v)(m_\infty(v) - m) \quad (2.2)$$

$$\partial n/\partial t = \gamma_n(v)(n_\infty(v) - n) \quad (2.3)$$

$$\partial h/\partial t = \gamma_h(v)(h_\infty(v) - h). \quad (2.4)$$

The right-hand side of (2.1) denotes the total membrane current. The unknown functions are $v(x, t)$, $m(x, t)$, $n(x, t)$, and $h(x, t)$ where $t \geq 0$ and $-\infty < x < \infty$. v is the potential difference across the membrane at time t and position x along the axon. The function $g_{Na}(m, h)$ represents the conductance of the membrane to sodium ions and was assumed by Hodgkin and

Huxley to have the form $g_{Na}(m, h) = \bar{g}_{Na} m^3 h$. Here $\bar{g}_{Na}$ is a constant, and the variables m and h control the sodium conductance activation and deactivation, respectively. Equations (2.2) and (2.4), which describe the change in time of m and h, were developed from experimental data. Similarly, $g_K(n)$ denotes the conductance of the membrane to potassium ions and has the form $g_K(n) = \bar{g}_K n^4$, a function of the single variable n, the so-called potassium activation variable. Equation (2.3) describes the rate of change of n with time and was also derived from experimental data. The chloride conductance $\bar{g}_l$ was found to remain at a constant level. The membrane capacitance c is given the experimentally determined value of 1 μF. Finally, the parameters v_K, v_{Na}, and v_l represent the equilibrium constants for potassium, sodium, and chloride, respectively. Through ingenious experimentation Hodgkin and Huxley measured all of the physical parameters in their model and the reader is referred to the original paper (Hodgkin and Huxley, 1952) for further details.

1. Numerical Computations

Numerical evidence indicates that the nerve impulse takes the form of a traveling wave solution of (2.1)–(2.4). That is, (v, m, n, h) is taken to be a function of the single variable, $\tau = (x/\theta) + t$, where θ is the speed of propagation. Substituting this single variable into (2.1)–(2.4), we obtain the system of first-order ordinary differential equations:

$$dv/d\tau = w \tag{2.5}$$

$$dw/d\tau = \theta[w + (g_K(n)(v - v_K) + g_{Na}(m, h)(v - v_{Na}) + g_l(v - v_l))] \tag{2.6}$$

$$dm/d\tau = \gamma_m(v)(m_\infty(v) - m) \tag{2.7}$$

$$dn/d\tau = \gamma_n(v)(n_\infty(v) - n) \tag{2.8}$$

$$dh/d\tau = \gamma_h(v)(h_\infty(v) - h). \tag{2.9}$$

Numerical computations show that (2.5)–(2.9) have one equilibrium state, $\pi_0 = (0, m_\infty(0), n_\infty(0), h_\infty(0))$. The existence of a single equilibrium solution is physically reasonable since the nerve axon appears to have only one rest state. A solitary traveling wave solution is a nonconstant bounded solution $\pi(\tau) = (v(\tau), m(\tau), n(\tau), h(\tau))$, of (2.5)–(2.9), that satisfies the boundary condition $\lim_{\tau \to \pm \infty} \pi(\tau) = \pi_0$ (i.e., the axon is at rest before and after the passage of an impulse). Such a solution is also called a homoclinic orbit for (2.5)–(2.9). Numerical computations by Fitzhugh and Antosiewicz (1959) and Huxley (1959) indicate that there are two values of θ, the wave speed, for which (2.5)–(2.9) has a homoclinic orbit. We discuss the notion of stability of these traveling wave solutions in Section IV.

2. The Space Clamped Axon

Much of the experimental data obtained by Hodgkin and Huxley resulted from an investigation of the space clamped axon. Briefly, a long thin silver wire (called the space clamp) is inserted into the interior of the axon and along its entire length. This eliminates the dependence of v upon x, and (2.1)–(2.4) are reduced to a system of ordinary differential equations, namely

$$dv/dt = -(g_K(n)(v - v_K) + g_{Na}(m, h)(v - v_{Na}) + \bar{g}_l(v - v_l)) \qquad (2.10)$$

$$dm/dt = \gamma_m(v)(m_\infty(v) - m) \qquad (2.11)$$

$$dn/dt = \gamma_n(v)(n_\infty(v) - n) \qquad (2.12)$$

$$dh/dt = \gamma_h(v)(h_\infty(v) - h). \qquad (2.13)$$

The numerical computations of FitzHugh (1960) show that (2.10)–(2.13) support a threshold of excitation, that is, there is a value $v^* > 0$ such that:

(i) if m, n, and h are initially at rest and $v(0) > v^*$, then v quickly rises to a peak value of approximately 100 mV before the system returns to equilibrium (Fig. 1).

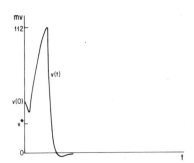

Fig. 1. When the initial depolarization in membrane potential exceeds its threshold value, v^*, an action potential occurs and v quickly attains a maximum value of approximately 112 mV before returning to rest.

(ii) if m, n, and h are initially at rest and $0 < v(0) < v^*$, then the system merely decays to rest and $v(t)$ does not exceed v^* (Fig. 2).

The spiked-shaped solution described above in (i) and shown in Fig. 1 is called an action potential. Note that the presence of the space clamp causes the action potential to occur simultaneously at all points x along the axon.

FitzHugh (1960) further investigated (2.10)–(2.13) and found that during an action potential the variables n and h change slowly with respect to v and m. Thus, setting $n \equiv n_\infty(0)$ and $h \equiv h_\infty(0)$, their resting values, FitzHugh

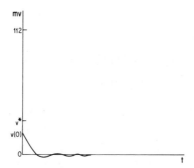

Fig. 2. An initial depolarization in membrane potential that does not exceed threshold quickly decays to rest and no action potential is produced.

investigated the reduced two-dimensional system

$$dv/dt = -(g_K n_\infty(0)(v - v_K) + g_{Na}(m, h_\infty(0))(v - v_{Na}) + \bar{g}_l(v - v_l)) \tag{2.14}$$

$$dm/dt = \gamma_m(v)(m_\infty(v) - m). \tag{2.15}$$

FitzHugh's machine calculations show that (2.14) and (2.15) have three equilibrium solutions, $(0, m_\infty(0))$, $(\mu_0, m_\infty(\mu_0))$, and $(\lambda_0, m_\infty(\lambda_0))$, where $0 < \mu_0 < \lambda_0$ (Fig. 3).

In Fig. 3 we plot the trajectories that solutions of Eqs. (2.14) and (2.15) follow as t increases. This is called the phase diagram of solutions. Note that there is a value $v^* > 0$ such that:

(i) if $m(0) = m_\infty(0)$ and $0 < v(0) < v^*$, then $(v, m) \to (0, m_\infty(0))$ as $t \to \infty$ and v does not exceed v^*;

(ii) if $m(0) = m_\infty(0)$ and $v(0) = v^*$, then $(v, m) \to (\mu_0, m_\infty(\mu_0))$ as $t \to \infty$;

(iii) if $m(0) = m_\infty(0)$ and $v(0) > v^*$, then $(v, m) \to (\lambda_0, m_\infty(\lambda_0))$.

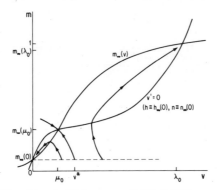

Fig. 3. For $n \equiv n_\infty(0)$ and $h \equiv h_\infty(0)$ the system (2.10)–(2.13) becomes the two-dimensional system (2.14) and (2.15). We plot the phase space of trajectories of solutions of Eqs. (2.14) and (2.15).

The value v^* acts as a threshold of excitation and the points $(0, m_\infty(0))$ and $(\lambda_0, m_\infty(\lambda_0))$ are called the rest state and excited state, respectively.

3. Existence of Traveling Wave Solutions

The numerical results obtained by FitzHugh for the space clamped axon have proved useful in the analysis of the unclamped model [Eqs. (2.1)–(2.4)]. Thus, using FitzHugh's computation as a guide, Hastings (1976) made several reasonable assumptions on the functions and parameters appearing in Eqs. (2.5)–(2.9). Under these assumptions he proved that the system has at least one homoclinic orbit. This solution represents a solitary traveling wave solution of Eqs. (2.1)–(2.4). Under a different set of assumptions on the functions and parameters in Eqs. (2.5)–(2.9), Carpenter (1976) proves that the system has periodic solutions as well as solitary traveling wave solutions with several spikes.

4. Stability of Traveling Waves

An important unsolved problem is to rigorously determine the stability of the traveling wave solutions found by Hastings and Carpenter. That is, does a solution of Eqs. (2.1)–(2.4) that is initially "close" to a traveling wave actually converge to a traveling wave solution as t increases? Mathematically, we formulate the problem as follows. We first specify the solution at time $t = 0$ by $(v(x, 0), m(x, 0), n(x, 0), h(x, 0)) = (v_0(x), m_0(x), n_0(x), h_0(x))$, where $v_0(x), m_0(x), n_0(x), h_0(x)$ are given functions of x corresponding to the initial state of the nerve. Next, let $(\bar{v}(\tau), \bar{m}(\tau), \bar{n}(\tau), \bar{h}(\tau))$ denote a known traveling wave solution of Eqs. (2.1)–(2.4) where $\tau = (x/\theta) + t$. Note that $(\bar{v}(\tau + l), \bar{m}(\tau + l), \bar{n}(\tau + l), \bar{h}(\tau + l))$ is also a traveling wave solution of Eqs. (2.1)–(2.4) for each real number l. We wish to show that if

$$\left| (v_0(x), m_0(x), n_0(x), h_0(x)) - \left(\bar{v}\left(\frac{x}{\theta}\right), \bar{m}\left(\frac{x}{\theta}\right), \bar{n}\left(\frac{x}{\theta}\right), \bar{h}\left(\frac{x}{\theta}\right) \right) \right|$$

is small, uniformly with respect to $x \in (-\infty, \infty)$, then there is a value l such that $(v(x, t), m(x, t), n(x, t), h(x, t))$ approaches

$$(\bar{v}(\tau + l), \bar{n}(\tau + l), \bar{n}(\tau + l), \bar{h}(\tau + l)) \qquad \text{as} \quad t \to \infty,$$

uniformly with respect to x. More precise definitions and elementary results on this problem may be found in Hastings (1975), Evans (1972a,b), and Sattinger (1977).

B. THE FITZHUGH–NAGUMO SIMPLIFICATION

Due to the complexity of the original Hodgkin–Huxley equations, Nagumo *et al.* (1962) and FitzHugh (1960) have developed a simpler system that abstracts the main qualitative features of (2.1)–(2.4). The so-called FitzHugh–Nagumo equations have the form

$$V_{xx} - V_t = F(V) + R \tag{2.16}$$

$$R_t = \varepsilon(V + a - bR) \tag{2.17}$$

where $V(x, t)$ denotes the potential difference across the membrane at time t and distance x along the axon and $R(x, t)$ represents the slowly changing variables n and h. The variable m in the Hodgkin–Huxley model has no counterpart here.

As with Eqs. (2.1)–(2.4), one investigates Eqs. (2.16) and (2.17) for the existence of traveling wave solutions by assuming that V and R are functions of the single variable $\tau = x - ct$. This reduces Eqs. (2.16) and (2.17) to the system of ordinary differential equation

$$V' = W \tag{2.18}$$

$$W' = F(V) + R - cW \tag{2.19}$$

$$R' = \varepsilon/c(bR - a - V). \tag{2.20}$$

Nagumo *et al.* (1962) took $F(V)$ to be cubic, set $b = 0$, and computed two values of the wave speed c for which Eqs. (2.18)–(2.20) have a homoclinic orbit. As with the earlier calculations for the Hodgkin–Huxley model, it appears that the faster traveling wave solution is stable as a solution of the partial differential equations while the slower pulse appears unstable.

FitzHugh (1961) considers the case $b \neq 0$, $F(V) = (V^3/3) - V$ and verifies the previous numerical results of Nagumo, Arimoto, and Yoshizawa.

Casten *et al.* (1975) use singular perturbation techniques in analyzing another form of Eqs. (2.16) and (2.17), namely,

$$\partial^2 V/\partial x^2 = (\partial V/\partial t) + V(V - a)(1 - V) + Z \tag{2.21}$$

$$\partial Z/\partial t = \varepsilon V. \tag{2.22}$$

Here $\varepsilon > 0$ and $0 < a < 1$. If W and Z are taken to be functions of the single variable $\tau = x + \theta t$, then Eqs. (2.21) and (2.22) become

$$V'' = V(1 - V)(V - a) + \theta V' \tag{2.23}$$

$$Z' = \varepsilon/\theta V. \tag{2.24}$$

Using formal asymptotic expansions, Casten, Cohen, and Lagerstrom construct homoclinic orbits of Eqs. (2.23) and (2.24) for two different values of the wave speed θ.

Rigorous proofs for the existence of homoclinic orbits of Eqs. (2.23) and (2.24) have recently been obtained. Conley (1975) has proved the existence of at least one solitary traveling wave as well as recurrent solutions. Subsequently, Hastings (1976b) proved that there must be a second wave speed for which Eqs. (2.23) and (2.24) have a homoclinic orbit, and for each wave speed between these two values there is a periodic solution.

1. A Piecewise-Linear Model

Replacing the cubic nonlinearity in Eqs. (2.21) and (2.22) with a Heaviside step function, Rinzel and Keller (1973) and Rinzel (1973) have investigated the simpler system

$$\partial V/\partial t = \partial^2 V/\partial x^2 - f(V) - W \tag{2.25}$$

$$\partial W/\partial t = bV. \tag{2.26}$$

Here $b > 0$ and $f(V) = V - H(V - a)$ where $0 < a < \frac{1}{2}$. The function $f(V)$ represents a piecewise-linear current–voltage relation.

Rinzel and Keller prove the existence of solitary traveling wave solutions of (2.25) and (2.26) for two different speeds of propagation. In addition, they find that for each wavelength greater than some minimum value there are two or more periodic solutions with different speeds. They also discuss the stability of these solutions.

C. Repetitive Firing

Repetitive firing has been observed in the space-clamped axon in response to an application of a constant external current to the membrane. This experimental preparation, called the current-clamped axon, is modeled with the system

$$dv/dt = I - (g_K(n)(v - v_K) + g_{Na}(m, h)(v - v_{Na}) + \bar{g}_1(v - v_1)) \tag{2.27}$$

$$dm/dt = \gamma_m(v)(m_\infty(v) - m) \tag{2.28}$$

$$dn/dt = \varepsilon\gamma_n(v)(n_\infty(v) - n) \tag{2.29}$$

$$dh/dt = \varepsilon\gamma_h(v)(h_\infty(v) - h). \tag{2.30}$$

Here I denotes the constant applied current. Following Hastings (1976a), we have inserted a small positive ε into Eqs. (2.29) and (2.30) to indicate that n and h change slowly with respect to v and m. The insertion of ε also makes the system more tractable to analytical treatment.

Equations (2.27)–(2.30) have been studied both numerically and analytically by several authors and we briefly describe these results below. Cole *et al.* (1955) consider a nerve axon initially at rest [i.e., $(v(0), m(0), n(0), h(0)) = (0, m_\infty(0), n_\infty(0), h_\infty(0))$]. They discontinuously raise I from 0 to 6.8 μA and compute what appears to be a large amplitude stable periodic solution. Physically, this corresponds to an infinite train of action potentials.

Cooley *et al.* (1966) follow a suggestion by FitzHugh that repetitive firing in the current clamped axon may be associated with the instability of the steady-state solution of Eqs. (2.27)–(2.30). For each $I \geq 0$ their calculations show that the system has a unique steady-state solution that we denote by $\pi_I = (V_I, m_\infty(V_I), n_\infty(V_I), h_\infty(V_I))$. Next, let A_I be the Jacobian of the right-hand side of Eqs. (2.27)–(2.30) evaluated at π_I. For $0 \leq I < 9.8$ all the eigenvalues of A_I have negative real parts and the steady-state solution is locally asymptotically stable. However, as I passes through 9.8 from below, the real parts of two of the eigenvalues become positive, the steady-state becomes unstable and small perturbations from π_I may grow with time and begin to oscillate. The steady state remains unstable for larger values of I until, at $I = 154\ \mu\text{A}$, the real parts of all the eigenvalues once again become negative. In addition, Cooley *et al.* (1966) have shown numerically that for each I in the interval $6.8 \leq I < 9.8$, where the steady state is stable, there is an unstable periodic solution of smaller amplitude than the earlier computed stable periodic orbit.

Sabah and Spangler (1970) completed the numerical computations. They found that as I increases from 9.8 to 154 the amplitudes of the large stable periodic orbits shrink until the oscillations disappear altogether at $I = 154\ \mu\text{A}$.

1. Simplification of the Current Clamped Model

FitzHugh (1961) simplified Eqs. (2.27)–(2.30) by introducing the system

$$du/dt = I + w - (u^3/3) + u \tag{2.31}$$

$$dw/dt = \rho(a - u - bw) \tag{2.32}$$

where

$$0 < b < 1, \qquad 1 - (2b/3) < a < 1, \qquad 0 < \rho < 1. \tag{2.33}$$

Here u denotes the potential difference at time t across the membrane of the current clamped axon, and w represents the slowly varying functions n and h. The restrictions on a and b guarantee that Eqs. (2.31) and (2.32) have exactly one rest state, (u_I, w_I), for each $I \in (-\infty, \infty)$.

It is apparent that Eqs. (2.31) and (2.32) closely resemble the Van der Pol oscillator (Hale, 1969). Thus one might apply standard phase plane techniques to obtain the existence of periodic solutions at values of I for which the steady-state solution is unstable. However, phase-plane techniques can be applied only to two-dimensional systems. Since our goal is to obtain information about the four-dimensional Hodgkin–Huxley model, we have chosen the more analytical approach of bifurcation theory that applies to arbitrarily many equations. In addition, much more detailed stability information can be obtained from the bifurcation approach than with phase-plane analysis. The theory of bifurcation of periodic solutions is a very powerful tool in analyzing systems of nonlinear differential equations. We present below a brief outline of the theory and then show how it easily applies to nerve conduction models.

2. The Hopf Theory of Bifurcation of Periodic Solutions

First we require a few definitions. Let $I^* \in (-\infty, \infty)$ be fixed, and let $\pi^* = (u_{I*}, w_{I*})$ denote the steady-state solution of Eqs. (2.31) and (2.32). We say that there occurs a bifurcation of small amplitude periodic orbits from the steady-state solution as I passes through I^* if there is a set of the form $\mathcal{U} = (I^*, I^* + \delta)$, or $\mathcal{U} = (I^* - \delta, I^*)$, or $\mathcal{U} = \{I^*\}$ such that for each $I \in \mathcal{U}$ there is a periodic solution $\pi_I(t) = (u_I(t), w_I(t))$ whose amplitude approaches zero as I approaches I^* through $\mathcal{U}$. If $\mathcal{U}$ is of the form $\mathcal{U} = (I^* - \delta, I^*)$, then we say that the direction of bifurcation is to the left of I^*, while if $\mathcal{U} = (I^*, I^* + \delta)$, then the direction of bifurcation is to the right of I^*. In the special case that $\mathcal{U} = \{I^*\}$, then the steady-state solution π^* is surrounded by a family of periodic solutions and π^* acts as a center for Eqs. (2.31) and (2.32). This notion of bifurcation of periodic orbits extends directly to a system of any dimension.

With this definition of bifurcation of periodic solutions in mind, we now present the Hopf theorem as stated by Hsu and Kazarinoff (1976). The Hopf theorem applies to an autonomous system of differential equations of the form

$$\dot{x} = A(\mu)x + F(x, \mu).$$

Here $x = \text{col}(x_1, \ldots, x_n)$, $F = \text{col}(f_1, \ldots, f_n)$, $F(0, \mu) \equiv 0$, and $F_x(0, \mu) \equiv 0$. Also, $A(\mu)$ is a real $n \times n$ matrix whose entries depend on the parameter μ. Suppose that $f_i (1 \le i \le n)$ is a real-valued analytic function on $G \times (-c, c)$ where $c > 0$ and G is an open connected domain in R^n. Also let $A(\mu)$ be analytic in μ for $\mu \in (-c, c)$ with exactly two purely imaginary eigenvalues $\pm i\beta_0$ at $\mu = 0$ and whose continuous extensions $\alpha(\mu) \pm i\beta(\mu)$ satisfy the conditions

$$\alpha(0) = 0, \qquad \beta(0) = \beta_0 > 0, \qquad d\alpha(\mu)/d\mu|_{\mu=0} \ne 0.$$

Under these assumptions we state the following theorem.

Theorem (*Hopf, 1942*): There is a value $\zeta_0 > 0$ such that for each $\zeta \in (-\zeta_0, \zeta_0)$ there is a periodic solution $p(t, \zeta)$, with period $T(\zeta)$, of the equation $\dot{x} = A(\mu)x + F(x, \mu)$ where the parameter ζ is related to μ by a functional relation $\mu = \mu(\zeta)$ such that $\mu(0) = 0$, $p(t, 0) \equiv 0$, and $p(t, \zeta) \ne 0$ for all sufficiently small $\zeta \ne 0$. Moreover, $\mu(\zeta)$, $p(t, \zeta)$, and $T(\zeta)$ are analytic at $\zeta = 0$, and $T(0) = 2\pi/\omega_0$. These periodic solutions exist for exactly one of three cases: either only for $\mu > 0$, or only for $\mu < 0$, or only for $\mu = 0$. Furthermore, for each $L > T(0)$ there exist $a > 0$, $b > 0$ such that if $|\mu| < b$ there is no nonconstant periodic solution of period less than L, besides the bifurcating periodic solutions $p(t, \zeta)$ that lie entirely within $\{x \mid |x| < a\}$.

3. Application of Bifurcation Theory to Nerve Models

In order to apply the Hopf theorem to FitzHugh's model we first compute the Jacobian of Eqs. (2.31)–(2.32) at the steady-state solution (u_I, w_I) and show that the matrix has purely imaginary eigenvalues for some critical value of I.

The Jacobian is given by

$$B_I = \begin{pmatrix} -u_I^2 + 1 & 1 \\ -\rho & -b\rho \end{pmatrix} \tag{2.34}$$

for each $I \in (-\infty, \infty)$. Necessary and sufficient conditions that B_I have purely imaginary eigenvalues are that

$$\text{tr}(B_I) = 0 \quad \text{and} \quad \det(B_I) > 0. \tag{2.35}$$

From (2.33) and (2.35) it follows that

$$\text{tr}(B_I) = 0 \Leftrightarrow u_I = \pm(1 - b\rho)^{1/2}. \tag{2.36}$$

It is not difficult to show that u_I is a strictly decreasing function of I, and that $\lim_{I \to \pm\infty} u_I = \mp\infty$. Thus there are exactly two values, $I_1 < I_2 < 0$ such that $u_{I_1} = -(1 - b\rho)^{1/2}$, $u_{I_2} = (1 - b\rho)^{1/2}$ and therefore $\text{tr}(B_{I_1}) = 0$ and $\text{tr}(B_{I_2}) = 0$. Next, from (2.33) and (2.34) it follows that $\det(B_I) = u_I^2 b\rho + \rho(1 - b) > 0$ for each $I \in (-\infty, +\infty)$. The conditions in (2.35) are satisfied if $I = I_1$ or $I = I_2$. Thus the matrices B_{I_1} and B_{I_2} have purely imaginary eigenvalues $\pm i\omega_1$ and $\pm i\omega_2$, respectively, for some nonzero values of ω_1 and ω_2.

Next, let $\alpha_1(I) \pm i\omega_1(I)$ denote the unique C^1 extensions of $\pm i\omega_1$ to the eigenvalues of B_I for values of I in a small open interval containing I_1. Similarly, let $\alpha_2(I) \pm i\omega_2(I)$ denote the unique C^1 extensions of $\pm i\omega_2$. From the equation

$$\det\left(B_I - (\alpha_1(I) + i\omega_1(I))\begin{pmatrix} 1 & 0 \\ 0 & 1 \end{pmatrix}\right) = 0$$

it easily follows that

$$d\alpha_1/dI|_{I=I_1} \neq 0 \quad \text{and} \quad d\alpha_2/dI|_{I=I_2} \neq 0.$$

In order to place Eqs. (2.31) and (2.32) into the setting required for an application of Hopf's theorem we let $I = \mu + I_1$ (or $I = \mu + I_2$). Then as μ

passes through 0 it is clear from the above analysis that the hypotheses of Hopf's theorem are satisfied and we obtain Theorem 1.

Theorem 1: From each of the steady-state solutions (u_{I_1}, w_{I_1}) and (u_{I_2}, w_{I_2}) there occurs a bifurcation of periodic solutions of Eqs. (2.31) and (2.32).

Poore (1973) has recently extended Hopf's theorem and has developed an algebraic criterion for the direction of bifurcation and stability of the bifurcating periodic solutions. Since the development and statement of Poore's criterion is rather lengthy we refer the reader to Poore's paper for the details and state here the results of an application of his criterion to our problem.

Theorem 2 (*Direction of Bifurcation*): Let $b \in (0, \frac{1}{2})$. Then there exist $\gamma_1 > 0$ and $\gamma_2 > 0$ such that the bifurcation of periodic orbits from (u_{I_1}, w_{I_1}) occurs on $(I_1, I_1 + \gamma_1)$ while the bifurcation from (u_{I_2}, w_{I_2}) occurs on $(I_2 - \gamma_2, I_2)$.

Note: An interesting phenomenon occurs if $\frac{1}{2} < b < 1$. It was proved by Troy (1974), and subsequently by Hsu and Kazarinoff (1976), that if $b \in (\frac{1}{2}, 1)$ and $\rho > 0$ is sufficiently small, then the bifurcation from (u_{I_2}, w_{I_2}) occurs to the right of I_2 where the steady-state is stable. That is, for values of I in a small interval of the form $(I_2, I_2 + \gamma_2)$ the steady-state solution (u_I, w_I) is locally asymptotically stable and yet surrounded by an unstable periodic orbit.

Theorem 3 (*Stability*): Let $b \in (0, \frac{1}{2})$. Then γ_1 and γ_2 can be chosen such that the periodic orbits bifurcating from (u_{I_1}, w_{I_1}) are orbitally asymptotically stable for $I \in (I_1, I_1 + \gamma_1) \cup (I_2 - \gamma_2, I_2)$.

To obtain a global description for the behavior of the families of bifurcating periodic orbits for values of I between $I_1 + \gamma_1$ and $I_2 - \gamma_2$, Troy (1976) has applied the global bifurcation theory of Yorke and Alexander (1977) to Eqs. (2.31) and (2.32). The results indicate that as I increases from I_1 the amplitudes of the periodic orbits increase, become fairly large, then shrink until the periodic solutions disappear altogether at $I = I_2$ (Fig. 4).

The bifurcation results described in Theorems 1–3 for the FitzHugh model have acted as an excellent guide in our investigations of the more complex Hodgkin–Huxley current clamped system [Eqs. (2.27)–(2.30)]. Thus, under an appropriate set of assumptions on the functions and parameters in Eqs. (2.27)–(2.30), an application of the Hopf Theorem results in the following.

Theorem 4 (*Troy, 1975*): There is an $\varepsilon_1 > 0$ and for each ε in $(0, \varepsilon_1)$ there exist positive values I_ε and I^ε with $0 < I_\varepsilon < I^\varepsilon$ such that as I passes through I_ε (or I^ε) there occurs a bifurcation of periodic solutions from the steady-state solution π_{I_ε} (or π_{I^ε}).

The direction of bifurcation and the stability of the periodic orbits have not yet been determined. However, Hassard (1978) has done some numerical

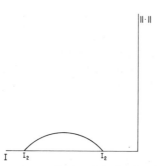

Fig. 4. Solid curve represents the amplitudes of the family of stable periodic solutions that bifurcate from the steady state as the current parameter I passes through I_1 from above. Note the growth and subsequent shrinkage of the orbits as I continues to decrease towards I_2, the second bifurcation value.

computations that indicate that the bifurcation is to the left from I_ε and I^ε for $\varepsilon = 1$, $I_\varepsilon = 9.8$, and $I^\varepsilon = 154$. Thus there is a bifurcation of small amplitude periodic orbits on $(I_\varepsilon - \delta, I_\varepsilon)$ for some $\delta > 0$. For these values of I the steady-state solution is stable and we conjecture (Fig. 5) that the small periodic orbits are unstable. These results were anticipated by the numerical computations of Sabah and Spangler (1970) and are in full agreement with our analytical investigations of FitzHugh's model [Eqs. (2.31) and (2.32)].

D. Oscillations in Cardiac Tissue

Another possible application of the Hopf bifurcation theory may give new insight into a cardiac Purkinje fiber problem. In their experiments on cardiac Purkinje fibers, Noble *et al.* (1969) have observed unexpected small

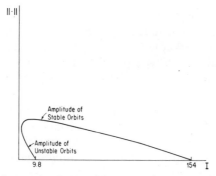

Fig. 5. Bifurcation diagram for the Hodgkin–Huxley equations. Note that for I slightly less than 9.8 the equations have two periodic solutions, a large amplitude stable orbit, and a smaller amplitude unstable periodic solution. This unexpected "backwards" bifurcation also is observed in FitzHugh's model Eqs. (2.31) and (2.32) for $b \in (\frac{1}{2}, 1)$ and $\rho > 0$ sufficiently small.

amplitude oscillations in the potential difference across the fiber membrane. These oscillations occur in response to the application of a constant current to the membrane. On the basis of experimenal data Noble *et al.* (1975) have modified the original Hodgkin–Huxley equations into a system that models the electrical activity in the membrane of the Purkinje fiber. Their model has the form

$$\dot{V} = I + f(V, W_1, \ldots, W_9) \qquad (2.37)$$

$$\dot{W}_i = g_i(V, W_i) \qquad (i = 1, \ldots, 9) \qquad (2.38)$$

where $V(t)$ represents the potential difference across the membrane of the fiber, and $W_i(t)$ $(i = 1, \ldots, 9)$ denote auxiliary variables that control the conductances of the same membrane currents. We conjecture that the bifurcation approach which proved useful in attaining the results in Theorems 1–4 may also show that there are small amplitude periodic solutions of Eqs. (2.37) and (2.38) that correspond to the small oscillations detected experimentally.

III. The Belousov–Zhabotinskii Reaction

A. INTRODUCTION

In this section we describe a mathematical model of a chemical system in which excitability interacts with diffusion to cause the formation and propagation of nonlinear waves of chemical activity in much the same manner as the electrical nerve impulse is formed and transmitted along the axon.

This "excitable" reaction was discovered by Belousov (1959) who observed *temporal* oscillations in the ratio [Ce(IV)/Ce(III)] during the oxidation of malonic acid by bromate in the presence of a cerium ion catalyst. Zhabotinskii (1964) later verified Belousov's observations and extended the scope of the reaction to include a few other metal ion catalysts and organic acids. Even more remarkable *spatial* structure may be observed when the reagent is unstirred. Thus Busse (1969) first noticed localized areas of high [Ce(IV)/Ce(III)] moving through the Belousov–Zhabotinskii reagent. In Busse's experiment horizontal bands of high [Ce(IV)/Ce(III)] propagate perpendicular to the axis of a vertical cylinder containing the reaction mixture. Zhabotinskii and Zaikin (1970) replaced cerium with ferroin as metal ion catalyst. Ferroin consists of the coordination complex of Fe(II) with three molecules of O-phenanthroline and has an intense red color. When Fe(II) is oxidized to Fe(III) the resulting complex, called ferroin, is blue. Zhabotinskii and Zaikin (1970) spread this reagent in a thin layer on the bottom of a Petri dish. In their preparation the reagent is initially red. However, at various points blue waves of high ferroin concentrations (i.e., a blue band) may form and propagate through the medium. It is evident that these traveling waves of chemical activity result from the same chemistry causing the temporal oscillations in the stirred reagent. An easy recipe for the reaction may be found in Tyson (1976).

1. Chemistry of the Temporal Oscillations

Field *et al.* (1972) proposed a detailed mechanism for the temporal oscillations observed in the Belousov–Zhabotinskii reaction. Subsequently, Field and Noyes (1974a) presented a simpler model that retains the main qualitative features of the earlier model. Their simplification consists of the following five step mechanism:

$$A + Y \rightarrow X + P \tag{3.1}$$

$$X + Y \rightarrow P \tag{3.2}$$

$$B + X \rightarrow 2X + Z \tag{3.3}$$

$$X + X \rightarrow B + P \tag{3.4}$$

$$Z \rightarrow fY \tag{3.5}$$

where $A \equiv B \equiv [BrO_3^-] = 0.06\ M$, $P \equiv$ inert products, $X \equiv [HBrO_2]$, $Y \equiv [Br^-]$, $Z \equiv$ [oxidized form of the metal ion catalyst], and f is the so-called stoichiometric factor. By analogy with the chemistry the rate constants for reactions (3.1)–(3.5) are assigned the values $k_1 = 2.1 \times [H^+]^2$, $k_2 = 2 \times 10^9 \times [H^+]$, $k_3 = 1 \times 10^4 * [H^+]$, $k_4 = 4 \times 10^7 \times [H^+]$, all in units of $M^{-1}\ sec^{-1}$, and $[H^+] = 0.8\ M$. The rate constant k_5 and the stoichiometric parameter f are considered variable parameters.

The character of the temporal oscillations in a stirred solution with ferroin as the metal ion catalyst is as follows. When $[Y](Br^-)$ is high and $[Z]$ (ferriin) is low the reagent remains red. During this time Y is removed by reactions (3.1) and (3.2). When $[Y]$ is high, $X(HBrO_2)$ nearly always reacts with Y by reaction (3.2) rather than with $B(BrO_3^-)$ by reaction (3.3). Reactions (3.1) and (3.2) are dominant. When Y has been sufficiently depleted, however, reaction (3.3) becomes competitive with reaction (3.2) as a fate of X, ferroin is oxidized to ferriin and *two* molecules of X are produced. Reaction (3.3) is *rapid and autocatalytic*. Reactions (3.4) and (3.5) become dominant as $[Y]$ is driven to very low values. Thus, in the reagent, the transition from red to blue is *sudden and nearly discontinuous*. When Z(ferriin) becomes an important component of the reaction mixture, reaction (3.5) starts producing Y and, if k_5 and f have appropriate values, then sufficient Y quickly accumulates to suppress reaction (3.3) in favor of reaction (3.2). Control of the system returns to reactions (3.1) and (3.2). The blue color of the reagent slowly fades to red as ferriin (Z) is slowly reduced to ferroin with simultaneous production of Y. The reagent remains red as reactions (3.1) and (3.2) deplete the Y produced by reaction (3.5). Eventually [Y] again decreases to the point where reaction (3.3) suddenly becomes the dominant fate of X and the very sudden transition from red to blue reoccurs. The steep concentration gradients that develop during the red-to-blue transition indicate that *diffusion* may become important in an unstirred reagent.

2. Two Types of Traveling Waves

When the Zaikin–Zhabotinskii reagent is spread in a thin layer (~ 2 mm) on the bottom of a Petri dish, the dominant form of the catalyst initially is as ferroin and the reagent appears red. However, at certain points in the reagent, called pacemaker centers by Winfree (1972), the ferroin is suddenly and autocatalytically oxidized to ferriin. A blue spot develops at that point and grows larger. As expected from the chemistry modeled by reactions (3.1)–(3.5), shortly after the reagent turns blue at a particular point, the color starts to slowly fade to red as the ferriin is reduced back to ferroin. Thus a blue wave is initiated around the pacemaker center. The change from dark red to blue at the leading edge of a wave is very sharp.

In an elegant series of experiments Winfree (1972, 1973, 1974a,b) devised reagents in which the traveling waves are particularly well-behaved, and he carefully characterized their properties. Winfree demonstrated experimentally that there are *two quite different varieties of traveling waves* that might appear in the Zaikin–Zhabotinskii–Winfree reagents. He calls these phase waves and trigger waves.

a. Phase Waves. Phase waves occur only in a reagent subject to temporal oscillation. When such a reagent is poured into a Petri dish it may continue to oscillate. Phase waves develop because it is possible that adjacent regions of the reagent may get out of phase with each other. If a continuous phase gradient is established, then sharp-edged wave fronts will *appear* to travel through the reagent as adjacent areas reach the point in their oscillatory cycle at which the sudden shift from red to blue occurs. One mechanism for the establishment of a continuous phase gradient is the passage of a trigger wave that we describe below. Phase waves reflect only apparent movement and are analogous to the spot of light seen moving around a theatre marquee. Movement is apparent because each light, which has its own on–off cycle, is slightly out of phase with the adjacent light. Thus one would expect phase waves to pass through physical barriers. Indeed, using a piece of plexiglass as a physical barrier, Howard and Koppel (1973) demonstrated that the traveling waves in Busse's experiment were actually phase waves that appear to pass through the barrier.

b. Trigger Waves. Trigger waves may occur in either an oscillatory or nonoscillatory reagent. Winfree adjusted reactant concentrations to obtain a reagent in which the temporal oscillations were suppressed when the mixture is spread in a thin layer in the Petri dish. He found that when he touched a heated needle to the reagent a single blue wave formed and propagated through the red medium. Field and Noyes (1974b) also made this observation. This type of wave, called a *trigger wave* by Winfree, must propagate by a reaction-diffusion mechanism in a medium stable to temporal oscillation but showing excitability. In this mode wave propagation proceeds thusly; interactions in the vicinity of a pacemaker perturb the stable steady state such that reactions (3.3) and (3.4) are temporarily dominant and large quantities of $X(HBrO_2)$ are autocatalytically and rapidly formed. A very steep concentration gradient of X develops at the edges of the excited area. The flux of X diffusing across this gradient and into the adjacent area interacts autocatalytically with $B(BrO_3^-)$ " triggering " this area to undergo a similar excursion in X and ferrin. This mechanism of wave propagation relies totally upon the excitability of the medium and subsequent action of diffusion across a sharp concentration gradient.

Schowalter and Noyes (1976) have demonstrated that trigger waves can also be initiated in the excitable reagent by the application of a 2.0-V positive square wave pulse to an Ag electrode universal in the reagent. A 0.6-V negative bias on the electrode suppresses trigger wave initiation. Thus they have produced an easily controlled pacemaker center.

B. The Field–Noyes Model

Applying the law of mass action to reactions (3.1)–(3.5), Field and Noyes (1974a) obtained the system

$$dx/dt = s(y - xy + x - qx^2) \tag{3.6}$$

$$dy/dt = (1/s)(-y - xy + fz) \tag{3.7}$$

$$dz/dt = w(x - z) \tag{3.8}$$

where $s = 77.27$, $q = 3.7 \times 10^{-6}$, $f > 0$, $w > 0$, and $x\alpha[HBrO_2]$, $y\alpha[Br^-]$, $z\alpha[Ce(IV)]$.

The steady-state solution of Eqs. (3.6)–(3.8) is found by setting the right-hand side of Eqs. (3.6)–(3.8) equal to zero. Solving the resultant system of algebraic equations, Field and Noyes (1974b) found that there is exactly one solution contained in the region $x > 0$, $y > 0$, $z > 0$. This solution, which we denote by $\pi_0 = (x_0, y_0, z_0)$, is given by

$$x_0 = ((1 - f - q) + [(1 - f - q)^2 + 4f(q + 1)]^{1/2})/2q \tag{3.9}$$

$$y_0 = fx_0/(1 + x_0) \tag{3.10}$$

$$z_0 = x_0. \tag{3.11}$$

Of special interest is the behavior of small perturbations from π_0. In order to investigate this problem one first evaluates the Jacobian matrix of the right-hand side of Eqs. (3.6)–(3.8) at the steady state. If all three eigenvalues have negative real parts then small perturbations from π_0 will decay exponentially back to π_0. However, if one or more of the eigenvalues has positive real part then a small perturbations from the steady state may begin to oscillate and grow in time. Field and Noyes (1974b) did a complete numerical analysis of the Jacobian matrix for positive values of f and w and obtained the steady-state diagram in Fig. 6.

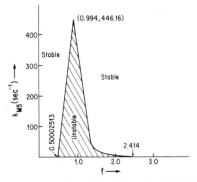

Fig. 6. Shaded region represents values of f and k_{M5} for which the steady state is unstable while unshaded region denotes values of f and W for which the steady state is stable.

Due to the amazing variety of experimental and numerical results described above, applied mathematicians became interested in studying the Field–Noyes model with the goal of obtaining more detailed information about the qualitative behavior of solutions. Thus Murray (1974) proved that any solution of Eqs. (3.6)–(3.8) with initial value in the positive octant must eventually enter the rectangular region

$$B = \left\{ (x, y, z) \mid 1 \le x, z \le \frac{1}{q}, \frac{fq}{1+q} \le y \le \frac{f}{2q} \right\}.$$

In addition, since the vector field points into the interior of B along its boundary then any solution that enters B cannot leave B. Subsequently Murray and Hastings (1975) investigated Eqs. (3.6)–(3.8) for values of f and w for which the steady state is unstable. For such values of f and w, they proved that Eqs. (3.6)–(3.8) have at least one periodic solution. The uniqueness and stability of this periodic orbit have not yet been determined.

Hsu and Kazarinoff (1975) continued the analysis of Eqs. (3.6)–(3.8) by investigating the behavior of solutions over intervals of f and w in which the stability of the steady state changes. They proved that for a fixed $f \in (0.5,$

2.412) there is a critical value $w_f > 0$ such that as w passes through w_f from above, the steady-state solution, π_0, loses stability and there occurs a bifurcation of small amplitude periodic solutions. They were also able to demonstrate the direction of bifurcation and the stability of the bifurcating periodic orbits. The principle tool which they used consists of the Hopf bifurcation theory described in Section II.

1. The Threshold of Excitation

The Field–Noyes model exhibits an interesting excitability phenomenon over a range of values of f and w where the steady-state solution is stable to small perturbations. When the Field–Noyes equations are incorporated into a spatial model for the traveling wave solutions, the threshold of excitation plays an important role in the formation and propagation of the wave.

Thus Troy and Field (1977) and Troy (1977a) investigated Eqs. (3.6)–(3.8) further and found that there is an interval $(f_1, f_2) \subset (2.412, +\infty)$ such that if $f \in (f_1, f_2)$ then:

(i) each solution with initial value in the positive octant must approach π_0 as $t \to \infty$; that is, the steady state is globally asymptotically stable;

(ii) there is a value $x^* > x_0$ such that if $y(0) = y_0$, $z(0) = z_0$ and $x(0) > x^*$, then for small $w > 0$ there is a rapid increase in x followed by the return of (x, y, z) to π_0 as $t \to \infty$ (Fig. 7);

(iii) if $x_0 < x(0) < x^*$ and $y(0) = y_0$, $z(0) = z_0$, then for sufficiently small $w > 0$ the solution decays to π_0 as $t \to \infty$ and x does not exceed x^* (Fig. 8).

As is evident from (ii) and (iii), x^* acts as a threshold of excitation for the system. We present below an outline of the proof of parts (ii) and (iii). We

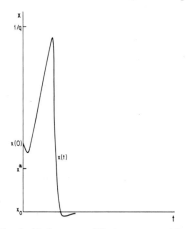

Fig. 7. If $x(0)$ exceeds threshold, then x rapidly increases and forms a spike similar to that observed in the nerve axon when an initial depolarization exceeds threshold.

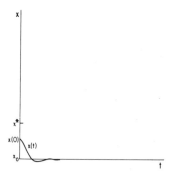

Fig. 8. If $x(0)$ does not exceed threshold, then x decays back to rest.

then compare the threshold phenomenon of Eqs. (3.6)–(3.8) with that of the Hodgkin–Huxley model.

First we set $w = 0$, which is equivalent to keeping z fixed. For convenience we set $z = z_0$, its steady-state value, and consider the reduced system

$$dx/dt = s(y - xy + x - qx^2) \tag{3.12}$$

$$dy/dt = (1/s)(-y - xy + fz_0). \tag{3.13}$$

A standard phase plane analysis of Eqs. (3.12) and (3.13) results in the phase portraits of solutions given in Fig. 9 below.

The function $y = h(x)$ in Fig. 9 represents the isocline $\dot{x} = 0$ while the function $y = k(x, z_0)$ represents the curve $\dot{y} = 0$. It is evident that there is a value $x^* > x_0$ that acts as a threshold of excitation in the following manner:

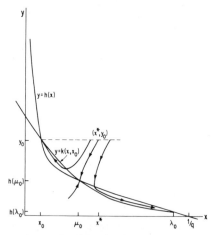

Fig. 9. A phase plane analysis of (3.12) and (3.13) shows that if $y(0) = y_0$ and $x(0) < x^*$, then $(x, y) \to (x_0, y_0)$ as $t \to \infty$. If $y(0) = y_0$ and $x(0) > x^*$, then $(x, y) \to (\lambda_0, h(\lambda_0))$ as $t \to \infty$.

(iv) if $y(0) = y_0$ and $x(0) > x^*$, then $\lim_{t \to \infty} (x(t), y(t)) = (\lambda_0, h(\lambda_0))$, the "excited state," while

(v) if $y(0) = y_0$ and $x_0 < x(0) < x^*$, then $\lim_{t \to \infty} (x(t), y(t)) = (x_0, y_0)$ and $x(t) < x^* \ \forall t \geq 0$.

Using continuity of solutions with respect to parameters together with some further analysis, it is not difficult to obtain the threshold phenomenon described in parts (ii) and (iii) above for the full system [Eqs. (3.6)–(3.8)].

2. Comparison with the Hodgkin–Huxley Model

It is interesting to note the similarities between the threshold of excitation in the Field–Noyes equations and the threshold phenomenon in the Hodgkin–Huxley model. Recall that in the Hodgkin–Huxley system the space variable is eliminated by insertion of a silver wire into the axon and the resulting temporal system consists of Eqs. (2.10)–(2.13). In this system a perturbation in v that exceeds threshold causes v to rise very rapidly to a peak value. Subsequently the slowly changing variables n and h cause v to decrease back to its rest state. Analogously, in the Field–Noyes model if an initial perturbation in x exceeds its threshold, x will rise quickly to a peak value. Then the slowly varying function $z(t)$ causes x to decay back to rest. Thus z plays a "recovery" role analogous to that of n and h.

Comparing Figs. 3 and 9 we see that the phase portrait of solutions of the reduced Hodgkin–Huxley model (i.e., $\varepsilon = 0$) is qualitatively the same as that of the reduced Field–Noyes system (i.e., $w = 0$). These basic similarities between the reduced systems then allow us to prove that for $\varepsilon > 0$ and $w > 0$ sufficiently small the global behavior of solutions of both system is exceedingly similar.

C. A SPATIAL MODEL

Using the excitability properties of the Field–Noyes equations as a guide, Field and Troy (1977) have investigated a spatial model for the existence of solitary traveling waves. Recall that in the vicinity of the stimulus (e.g., a heated needle) a local reaction takes place that causes a large buildup in $x(\text{HBrO}_2)$ and a steep concentration gradient in x quickly forms. Subsequently the natural action of diffusion causes x to spread into adjacent regions, triggering these regions to undergo a similar rapid build-up in x. Thus, since x appears to play the most important role in the diffusion process, Field and Troy add a diffusion term to the equation for x, and study the system

$$\partial x/\partial t = s(y - xy + x - qx^2) + D(\partial^2 x/\partial \zeta^2) \tag{3.14}$$

$$\partial y/\partial t = (1/s)(-y - xy + fz) \tag{3.15}$$

$$\partial z/\partial t = w(x - z) \tag{3.16}$$

where $D > 0$ denotes the diffusion coefficient and ζ is a one-dimensional space variable.

1. Statement of Results

In order to investigate Eqs. (3.14)–(3.16) for the existence of solitary traveling wave solutions we assume that x, y, and z are functions of the single variable $\tau = (x/\alpha) + t$. Substitution of this single variable into Eqs. (3.14)–(3.16) results in the system of first-order ordinary differential equations:

$$dx/d\tau = v \tag{3.17}$$

$$dV/d\tau = \theta[v - s(y - xy + x - qx^2)] \tag{3.18}$$

$$dy/d\tau = (1/s)(-y - xy + fz) \tag{3.19}$$

$$dz/d\tau = w(x - z) \tag{3.20}$$

where $\theta = \alpha^2/D$.

It is not difficult to show that there is exactly one physically reasonable steady-state solution of Eqs. (3.17)–(3.20), which we denote by $\pi_0 = (x_0, 0, y_0, z_0)$. Also, for ease of notation, we denote a solution of Eqs. (3.17)–(3.20) by $\pi(t) = (x(t), v(t), y(t), z(t))$.

As shown in the previous section, there is an interval $(f_1, f_2) \subset (2.412, +\infty)$ such that if $f \in (f_1, f_2)$ then the temporal system, [Eqs. (3.6)–(3.8)], has no periodic solution. We now state the following theorems.

Theorem 5 (Existence of Solitary Traveling Waves): There is an interval $(f_3, f_4) \subset (f_1, f_2)$ such that if $f \in (f_3, f_4)$ there exists $w_f > 0$ such that if $0 < w < w_f$, then there is at least one value $\theta^* = \theta^*(w, f)$ for which Eqs. (3.17)–(3.20) have a nonconstant solution $\pi^*(\tau)$ that satisfies

$$1 < x(\tau), \quad z(\tau) < \frac{1}{q}, \quad \frac{fq}{1+q} < y(\tau) < \frac{f}{2q} \quad \forall \tau \in (-\infty, \infty)$$

and

$$\lim_{\tau \to \pm \infty} \pi^*(\tau) = \pi_0.$$

Theorem 6 (Disappearance of Traveling Waves): There is an interval $(f_5, f_6) \subset (f_4, +\infty)$ such that if $f \in (f_5, f_6)$ there exists $w_f > 0$ such that if $0 < w < w_f$, then Eqs. (3.17)–(3.20) have no nonconstant solution whose trajectory is entirely contained in the region $1 < x, z < 1/q$, $[fq/(1 + q)] < y < (f/2q)$ and satisfies $\lim_{\tau \to \pm \infty} \pi(\tau) = \pi_0$.

2. Outline of Proofs of Theorems 5 and 6

We use a shooting technique to prove the existence of solitary traveling wave solutions. The first step is to compute the Jacobian matrix at π_0. It is not difficult to show that there is an interval $(f_3, f_4) \subset (f_1, f_2)$ such that if $f \in (f_3, f_4)$ and $w > 0$ is small, then one eigenvalue of the Jacobian is positive and the other three eigenvalues have negative real parts. Thus there is a one-dimensional unstable manifold $\mathscr{U}$ of solutions that tend to π_0 as $\tau \to \infty$. Therefore we assume that $\pi(0) \in \mathscr{U}$. Next, a standard analysis shows that (f_3, f_4) can be chosen such that if $f \in (f_3, f_4)$ is fixed, then for large θ the function $x(\tau)$ crosses the line $x = 1/q$ and $\lim x(\tau) = +\infty$ while if θ is small the solution $\pi(\tau)$ enters the physically meaningless region $x < 0$. A topological argument shows that for some intermediate value θ^*, $\pi(\tau)$ must remain in the bounded region

$$1 < x, \qquad z < \frac{1}{q}, \qquad \frac{fq}{1+q} < y < \frac{f}{2q} \qquad \text{for all } \tau \in (-\infty, \infty).$$

From a careful analysis it then follows that $\lim_{\tau \to +\infty} \pi(\tau) = \pi_0$.

To prove Theorem 6 we use the "energy" function

$$E = \frac{v^2}{2} + \theta s \int_{x_0}^{x} (1 - u) \left(\frac{fx_0}{1+u} - \frac{qu^2 - u}{1 - u} \right) du.$$

It is fairly simple to show that there is an interval $(f_5, f_6) \subset (f_4, +\infty)$ such that if $f \in (f_5, f_6)$, then

$$\int_{x_0}^{x} (1 - u) \left(\frac{fx_0}{1+u} - \frac{qu^2 - u}{1 - u} \right) du < 0. \tag{3.21}$$

Condition (3.21) guarantees that the curve $E = 0$ lies entirely above the x-axis in the region $x > x_0$ (Fig. 10).

An algebraic manipulation shows that the projection of $\mathscr{U}$, the unstable manifold, onto the (x, v)-plane points into the region $E > 0$. Thus $\pi(0)$ lies in the region $E > 0$, and along the solution $\pi(\tau)$, $E(\tau) > 0$ on a maximal interval of the form $(-\infty, T)$. If $T < \infty$, then $E(T) = 0$, and since $E(\tau) > 0$ on

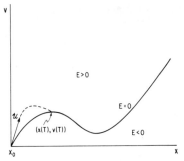

Fig. 10. In the (x, v) plane the unstable manifold initially lies above the curve $E = 0$. Dotted curve represents a trajectory of $\mathscr{U}$ which leads to a contradiction.

$(-\infty, T)$ we conclude that

$$\dot{E}(T) \leq 0. \tag{3.22}$$

However from Eqs. (3.17)–(3.20) it easily follows that $\dot{E}(T) > 0$, contradicting (3.22). Therefore $E(\tau) > 0$ for all $\tau \in (-\infty, \infty)$ and we conclude that $\lim_{\tau \to \infty} x(\tau) = +\infty$ for each $\theta > 0$. Thus there can be no bounded homoclinic orbit.

Remark: The methods used in proving Theorems 5 and 6 also apply directly to the Hodgkin–Huxley model and the FitzHugh–Nagumo equations. See, for example, Hastings (1976a,b) and Conley (1975). Further details concerning the proofs of Theorems 5 and 6 can be found in Field and Troy (1979) and Troy (1977c).

D. OTHER MODELS

Various authors have proposed other models for the Belousov–Zhabotinskii reaction, each of which appears to retain some of the basic qualitative features observed in the Belousov–Zhabotinskii reaction.

Hassard, Hastings, and Greenburg (1977) have proposed a discrete model and proved that their system has solitary, spiral, and periodic wave solutions as well as a threshold of excitation.

Koppell and Howard (1977) investigate a fairly general reaction diffusion equation for which the kinetics exhibit an orbitally asymptotically stable periodic solution. They then show that the diffusion term causes slowly rotating waves to appear. They also prove that their model has shock solutions similar to those observed in the Petri dish.

In private communications to the author R. J. Field has reported that he has added diffusion terms to each of the equations in the Field–Noyes kinetic model and he numerically obtains solitary traveling waves in good agreement with those observed experimentally.

It appears to be an interesting and very difficult problem to further extend the model to three dimensions and prove that there are scroll waves analogous to those observed by Winfree.

Another unsolved and very important problem is to determine the stability of each of the wave solutions described above.

References

Belousov, B. P. (1959). *Sb. Ref. Rad. Med.* (*Moscow*), 145.
Busse, H. (1969). *J. Phys. Chem.* **73**, 750.
Carpenter, G. (1976). Ph.D. Dissertation, Univ. of Wisconsin, Madison, Wisconsin.
Casten, R., Cohen, H., and Lagerstrom, P. (1975). *Q. J. Appl. Math.* **32**, 365–402.
Cole, K. S., Antosiewicz, H. A., and Rabinowitz, P. (1955). *SIAM J. Appl. Math.* **3**, 153–172.
Conley, C. C. (1975). *Proc. Tulane Conf. Partial Differential Equations.*
Conley, J. W., Dodge, F. A., and Cohen, H. (1966). *J. Cell. Comp. Physiol.* **66**, 99–110.

Evans, J. (1972a). *Ind. Univ. Math. J.* **22**, 75–90.
Evans, J. (1972b). *Ind. Univ. Math. J.* **22**, 577–593.
Field, R. J., Korös, E., and Noyes, R. M. (1972). *J. Am. Chem. Soc.*, 8649–8664.
Field, R. J., and Noyes, R. M. (1974a). *J. Chem. Phys.* **60**, 1877–1884.
Field, R. J., and Noyes, R. M. (1974b). *J. Am. Chem. Soc.* **96**, 2001–2006.
Field, R. J., and Troy, W. C. (1979). *SIAM J. Appl. Math.* (to be published).
FitzHugh, R., and Antosiewicz, H. A. (1959). *SIAM J. Appl. Math.* **7**, 447–458.
FitzHugh, R. (1960). *J. Gen. Physiol.* **43**, 867–896.
FitzHugh, R. (1961). *Biophys. J.* **1**, 445–466.
Hale, J. (1969). "Ordinary Differential Equations." Wiley (Interscience), New York, Ch. 2.
Hassard, B. (1978). *J. Theor. Bio.* (in press).
Hassard, L. N., Hastings, S., and Greenburg, J. (1978). *SIAM J. Appl. Math.* (in press).
Hastings, S. P. (1975). *Amer. Math. Monthly* **82**, 881–895.
Hastings, S. P. (1976a). *Arch. Rat. Mech. Anal.* **60**, 229–258.
Hastings, S. P. (1976b). *Quart. J. Math.* **27**, 123–134.
Hodgkin, A. L., and Huxley, A. F. (1952). *J. Physiol.* **117**, 500–544.
Hopf, E. (1942). *Sb. Math. Phys. Kh. Sächs. Akad. Wiss. Leipzig* **94**, 1–22.
Howard, L. N., and Kopell, N. (1977). *Stud. Appl. Math.* **LVI**, no. 2.
Hsu, I. D., and Kazarinoff, N. D. (1976). *J. Math. Anal. Appl.* **55**, 61–89.
Huxley, A. F. (1959). *Ann. N.Y. Acad. Sci.* **81**, 221–246.
Koppel, N., and Howard, L. (1973). *Stud. Appl. Math.* **52**, 291–328.
Koppel, N., and Howard, L. (1977). *Stud. Appl. Math.* **LVI**, 95–146.
Murray, J. (1974). *J. Chem. Phys.* **61**, 3610–3613.
Murray, J., and Hastings, S. P. (1975). *SIAM J. Appl. Math.* **28**, 678–688.
Nagumo, J., Arimoto, S., and Yoshizawa, S. (1962). *Proc. IRE* **50**, 2061–2070.
Noble, D., Hausworth, O., and Tsien, R. (1969). *J. Physiol.* **200**, 255–265.
Noble, D., McAllister, R. E., and Tsien, R. (1975). *J. Physiol.* **251**, 1–59.
Poore, A. B. (1973). *Arch. Rat. Mech. Anal.* **52**, 358–388.
Rinzel, J. (1973). Ph.D. Dissertation, New York University.
Rinzel, J., and Keller, J. (1973). *Biophys. J.* **13**, 1313–1330.
Sabah, N. H., and Spangler, R. A. (1970). *J. Theor. Bio.* **29**, 155–171.
Sattinger, D. H. (1977). *Adv. Math.* **22**, 312–355.
Schowalter, K., and Noyes, R. M. (1976). *J. Amer. Chem. Soc.* **98**, 3730–3731.
Turing, A. (1952). *Philos. Trans. R. Soc. London* **237**, 37–72.
Troy, W. C. (1974). Ph.D. Dissertation, State University of New York at Buffalo.
Troy, W. C. (1975). *Proc. R. Soc. Edinburgh* **74A**, 299–310.
Troy, W. C. (1976). *J. Math. Anal. Appl.* **54**, 678–690.
Troy, W. C. (1977a). *J. Math. Anal. Appl.* **58**, 233–248.
Troy, W. C. (1977b). *Arch. Rat. Mech. Anal.* **65**, 227–247.
Troy, W. C. (1977c). *Rocky Mount. J. Math.* **7**, 467–478.
Troy, W. C., and Field, R. J. (1977). *SIAM J. Appl. Math.* (in press).
Tyson, J. (1976). "Lecture Notes in Biomathematics." Springer-Verlag, Berlin and New York.
Winfree, A. T. (1972). *Science* **175**, 634–636.
Winfree, A. T. (1973). *Science* **181**, 937–939.
Winfree, A. T. (1974a). "Science and Humanism: Partners in Human Progress" (H. Mel, ed.). Univ. California Press, Berkeley.
Winfree, A. T. (1974b). "Lecture Notes in Biomathematics" (P. Van den Driesche, ed.). Springer-Verlag, Berlin and New York.
Yorke, J. A., and Alexander, J. C. (1977). *Ann. Math.* (in press).
Zhabotinskii, A. M. (1964). *Dokl. Akad. Nauk. SSSR* **157**, 362–365.
Zhabotinskii, A. M., and Zaikin, A. N. (1970). *Nature* **225**, 535–537.

Oscillating Enzyme Reactions

Benno Hess

Max Planck Institute für Ernährungsphysiologie
Dortmund, Federal Republic of Germany

Britton Chance

Johnson Research Foundation,
University of Pennsylvania Medical School, Philadelphia, Pennsylvania

I. Introduction

A. Historic

The occurrence of oscillating reactions in enzyme systems was first suggested with the recognition of overshoot phenomena in glycolysis, and soon followed by a direct demonstration of oscillations of the concentrations of intermediates of the photosynthetic cycle in the dark and of glycolysis in yeast cells. These observations coincided with the discovery of oscillating ion motions in mitochondria and of the oscillation of reactions catalyzed by horseradish peroxidase and lactoperoxidase [for a review see Hess and Boiteux (1971) and Chance et al. (1973a)].

The rapidly growing interest in oscillating enzyme reactions was enhanced by the recognition of the general role of feedback control in biochemical and

159

biological systems, and furthered by the thermodynamic concept of "dissipative structures" predicting the occurrence of unstable dynamic states in open systems. Moreover, the possible link between oscillating enzyme reactions and the mechanism of long-range periodicities occurring at all levels of biological organizations, such as mitotic cycles and differentiation or the biological clock, logically followed and opened a new broad field of investigations (Hess, 1977).

Two types of oscillating enzyme reactions can be distinguished and were defined as epigenetic and metabolic oscillations, respectively (Hess and Boiteux, 1971; Dahlem Workshop on the Molecular Basis of Circadian Rhythms, 1976). The first type occurs as a result of control mechanisms exerted at the level of transcription and/or translation and leads to a periodic change of the structure and/or the amount of an enzyme. The second type is a consequence of the periodic change of the direct regulatory interactions leading to a periodicity in enzyme activity. This chapter deals only with the latter type of oscillating reactions. It summarizes a variety of enzymic reactions and presents in more detail the current views on the mechanism of glycolytic oscillations in yeast as a general example of an oscillating enzyme system.

B. NONEQUILIBRIUM DYNAMICS

Oscillatory enzyme reactions have been found in single isolated enzymes as well as multiple enzyme systems analyzed either in a cell-free state or as an intact cellular system. In all cases described so far the oscillatory state is only one of a variety of dynamic states that have been demonstrated in theory and experiments, such as multiple steady states defined by a stable singular point, oscillations in terms of a rotation of a limit cycle around an unstable singular point (upon coupling to transport processes resulting in chemical waves), bistability phenomena, and chaotic behavior.

The generation of any of these dynamic states is a function of the rate of supply of a substrate to an enzyme or enzyme system, the rate law and the rate at which the product of the system leaves towards the sink. Thus, the system must be open and maintained far from its thermodynamic equilibrium. The occurrence of these states apart from simple steady states is further bound to the condition that the underlying kinetics must be nonlinear, a property which is indeed a general feature of biological systems.

In enzyme systems the kinetic nonlinearity arises from a number of different features. There are multiple types of positive or negative interactions and many parts of metabolism occur in form of enzymic cycles. Enzymic processes in all living systems are controlled by allosteric enzymes that respond to small changes of substrates, products, and controlling ligands in a highly nonlinear function. Finally the coupling of enzymic reactions to transport processes, not only by free diffusion but to a large extent by active transport through transmembrane process, must be considered (Faraday Symposia of the Chemical Society, 1974; Boiteux et al., 1977; Hess et al., 1978).

C. SUMMARY OF OSCILLATORY ENZYMIC SYSTEMS

The analysis of the oscillatory reaction mechanism of soluble horse radish peroxidase and lactoperoxidase reveals a highly complex dynamic behavior quite different from the Michaelis–Menten behavior of their peroxidatic function. Indeed, all four states mentioned above have been demonstrated in the case of horse radish peroxidase under the conditions of constant stirring to avoid interference with diffusion processes. The source of the nonlinearity in the kinetics of peroxidases is the autocatalytic character of the oxidation of NADH resulting from significant concentrations of a variety of radical species (Chance et al., 1973a).

The oscillatory state of simple membrane bound enzyme is exemplified in the case of the artificial papain membrane oscillator. This enzyme was experimentally immobilized in a membrane. It hydrolyzes benzoyle-L-arginine-ethyl ester in a classical reaction pathway. The overall reaction leads to an acidification of the system upon hydrolysis. The enzyme activity responses to changes of pH with a classical bell-shaped curve decreasing its activity on the acid as well as the alkaline side. If the activity is analyzed on the alkaline side of its pH-activity profile, the rate of the reaction is effected by the acid released upon hydrolysis that activates the enzyme activity. If the enzyme is bound to a membrane this activating effect is strongly enhanced because a membrane-dependent diffusion process is imposed on the system, which favors a local accumulation of protons and therefore an autocatalytic effect. Whenever the rate of substrate supplied to the enzyme through the membrane is maintained at a constant rate to a given domain the rate of product formation oscillates between diffusion and enzyme activity-limited states. The system has been described in approximation as a limit-cycle phenomenon (Goldbeter and Caplan, 1976).

In contrast to these two single enzyme reactions serving as simple models of soluble and membrane bound enzymes, the oscillations of multienzyme systems are much more complex. It is of interest to note that oscillations have been found in all pathways of energy metabolism: photosynthesis, mitochondrial respiration, and glycolysis. Furthermore, recently periodic functions have been identified during morphogenesis of the slime mould *Dictyostelium discoideum*.

The enzymatic mechanism of the oscillatory transient of the dark reactions of photosynthesis is not known, although a kinetic theory has been presented on the basis of a system of chemical kinetic equations, describing the overall reactions of the Calvin cycle. No detailed stability analysis of the kinetic properties of the individual enzymes involved is at hand and the primary source of the oscillation is not detected (Hess and Boiteux, 1971).

The observation of oscillation of mitochondrial respiration in isolated mitochondria as well as in intact cells attracted considerable attention and the mechanism of its generation is associated with the membrane transient phenomena. Simultaneous oscillations of pyridinnucleotide, flavoproteins, cytochromes, and oxygen uptake of ATP and AEP as well as of the transport of protons and potassium ions can be recorded and are coupled to osmotic swelling and shrinking. The oscillations were discovered during a study of the influence of the effect of ionophores of mitochondria, however, can also be observed in their absence (Hess and Boiteux, 1971).

Figure 1 illustrates a cutout of a long train of undamped oscillations of

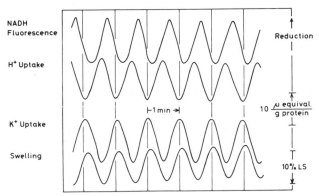

Fig. 1. Simultaneous record of NADH fluorescence, H^+ uptake, K^+ uptake, and swelling in oscillating mitochondria (from Boiteux and Hess, 1974).

simultaneously recorded NADH fluorescence, H^+-uptake, K^+-uptake, swelling, and shrinking indicative of water transport in oscillating mitochondria (Chance *et al.*, 1973a; Boiteux and Hess, 1974). The figure shows appreciable phase shifts between NADH and the uptake of protons as well as of potassium ions, the latter two components oscillating with an opposite flux direction and a varying stoichiometry of K^+/H^+ up to ten depending on the experimental conditions. Phase-plane plots reveal a stable limit cycle of the components with a frequency of approximately 1 min^{-1}. A detailed phase analysis of the components and the rate of oxygen uptake indicate a tight coupling between the rate of redox reactions of the respiratory pathway and the rate of ion fluxes as a function of the electrochemical gradients of protons and other charged components. Correlating studies showed the additional coupling between these latter processes with the state of the mitochondrial ATP/ADP system as expected in terms of current theories of oxidative phosphorylation (*Annu. Rev. Biochem.*, 1977). Thus, the occurrence of mitochondrial oscillation illustrates the delicate coupling between all redox reactions, phosphorylating reactions, ion and charge translocation processes.

Glycolytic oscillations have first been observed in intact yeast cells and yeast cell extracts and later also in the extracts of beef hearts. Although the general observation in both systems indicates that the primary source of the oscillation is the enzyme phosphofructokinase, the mechanisms of oscillation in both systems are different and only the first system has been analyzed in great detail, which will be presented below (Hess and Boiteux, 1971).

The oscillation of enzyme activities and transport processes of the slime mould *D. discoideum* during morphogenesis plays a significant physiological role in the life cycle of this species. The oscillations are integral part of the mechanism of aggregation. In a study of the mechanism of aggregation spontaneous oscillations in the minute period range of light scattering changes of a suspension of the amoeba during its interphase state were discovered. Later on it was shown that these oscillations can be initiated by extracellular addition of pulses of cyclic AMP. Furthermore, the phase of the oscillation was found to be sensitive to pulses of this compound. Titration experiments indicated that a response of the cellular system was still observed when the molecule to cell ratio was 3×10^3, indicating that cyclic AMP serves as an intercellular transmitter signalling the dynamic state within the amoeba population. These observations were extended by the direct demonstration of intra- and extracellular oscillations of cyclic AMP corresponding to the light scattering signal. Finally, it could be shown that each cyclic AMP pulse received by a cell population led to an intracellular up to 100-fold synthesis of new cyclic AMP, which was then released again in pulses into the cellular environment. The intracellular events are also coupled to a chemotactic motion in the contradirection of the arriving cyclic AMP pulse.

In the slime mould a number of periodic cellular functions are perfectly ordered in time: a periodic production and ejection of cyclic AMP leading to extracellular cyclic AMP waves, which are received by neighboring cells over a large distance. Each cell responds with a chemotactic motion as well as a new cyclic AMP synthesis serving as a cyclic AMP relay station. The oscillations are furthermore coupled to cyclic GMP-pulses that control in part the breakdown of cyclic AMP by phosphodiesterase and to oscillations of proton fluxes through the cellular membrane as well as oscillations of cytochrome *b*.

The evidence for the periodic synthesis of cyclic AMP suggested that the control of adenylate cyclase by extra- and intracellular interaction might be the primary source of the oscillatory phenomena. Indeed, a mathematical model of an allosteric control mechanism of the membrane bound adenylate cyclase could explain the periodic synthesis of cyclic AMP [the structure of the model is very similar to the structure of the model of the phosphofructokinase oscillator in glycolysis (see below)], although a detailed biochemical mechanism of the adenylate cyclase as well as of its interaction with the cyclic AMP receptor at the membrane and other periodic functions in the cell is not known (Hess *et al.*, 1978; Boiteux and Hess, 1974; Gerisch and Hess, 1974; Gerisch *et al.*, 1978).

II. Glycolysis

A. GENERAL OBSERVATIONS

Metabolic oscillations have been observed in many cellular systems (Hess and Boiteux, 1971; Chance *et al.*, 1973a). Glycolytic oscillations were first observed in a suspension of yeast cells of *Saccharomyces carlsbergensis* and later also in many other yeast strains. It was soon proposed that the observation of periodic phenomena in a suspension of cells or cultures of cells implies that glycolysis of all individual cells in the whole population oscillates and a full synchronization of the metabolic states in time of all cells is achieved. Indeed, oscillations of NADH as a parameter of the glycolytic state of one single yeast cell could be demonstrated to have a damping factor slightly less than that of the total population and a synchronization phenomenon was suggested (see Fig. 2) (Chance *et al.*, 1973b).

The observation of glycolytic oscillations in yeast cells is not dependent upon their biological state. Oscillations were found to occur in yeast cells grown on different substrates, aerobically or anaerobically, in their stationary or logarithmic phase (Hess and Boiteux, 1968a). If harvested and analyzed in suspension, the oscillations occur spontaneously or may be induced by single or continuous additions of substrates. The latter technique

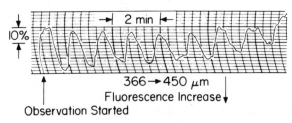

Fig. 2. An example of a series of continuous oscillations of cytosolic NADH in a single yeast cell recorded with a microfluorometer (from Chance *et al.*, 1973b).

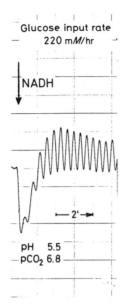

Fig. 3. Oscillations of NADH in yeast cell suspensions grown anaerobically for 24 hours (*S. carlsbergensis*). Glucose was injected at a constant rate of 220 mM/hr (from Hess and Boiteux, 1968a).

allows the maintenance of a steady state that is only dependent on the rate of substrate supply. A typical case is shown in Fig. 3, demonstrating a full cycle of the transient response of NADH as a parameter of glycolysis leading to sustained oscillations in a suspension of anaerobically grown yeast cells. The record is a cut out of a series of over 65 cycles. The reaction was initiated by starting a constant supply of glucose with a rate of 220 mmoles $1^{-1} h^{-1}$ to the suspension. The initial transient response is an overshoot phenomenon. After a series of cycles the system oscillates with a rather constant amplitude, a period of 19.6 seconds with a standard deviation of $\pm 3.5\%$ ($n = 15$). In such experiments damping factors up to 1000 have been observed. The waveform of the NADH oscillation in intact yeast cells is sinusoidal with a narrow bandwidth of a frequency in the range of 1–3 min^{-1}, depending on the substrate and its rate of addition. In the range between 19° and 40° a temperature dependency of the frequency of oscillation in glycolyzing yeast extract has been found with a frequency increasing by a factor of 4–2.2 per 10° (Hess *et al.*, 1966).

B. GLYCOLYTIC OSCILLATIONS IN YEAST EXTRACTS

The demonstration of glycolytic oscillations in highly concentrated cell-free extracts of yeast cells allowed a detailed analysis of the dynamics and mechanism of glycolytic oscillations (Chance *et al.*, 1973a; Hess *et al.*, 1966).

The study of the phenomenon in extracts greatly facilitated the selection of suitable experimental conditions and was complementary to the studies carried out in whole cell suspensions. Although there are differences in the range and greater sensitivity towards perturbations of glycolytic oscillations in the extract compared to yeast cells, no principle differences in the mechanism of the glycolytic oscillations have been observed between both states.

All glycolytic intermediates oscillate in the range of 10^{-5}–10^{-3} M. The concentration of one of the components, NADH, can be continuously monitored by fluorescence or absorption spectroscopy and the pulsed production of protons and carbon dioxide has been analyzed directly by suitable electrodes. In experiments using a continuous substrate injection technique to induce dynamic states like those observed in classical stirred flow reactor systems, it was shown that the frequency and amplitude of glycolytic NADH oscillations in a given yeast extract is flux-rate dependent (Hess *et al.*, 1969). A critical flux range (see Table I) was found extending in the extract over approximately one order of magnitude, well within the physiological flux range of yeast glycolysis. In addition, the waveform, frequency, and amplitude are flux dependent (see Table I). A large variety of sinusoidal, single and double pulses, strong asymmetric cycles, square-wave approximations, and spikes have been recorded depending on experimental conditions with frequencies between 0.05 and 0.5 min^{-1} and well over 160 continuous cycles (Hess and Boiteux, 1971; Hess *et al.*, 1966).

By periodic addition of substrate, the glycolytic system is readily entrained with the oscillation period of glycolysis synchronizing with a period of substrate supply. Furthermore, synchronization to a subharmonic of the driving frequency of the periodic substrate supply was recorded. This is observed if the driving frequency is near to an integral multiple of the frequency being recorded, if a continuous rate of substrate supply is used (defined as fundamental frequency). This phenomenon is also known as subharmonic resonance or frequency division. Figure 4 illustrates the entrainment by the 1/2-harmonic in a record of NADH-absorbance in a yeast extract supplied with the periodic glucose injection rate. The ranges of dynamic interactions of the glycolytic oscillator with a periodic source of substrate is summarized in Table II demonstrating the domains of coupling (for comparison with a model see below). The observation of subharmonic synchronization in glycolytic oscillation proves the nonlinear nature of the glycolytic oscillator, the mechanism of which has been analyzed to a considerable extend by biochemical experimentation leading to the notion that the "primary oscillophore" of the system is the enzyme phosphofructokinase (Boiteux *et al.*, 1975).

The influence of random perturbation of the glycolytic system was analyzed by means of stochastic addition of substrate to a glycolytic extract.

Table I

COMPARISON OF OSCILLATORY BEHAVIOR IN MODEL AND EXPERIMENT WITH A CONSTANT SOURCE OF SUBSTRATE[a]

	Model	Experiment
Sustained oscillations		
Oscillatory range of substrate input	$v_1 = 19–246 \text{ m}M \cdot \text{hr}^{-1}$	$v_1 = 20–160 \text{ m}M \cdot \text{hr}^{-1}$
Period	Of the order of minutes[b]	
Amplitude	In the range $10^{-5}–10^{-3}$ M; passes through maximum as v_1 increases	
Periodic change in phosphofructokinase activity	Minimum: 0.95% Activity maximum: 73% ($\%V_M$) mean: 17.5% Activation factor: 77	Minimum: 1% Activity maximum: 80% ($\%V_M$) mean: 16% Activation factor: 80
Phase-shift by ADP	Delay of 1.5 min upon addition of 0.7 mM ADP (14 units of γ) at the minimum of ADP oscillations	

[a] From Boiteux et al. (1975).
[b] Decreases by a factor of > 0.1 as v_1 increases.

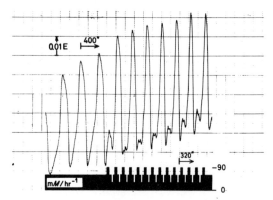

Fig. 4. NADH absorbance (upper trace) in a yeast extract entrained by the 1/2-harmonic of a periodic glucose injection rate (lower trace). Both $T_0 = 400$ sec, $T' = 160$ sec. Resulting period: $T = 2 \times T' = 320$ sec (from Boiteux *et al.*, 1975).

Table II

INTERACTION OF THE GLYCOLYTIC OSCILLATOR WITH A PERIODIC SOURCE OF SUBSTRATE[a]

Case	Relation between T' and T_0	Interaction
Glycolysis: experiment	$T'/T_0 \approx 1/n \ (n = 2, 3, \cdots)$	Entrainment by the 1/n-subharmonic of the input frequency
	$0.7 \leq T'/T_0 \leq 1.2$	Entrainment by the fundamental frequency of the input
	$1.2 \leq T'/T_0 \leq 1.6$	No entrainment
	$T'/T_0 > 3$	Double periodicity: separation of autonomous and input periodicities
Model	$T'/T_0 \approx 1/n \ (n = 2, 3)$	
	$0.89 \leq T'/T_0 \leq 1.11$	Entrainment

[a] From Boiteux *et al.* (1975).

In a record of NADH absorbancy changes, it was found that these conditions lead to periodic behavior within a narrow range around the period length that is observed if the continuous rate of substrate supply is used. This experiment demonstrates that the system reacts as a high "Q" filter centered at the mean fundamental frequency and keeping the period stable in spite of short time variations of the source rate (Boiteux *et al.*, 1975).

Although the analysis of glycolytic oscillations has been carried out thus far under homogeneous experimental conditions obtained by constant stirring, where transport processes can be neglected, it is expected that time-dependent spatial structures should occur whenever the evolution of enzymic processes is allowed to couple with transport via appropriate diffusion gradients. Recently the occurrence of macroscopic periodic structure formation synchronized with glycolytic oscillations have been recorded in cell-free extracts of yeast in accord with a kinetic model representing the evolution of spatio temporal patterns of glycolysis (Hess et al., 1975).

The evidence for the phosphofructokinase theory is based on a variety of experimental results and supported by computer studies of models of glycolytic oscillations (Hess and Boiteux, 1971; Chance et al., 1973a). The observation of the oscillation of all glycolytic intermediates and its quantitative analysis yielded a detailed balance of metabolites during one oscillatory cycle indicating that the time changes of the metabolite concentrations are coupled to the enzyme reaction rates by equations of the form

$$c_{(i)}(t) = v_{(i)+}(t) - v_{(i)-}(t).$$

This shows that the observed time course of metabolite i concentration reflects the imbalance between the source term $(v_{(i)+})$ and the sink term $(v_{(i)-})$ for this metabolite. The concentrations of the metabolites change during the oscillation with equal frequency, but different phase angles relative to each other. An analysis of the phase angle allows to locate the kinetic control points of the oscillation. A phase shift of 180° was found between the metabolites fructose-6-phosphate and fructose-1,6-bisphosphate, as well as between phosphoenolpyruvate and pyruvate, indicating the enzymes phosphofructokinase and pyruvate kinase as essential control points of the oscillation dynamics. A much smaller phase angle was found between the reaction components catalyzed by glyceraldehyde-3-phosphate dehydrogenase and phosphoglycerate kinase in the order of 80°. These results suggested that phosphofructokinase is indeed the primary source of the oscillation and kinetically coupled via the ATP/ADP-metabolites with pyruvate kinase and to a small degree with phosphoglycerate kinase. This view was further supported by the fact that injection of fructose-6-phosphate readily induces oscillation of glycolysis whereas injection of fructose-1,6-bisphosphate the product of the phosphofructokinase reaction does not lead to an oscillatory state of the residual glycolytic pathway. Further support of this mechanism was presented by an analysis of the phase sensitivity towards the addition of ATP/ADP/AMP. This clearly demonstrates a periodically changing affinity of the phosphotransferases toward the adeninnucleotides during oscillation. During a high-affinity phase towards ATP, the system responses to addition of ATP with the rapid phase shift, but does not react to additions either AMP of ADP, their binding sites being insensitive. In contrast, during the low-affinity phase towards ATP, a high sensitivity and phase response is obtained by addition of ADP or AMP, the latter two ligands reacting now with a high-affinity binding site of the enzymes. Here it is important to stress that whereas phosphofructokinase reacts directly with all three nucleotides, phosphoglycerate kinase and pyruvate kinase response only to ADP and ATP. Any reactivity of the latter two enzymes towards AMP is mediated via the adenylate kinase reaction (Boiteux and Hess, 1974; Hess et al., 1969).

The change of activity of phosphofructokinase, to a large extent, is controlled by the rate of generation and consumption of its second substrate ATP, its allosteric ligand AMP, and finally by the rate of generation of its first substrate fructose-6-phosphate. The reaction product ADP couples the activity state of phosphofructokinase with the other phosphotransferases of the glycolytic pathway, and all three nucleotides equilibrate via the adenylate kinase. Depending on the experimental condition (especially upon experimentation in a yeast extract), a propagation of the periodic activity change of phosphofructokinase by its second product fructose-1,6-bisphosphate can be neglected because its high accumulation saturating the residual enzymic pathway. Thus, the adenine nucleotide system controls the propagation of the pulsed production of intermediates along this enzymic chain and synchronizes the change of activity of phosphoglycerate kinase and pyruvate kinase with the primary change of the enzyme phosphofructokinase.

An analysis of the activity of phosphofructokinase in the extracts under oscillating conditions demonstrates a maximum activity in the order of 70% of V_{max}, a minimum activity in the order of 1% and a mean activity in the order of 16% during one oscillatory cycle (Hess et al., 1969; Boiteux et al., 1975). A static record of the relationship between the enzyme activity and the concentration of fructose-6-phosphate under oscillating condition in the extract is given in Fig. 5 for the boundary conditions of all three adenine nucleotides as found during the oscillatory state and the respective fructose-6-phosphate concentration. The characteristic curves (A and B) are given for the conditions of maximum and minimum concentrations of a controlling ligand and clearly illustrate the allosteric properties of the enzyme. The area on the left side between the two boundary characteristics indicate the change of activity during one oscillatory period (Boiteux and Hess, 1974).

Phosphofructokinase of yeast was isolated in a highly purified form. Its molecular weight was found to be 720,000. It is composed of at least four protomers $(n = 4)$ associating into eight subunits of two types with molecular weights of approximately 86,000 and 94,000. The kinetics of the enzyme follow an allosteric model with the homotropic effectors fructose-6-phosphate and ATP as substrates and a strong heterotropic activator AMP, indicating a highly cooperative response of the enzyme. It is interesting to note that the Hill coefficients up to 4.9 are pH dependent. The pH depen-

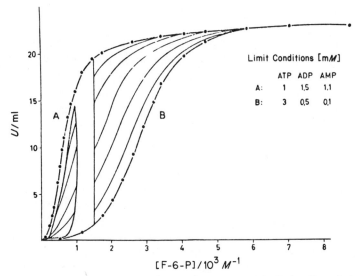

Fig. 5. Activity of phosphofructokinase under conditions of oscillation (from Boiteux and Hess, 1974).

dency of the allosteric properties of this enzyme coincide with the pH dependency of the glycolytic oscillation. The detailed structure analysis as well as the kinetic mechanism of the enzyme is still under investigation (Hess *et al.*, 1969; Tamaki and Hess, 1975).

In the context of this presentation, the function of the other enzymes, especially of the allosteric pyruvate kinase, cannot be discussed (see below), although it should be noted that the function in general is an integral part of the oscillating glycolytic system. Until now reconstitution of an oscillatory state of the total glycolytic system was achieved, although a reconstitution of a minimum system composed only of phosphofructokinase and pyruvate kinase coupled to adenylate kinase has not been observed experimentally and is under current investigation (for model studies see below) (Hess and Boiteux, 1968b).

C. MODELS OF GLYCOLYTIC OSCILLATIONS

The biochemical studies of the mechanistics of oscillations were complemented by analyses of various dynamic models allowing application of stability criteria and demonstration of domains of sustained oscillations. The earlier models were based on a phenomenological description of a Michaelis–Menten mechanism of phosphofructokinase activated by its reaction product in terms of a positive feed back reaction. This analysis was followed by a similar study demonstrating the range of the substrate injection rate critical for the occurrence of limit-cycle oscillations. Both models were simple, qualitative, and empirical and stress the importance of the autocatalytic feature of the model structure as the source of nonlinearity in its kinetic equations. Qualitatively, the models were in accord with the experimental results (Hess and Boiteux, 1971; Chance *et al.*, 1973a; Hess *et al.*, 1978; Boiteux *et al.*, 1975; Goldbeter and Nicolis, 1976).

With the recognition of the allosteric properties of phosphofructokinase of yeast, the source of the autocatalytic property was evidently not a simple feedback activation by the metabolic product or substrate, but a more complicated feedback involving alteration of the transfer function of the enzyme, which in biochemical terms means a change of the intrinsic properties of the oligomeric enzyme. These properties are defined as chemical conformation states determining the enzyme activity as a function of activating or inhibiting ligand interactions. Such a mechanism relies on the cooperation of several enzyme subunits and can be described on the basis of the concerted transition theory of Monod, Wyman, and Changeux or the sequential theory of Koshland, Nemethy, and Filmer (Hess *et al.*, 1978; Hess *et al.*, 1975; Goldbeter and Nicolis, 1976). In a series of studies the properties of a model based on the transition theory were analyzed and compared successfully to experimental results, although it should be noted that this model is not unique and comparable results were obtained upon application of the sequential theory. Furthermore, a multistate theory in general would also give the necessary change of the enzyme transfer function required for the oscillations (Hess *et al.*, 1978).

The concerted transition model of phosphofructokinase (see Fig. 6) is

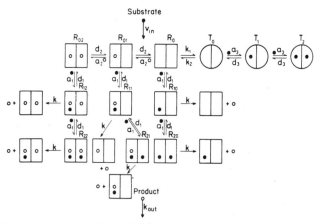

Fig. 6. Concerted model of a dimer allosteric enzyme activated by the reaction product. The substrate (●) enters at a constant rate (v_1) and the product (○) leaves the system at a rate proportional to its concentration (from Goldbeter and Lefever, 1972).

based on a simplified protein structure composed of two peptide chains with two binding sites for substrates $(n = 2)$ in dimer form and occuring in at least two conformation states that are designated (R, T). The two conformation states differ by their affinity towards the substrate ATP and the product ADP. Indeed the model simplifies the biochemical enzyme structure to quite an extent, mainly with respect to the neglection of the substrate and product fructose-6-phosphate and fructose-1,6-bisphosphate, respectively, of AMP, as well as the oligomeric state with $n = 8$ in the native structure. However, as shown elsewhere, these simplifications did not result in a loss of quantitative or qualitative properties of the model compared to experimental observations. The substrate in the scheme of Fig. 6 enters the system at a constant rate (v_{in}), the product leaves the system at a rate proportional to its concentration (for further designations see figure legends).

This model enzyme system can be described by a total of 13 kinetic equations and greatly simplified by assuming a quasisteady state for the enzymic forms. This assumption is based on the observation that the concentration of metabolites exceeds that of the enzymes sometimes by several orders of magnitudes. In the case of phosphofructokinase in yeast extract, the concentration of ATP and ADP are in the range of 10^{-4}–10^{-3} M, whereas the enzyme concentration is in the range of 5×10^{-7} M. It should be added here that this simplification is not a prerequisite for the periodic behavior in such systems. Computer simulations of a reaction system involving both enzymes and metabolites in principle also yield periodicities in metabolites as well as conformation changes (Hess *et al.*, 1978; Goldbeter and Nicolis, 1976; Hess, 1968).

Assuming a quasisteady state for the enzymic forms as well as neglecting diffusions, the time dependency of the normalized substrate $(\alpha = \text{ATP})$ and product $(\gamma = \text{ADP})$ concentrations are described by the following algebraic equations, with the nonlinear contribution given by the quotient (Φ):

$$d\alpha/dt = \sigma_1 - \sigma_M \Phi$$

$$d\gamma/dt = \sigma_M \Phi - k_s \gamma$$

where

$$\Phi = \frac{\left(\dfrac{\alpha}{\varepsilon + 1}\right)\left(1 + \dfrac{\alpha}{\varepsilon + 1}\right)(1 + \gamma)^2 + \text{L}\Theta \left(\dfrac{\alpha c}{\varepsilon' + 1}\right)\left(1 + \dfrac{\alpha c}{\varepsilon' + 1}\right)}{\text{L}\left(1 + \dfrac{\alpha c}{\varepsilon' + 1}\right)^2 + (1 + \gamma)^2 \left(1 + \dfrac{\alpha}{\varepsilon + 1}\right)^2}.$$

Normalization of the substrate ATP and the product ADP are obtained by division through the respective dissociation constants K_R. The production rates of the substrate (σ_1) and the rate constant (k_s) related to the sink of the product were also normalized [for further definitions see (Hess *et al.*, 1978; Goldbeter and Nicolis, 1976)].

The quotient is the rate function (v/V_{max}) of the phosphofructokinase. Its nonlinearity arises from the allosteric constant as well as the positive feed back term $(1 + \gamma)^2$. The latter term in connection with the allosteric constant yields the evolution of nonequilibrium instability states. The allosteric constant L itself, also still a source of nonlinearity, allows only the occurrence of a unique stable state, but no limit cycle and if L becomes small the system reduces to a simple Michaelis–Menten system (for limiting conditions see below) (Hess *et al.*, 1978; Goldbeter and Nicolis, 1976; Goldbeter and Lefever, 1972).

The stability analysis of the model yields the conditions of instability as a function of the various parameters. A detailed comparison between model and experiment shows remarkable quantitative agreement. It also demonstrates that the dynamic behavior of a complex system, such as glycolysis, can be reduced to the molecular properties of a single protein species operating as a master enzyme in a biochemical pathway. The comparison of the model and biochemical experiments for a constant source of substrate is illustrated in Table I. The oscillatory range of substrate injection rate, the period, amplitude, activity range of phosphofructokinase, and phase shift properties agree satisfactorily (Boiteux *et al.*, 1975).

The response of the enzyme towards a periodic source of substrate was analyzed in the model. The results are given in Fig. 7, where the domains of

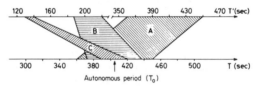

Fig. 7. Domains of entrainment of the enzyme model by the (A) fundamental frequency, (B) 1/2-harmonic and, (C) 1/3-harmonic of a sinusoidal source of substrate $\sigma_1 = [0.5 + 0.25 \sin(2\pi t/T')]$ sec^{-1}. In the given range no entrainment takes place outside these domains, which extends symmetrically around $T_0 = 406$ sec; T denotes the period after entrainment (from Boiteux *et al.*, 1975).

entrainment of the oscillatory enzyme model by (A) the fundamental frequency, (B) 1/2-harmonic, and (C) 1/3-harmonic of a sinusoidal source of substrate are illustrated. Entrainment domains extends symmetrically around the 406-second autonomous period of the oscillating enzyme. Furthermore, it is clearly seen that the range of coupling decreases with increasing driving frequency. A comparison between the results of the model and experiments for the periodic rate of substrate injection is represented in Table II. It shows that the domain of entrainment obtained for one case of the model is only slightly smaller than that obtained for the case in the experiment. This agreement is also found between subharmonic entrainment and the modulation of autonomous oscillation by a source of shorter period. In addition, the biochemical results obtained by stochastic substrate input were found to agree in the model and experiment (Boiteux *et al.*, 1975).

The frequency of the oscillation is also a function of the enzyme concentration. It was therefore of interest to examine the effect of periodic enzyme synthesis on limit-cycle behavior. It was found that the period of a possible epigenetic oscillation (although not detected in oscillating glycolysis) is usually larger in the model by one order of magnitude than the period of metabolic oscillation. The modulation of the periodicity was quite similar to that found for metabolic oscillations (Boiteux *et al.*, 1975).

The problems of the necessary conditions to generate spatiotemporal states of glycolysis were also investigated on the basis of the allosteric model of phosphofructokinase. An extension of the equations given above by the proper transport terms for ATP and ADP to allow for diffusion and neglecting possible translational motion of enzyme species was solved as a system of partial differential equations by finite differences in one dimension. These studies showed that with the domain of parameter given for the homogeneous case (above), depending on the fixed concentrations of the diffusing components (ATP, ADP) at the boundaries, the size of the system and the diffusion coefficient, time-independent regimes, as well as standing and propagating waves, are computed (Goldbeter, 1973).

The analysis of the relationship between the transport terms and the chemical reaction terms yield a critical length as the square root of the ratio of the transport term/kinetic term. Earlier, on the basis of a phenomenological model of oscillating glycolysis the critical length of 10^{-2}–10^{-4} cm was estimated (Glansdorff and Prigogine, 1971). On the basis of the allosteric model for given boundaries (see above) propagating waves were computed for a critical length of 0.3 cm, which is qualitatively well within the range of experimental observations (Hess *et al.*, 1975). The studies illustrate that the solution of kinetic equations describing the delicate balance between chemical kinetics and transport processes far from equilibrium depend on global features of the system such as size and geometry. These theoretical studies predict the occurrence of macroscopic spatiotemporal structure in glycolytic systems, which have been recorded recently in biochemical experiments using yeast extract under the conditions of oscillating glycolysis (Hess *et al.*, 1975).

The allosteric nature of phosphofructokinase as a major source of glycolytic oscillation as well as the result of the studies of the allosteric models simulating the dynamic states of glycolytic oscillations stress the importance of cooperative allosteric transitions as an essential property of periodic phenomena in biochemical nonequilibrium systems. Indeed, it has been suggested that this property observed at the genetic, enzymatic, and membrane levels might well represent the common molecular basis to many periodic phenomena in biology. This concept has also been recently invoked quite successfully in the analysis of the various periodic functions of the slime mould *D. discoideum* (Hess and Boiteux, 1971; Hess *et al.*, 1978).

The role of cooperativity in the generation of oscillations can be illustrated by an observation of the critical values of the Hill coefficient (n_H), the allosteric constant (L), and the nonexclusive binding coefficient (c). The relationship between the Hill coefficient and the allosteric constant under steady-state conditions is shown in Fig. 8 for different values of the non-

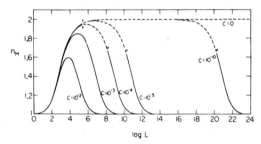

Fig. 8. Hill coefficient of the steady state and oscillatory domain in the dimer phosphofructokinase model as a function of the allosteric constant (L) and the nonexclusive binding coefficient c. The data obtain for a perfect K-system. The dashed line of each curve represents the domain of sustained oscillations (from Goldbeter, 1977).

exclusive binding coefficient c. The dashed line indicates the oscillatory domain on each curve. For binding coefficients larger than zero, a finite domain of allosteric constants allows the occurrence of oscillations with a lower limit of L and an upper limit of c and in addition a low limit of the Hill coefficient is indicated for a given number of protomers $(n = 2)$ (Hess *et al.*, 1978; Goldbeter, 1977).

The results indicate the importance of the cooperativity function of the allosteric enzyme in the generation of oscillatory states. In accordance with biochemical experiments, any destruction of the cooperativity property results in a disappearance of the oscillation. The figure also illustrates that the Hill coefficient during the oscillatory state is near its maximum value with a lower limit of 1.6 for the condition of constant maximum activity of the enzyme. A computer solution of the model described above with respect to the periodic change of the two metabolites and the Hill coefficient is shown in Fig. 9, indicating the Hill coefficient near its maximum value. Although the model is analyzed for a dimeric structure of the enzyme, correlating studies have shown that larger subunit numbers do not significantly change the oscillatory properties of the system (Hess *et al.*, 1978; Plesser, 1975).

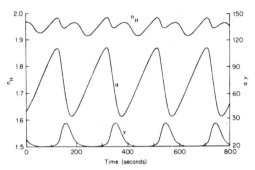

Fig. 9. Sustained oscillation of the Hill coefficient and of the metabolite concentrations in the dimer allosteric model for the phosphofructokinase reaction (from Goldbeter, 1977).

III. Physiology

The demonstration of the quantitative agreement between the oscillation analyzed in biochemical experiments and in the phosphofructokinase model leads to the conclusion that the dynamic behavior of the complex system can be reduced to the molecular property of a single protein species operating as a master enzyme in a biochemical pathway.

This view allows classification of the enzymes comprising the glycolytic pathway into various types of functions: 1) an input supplying the substrate, 2) the primary "oscillator" enzyme, 3) a feedback function, and 4) a sink.

Mechanistically, the functions are performed by two types of enzymes: allosteric enzymes with the potentiality of autonomous and stable oscillations such as phosphofructokinase and on the other hand enzymes such as Michaelis–Menten enzymes with kinetic properties insufficient to maintain an autonomous oscillatory state. In glycolysis it can be concluded that most of the enzymes of the Michaelis–Menten type are driven into an oscillatory state by induction through the autonomous activity changes of the phosphofructokinase propagated by periodically produced products like ADP. The sink and feedback function of the system are found in the activity of pyruvate kinase. However, it should be recognized that pyruvate kinase is also an allosteric enzyme and might follow its own autonomous periodicity for a given concentration range of phosphoenolpyruvate, but locks onto or synchronizes at the same frequency like phosphofructokinase because of being entrained by pulses of ADP as discussed above. Since the ATP pulses produced by pyruvate kinase feedback to phosphofructokinase, perfect synchronization is achieved. The sink and feedback function of allosteric pyruvate kinase might imply a strong influence of the enzyme on the oscillatory dynamics of glycolysis. Indeed, in model analysis it was shown that a period range of 2–70 min and a variety of waveforms, such as double period-icities are observed as a function of the concentration of pyruvate kinase (Plesser, 1975).

Oscillations are a predictable consequence of control mechanisms that involve gain and feedback. They may or may not be harmful. In the case of the slime mould *D. discoideum* the physiological function of the synchroniza-tion of several oscillating enzyme functions is well documented since it is a decisive part of the time-dependent aggregation mechanism within the life cycle of the organism (Hess *et al.*, 1978; Goldbeter and Caplan, 1976; Gerisch and Hess, 1974; Gerisch *et al.*, 1978). Although it has been observed that oscillation of glycolysis does occur in single yeast cells as well as large yeast cell population, its ·physiological function is not known. However, a significant influence on the ATP utilizing pathways by modulation of the adenylate energy charge has been suggested, a matter which is still under investigation (Goldbeter and Caplan, 1976).

The experimentally observed synchronization of glycolytic oscillations at fundamental and subharmonic frequencies illustrate how oscillators of dif-ferent periods in metabolism might couple synchronously. Thus, very low frequencies might be generated from higher frequency oscillators. Synchron-ization in both glycolytic and circadian oscillators have been observed. Synchronization of circadian rhythms by illumination and temperature cycles are well known, both in unicellular and multicellular systems. The range of entrainment of these rhythms with the driving frequency extends from 18 to 30 hours. These limits $(0.75 \leq T'/T_0 \leq 1.25)$ correspond to the domains observed in glycolytic oscillations (see Table II) (Dahlem Work-shop on the Molecular Basis of Circadian Rhythms, 1976; Boiteux *et al.*, 1975).

The possibility of frequency diminution in enzymic oscillators to very low frequency of epigenetic or even circadian rhythms has been suggested (Dahlem Workshop on the Molecular Basis of Circadian Rhythms, 1976). This may be done by beating mechanism, frequency division, or decreasing the oscillatory frequency by fitting source rate, sink, and enzyme activity appropriately. By suitable adjustment of the allosteric constant and turn-over numbers of the allosteric enzyme of a three-enzyme model composed of phosphofructokinase, pyruvate kinase and adenylate kinase for given input conditions periods of one hour have been computed in model studies (Plesser, 1975). Even with a simple model described above 24-hour periods have been obtained without a frequency division. Summarizing, it may be stated that enzymic oscillations might well be of physiological significance in the mechanism of a large number of biological phenomena, such as syn-chronization of cell metabolism, signal transmission, cell differentiation, as well as cell division and clock phenomena.

References

Annu. Rev. Biochem. (1977).

Boiteux, A., and Hess, B. (1974). *Faraday Symp. Chem. Soc.* **9**, 202–214.

Boiteux, A., Goldbeter, A., and Hess, B. (1975). *Proc. Natl. Acad. Sci. U.S.A.* **72**, 3829–3833.

Boiteux, A., Hess, B., Plesser, Th., and Murray, J. D. (1977). *FEBS Lett.* **75**, 1–4.

Chance, B., Pye, E. K., Ghosh, A. K., and Hess, B. (eds.) (1973a). "Biological and Biochemical Oscillators." Academic Press, New York.

Chance, B., Williamson, G., Lee, I. Y., Mela, L., and DeVault, D. (1973b). *In* "Biological and Biochemical Oscillators" (B. Chance, E. K. Pye, A. K. Ghosh, and B. Hess, eds.), pp. 285–300. Academic Press, New York.

Dahlem Workshop on the Molecular Basis of Circadian Rhythms (1976). "The Molecular Basis of Circadian Rhythms," Life Sciences Research Reports, no. 1.

Faraday Symp. Chem. Soc. (1974). "Physical Chemistry of Oscillatory Phenomena," no. 9. University Press, Aberdeen.

Gerisch, G., and Hess, B. (1974). *Proc. Natl. Acad. Sci. U.S.A.* **71**, 2118–2122.

Gerisch, G., Hülser, D., Malchow, G., and Wieck, M. (1975). *Philos. Trans. R. Soc. London B.* **272**, 181–192.

Glansdorff, P., and Prigogine, I. (eds.) (1971). "Thermodynamic Theory of Structure, Stability and Fluctuations." Wiley (Interscience), New York.

Goldbeter, A. (1973). *Proc. Natl. Acad. Sci. U.S.A.* **70**, 3255–3259.

Goldbeter, A. (1977). *Biophys. Chem.* **6**, 95–99.

Goldbeter, A., and Caplan, S. R. (1976). *Annu. Rev. Biophys. Bioeng.* **5**, 449–476.

Goldbeter, A., and Lefever, R. (1972). *Biophys. J.* **12**, 1302–1315.

Goldbeter, A., and Nicolis, G. (1976). *Prog. Theor. Biol.* **4**, 65–160.

Hess, B. (1968). *Nova Acta Lepold.* **33**, 195–230.

Hess, B. (1977). *TIBS* **2**, 193–195.

Hess, B., and Boiteux, A. (1968a). *Hoppe-Seyler's Z. Physiol. Chem.* **349**, 1567–1574.

Hess, B., and Boiteux, A. (1968b). *In* "Regulatory Function of Biological Membranes" (J. Järnefeldt, ed.), pp. 148–162. Elsevier, Amsterdam, The Netherlands.

Hess, B., and Boiteux, A. (1971). *Annu. Rev. Biochem.* **40**, 237–258.

Hess, B., Boiteux, A., and Krüger, J. (1969). *Adv. Enzyme Regul.* **7**, 149–167.

Hess, B., Brand, K., and Pye, E. K. (1966). *Biochem. Biophys. Res. Commun.* **23**, 102–107.

Hess, B., Goldbeter, A., and Lefever, R. (1978). *Adv. Chem. Phys.* (in press).

Hess, B., Boiteux, A., Busse, H. G., and Gerisch, G. (1975). *Adv. Chem. Phys.* **20**, 137–168.

Plesser, Th. (1977). "Dynamic States of Allosteric Enzymes." *In* Proc. 7th Int. Conf. Nonlinear Oscillations, Vol. 2, pp. 273–280. Akademie der Wissenschaften, Berlin.

Tamaki, N., and Hess, B. (1975). *Hoppe-Seyler's Z. Physiol. Chem.* **356**, 399–415.

Oscillatory Properties and Excitability of the Heart Cell Membrane

Robert L. DeHaan and Louis J. DeFelice

Department of Anatomy,
Emory University, Atlanta, Georgia

I. Introduction

Toward the end of the nineteenth century, Hermann (1899) applied the theory of current flow, initially developed for submarine cables by Lord Kelvin (1892), to the cylindrical excitable cells that characterize nerve and muscle. The Kelvin cable was generalized by Heaviside (1893), who included terms for self-induction to help support the signal over long distances and reduce their distortion during propagation (see also DeFelice, 1978). Hermann suggested that current flow described by cable theory may also repre-

181

sent the mechanism underlying propagation of nerve impulses. Only a few years later the concept that the electrical properties of such excitable cells may reside primarily in the cell membrane was introduced by Bernstein (1902, 1912), and an electrical equivalent circuit for the cell membrane was developed by Fricke (1925) and Cole (1928). These workers applied sinusoidal current analysis to suspensions of erythrocytes and other cell types and concluded that the cell membrane could be represented by a parallel resistance–capacitance (RC) circuit. Their estimates of membrane capacity, on the order of 1 μF/cm^2, have since been corroborated by modern electrophysiological and ultrastructural techniques with remarkable consistency.

The concept of the membrane as an RC system was strengthened when Hodgkin and Rushton (1946) published an experimental and theoretical analysis of the subthreshold responses of nerve axons to locally applied currents. This work emphasized the purely passive ("electronic") response of the membrane in the voltage region near rest, but also yielded important clues to its nonlinear properties. Real understanding of the nonlinear behavior of excitable cells was made possible when the "voltage clamp" technique, introduced by Cole and Marmont (1949), was exploited in the monumental studies of Hodgkin and Huxley (1952) in which they devised a set of semiempirical equations for describing the time- and voltage-dependence of membrane current.

Hodgkin and Huxley focused mainly on a description of the "macroscopic"† action potentials of the axon. However, previous work had already demonstrated that even in the subthreshold voltage range near rest, the view of the membrane as an RC circuit was too simple. Cole and Baker (1941) had applied ac impedance measurements to the squid giant axon, and discovered that the preparation behaved as if it contained an inductive element. At certain frequencies (around 100 Hz) voltage changes led current changes. The frequency at which this occurs is known as the resonant frequency. At the resonant frequency the impedance of the membrane has its maximum value; the impedance falls rapidly at frequencies on either side of resonance. The RLC circuit is said to be tuned at the resonant frequency. The excitable membrane can also be tuned to different frequencies by modifying the voltage across the membrane or other parameters that influence the R, L, or C elements of the circuit. Voltage is effective in tuning because the values of the resistive and inductive components in the membrane are voltage-dependent. Cole and Baker termed this behavior "anomolous

---

† We will distinguish macroscopic electrical activity as that which involves voltage displacements in the range of 10–100 mV, from "microscopic" or "subthreshold" events, which range from microvolts to a few millivolts.

impedance." More recent workers that have investigated membrane impedance by means of small-current perturbations have referred to the "phenomenological inductance" of excitable membranes (Mauro *et al.*, 1970), and have shown that inductive behavior arises from time- and voltage-dependent changes in membrane conductance (Hodgkin and Huxley, 1952; Chandler, Fitzhugh, and Cole, 1962; Mauro *et al.*, 1970). The impedance is usually measured under current clamp. It is also possible to measure the membrane admittance under voltage clamp conditions, where small transient currents are recorded in response to known input voltage changes. Single-value command potentials were replaced by clamp signals that varied with time (Palti and Adelman, 1969) as an effective method for measuring ionic conductance and membrane capacitance of the squid giant axon.

The small-signal response obtained by any of these methods provides a measure of membrane impedance and thus represents the transfer function of the system. This is equally true for the membrane response to its inherent stochastic conductance fluctuations recorded as electrical noise. Wanke *et al.* (1974) showed experimentally that the expected relationship between voltage noise, current noise, and impedance required an equivalent circuit model containing passive resistive and capacitive components as well as time-varying voltage-dependent conductances to provide inductive properties.

An important property of circuits that contain capacitive and inductive elements (RLC circuits) is that they tend to oscillate when perturbed and exhibit a resonant frequency (Chandler *et al.*, 1962; Mauro *et al.*, 1970). One of the simplest mathematical descriptions of oscillatory systems was first developed by van der Pol (1926), who devised an equation for a diode oscillator with a nonlinear relation between anode current and grid voltage. The van der Pol relationship is of the same class as an earlier equation given in 1883 by Lord Rayleigh for vibrating systems subject to internal dissipative forces. The voltage-time curve given by the solution of the van der Pol equation resembled the action potentials and pacemaker potentials observed in cardiac muscle fibers (van der Pol and van der Mark, 1928; see also Davis, 1962). It has since been shown that the van der Pol equation may be derived from a simple extension of the H-H model, and that with minor parametric adjustments, the equation yields low amplitude sinusoidal solutions similar to those for a simple harmonic oscillator (see Jack *et al.*, 1975, pp. 325–330).

Even the earliest extracellular electrical recordings from excitable cells revealed spontaneous, subthreshold oscillatory voltage fluctuations (Arvanitaki, 1939; see Guttman, 1969, for review of the older literature and Weidmann, 1971, for an historical approach). Less than a year after Ling and

Gerard (1949) discovered how to pull intracellular glass microelectrodes, the first transmembrane electrical recordings were made from heart cells (Woodbury *et al.*, 1950), revealing both macroscopic and microscopic oscillatory voltage changes. Some examples of such behavior are shown in Fig. 1.

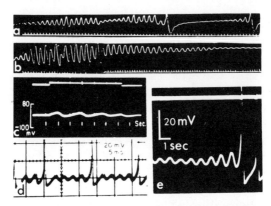

Fig. 1. Examples of spontaneous voltage oscillations recorded from various excitable preparations. (a) Extracellular recording from turtle sinus venosus after a brief exposure to 45 mM KCl. Temperature, 26°C; time marker, seconds; vertical scale, 0.4 mV (Bozler, 1943). (b) Extracellular record from cat atrial strip bathed in normal Ringer's solution. Temperature, 29°C; time, seconds; vertical scale, 0.2 mV (Bozler, 1943). (c) Transmembrane voltage oscillations in a quiescent sheep Purkinje fiber in response to a sustained weak depolarizing current, Tyrode's solution, temperature, 37°C (Trautwein and Kassebaum, 1961). (d) Squid axon exposed to low-calcium seawater and a constant weak depolarizing current; 25°C (Guttman and Barnhill, 1970). (e) Intracellular record of spontaneous underdamped oscillations recorded from a cluster of 6-day-old chick embryo heart cells in culture for 24 hours; 21°C (Fänge, Persson, and Thesleff, 1956).

In the present article we wish to consider the relationship between the spontaneous low-amplitude voltage fluctuations of the heart-cell membrane and its macroscopic rhythmic properties. We will examine the proposition that the rhythmic firing of heart cells, and that of many neural systems as well, arises from two distinguishable membrane properties: the RLC impedance characteristic that causes the membrane to respond to voltage or current perturbation in an oscillatory fashion, and an intrinsic stochastic, current noise generator that results from the random opening and closing of individual conductance channels, and that continuously activates the oscillatory response. To test these ideas experimentally, we have employed a model heart tissue system composed of spheroidal aggregates of embryonic heart cells prepared according to the methods of Sachs and DeHaan (1973). We will describe the electrical properties of this system in some detail.

II. The Cardiac Myocyte

A. CELL STRUCTURE

Heart tissue is composed primarily of discrete cardiac muscle cells, or myocytes, joined to neighbors by specialized junctions that provide adhesive strength and are responsible for the behavior of the tissue as an electrical syncytium. Cardiac myocytes in the adult mammal are generally irregularly shaped cylinders, containing a well-organized myofibrillar system of contractile proteins, and prominent elongate mitochondria. Myofibrils make up about 50% of the total cell volume; mitochondria about 32% (McAllister and Page, 1973). The myocytes branch and unite with neighboring cells to form a three-dimensional network. At magnifications of 1000 × or less (in the light microscope) it is difficult to tell whether heart muscle is composed of single cells or has a syncytial structure. With appropriate stains, periodic dark bands are seen running transversely across the myofibers in register with the z bands of the myofibrillar striations. These are the intercalated disks. In electron micrographs, the intercalated disks are seen to be highly specialized regions of end-to-end apposition of discrete heart cells, in which several types of intercellular junctions are most frequently seen. The disk is composed of the two membranes of the apposed cells bounded by dense granular material and separating an intercellular space of 2–3 nm. The parallel membranes within the disk are thrown into folds and fingerlike interdigitations. Most disks are of the "step-type" (Spira, 1971) with transverse segments located at different sarcomeric levels, joined by longitudinally oriented regions of membrane. A heart muscle fiber then consists of chains of myocyes a few cells thick joined end to end across intercalated disks.

Mammalian heart cells are remarkably uniform in size within each region of the heart, although they vary in size from region to region. Spira measured the length and diameter of 96 atrial cells sampled from three dog hearts as 92.2 × 10.2 μm with the standard error of the length being only 2% of the mean. Cat ventricular cells have been estimated as 9–12 × 125 μm, while cat atrial cells are about 6 μm in diameter (Weidmann, 1966; McNutt and Fawcett, 1969). Bonke (1973b) reports that cells of the SA node are much smaller, and spindle-shaped (length 15–20 μm, diameter in the nuclear region 5–8 μm and near the ends 2–3 μm). Myocardial cells of the domestic chicken embryo average 3–7 μm (Hirakow, 1970). Rat ventricle myocytes of about the same overall size as those of the cat have a sarcolemmal surface/cell volume ration of $0.35 \pm 0.028 \ \mu\text{m}^2/\mu\text{m}^3$ (McCallister and Page, 1973).

The cardiac myocyte membrane invaginates at frequent intervals in most mammalian hearts to form an extensive tubular network that penetrates deep into the volume of the cell to form the " T system." T tubules surrounding myofilament bundles, generally near z bands, serve to carry the extracellular milieu into close proximity with the innermost contractile material. The T tubules are 500–3000 nm in diameter and usually contain the same glycoprotein coat found on the external sarcolemma. Forssmann and Girardier (1970) have shown that the tracer molecule, horseradish peroxidase (MW = 40,000), can pass from the extracellular medium into the T system and penetrate even the smallest T tubules. The surface to volume ratio of rat ventricle cells calculated to include both sarcolemma and T tubular surface is about 12% greater than that determined with the sarcolemma alone (McCallister and Page, 1973). The amount of T system membrane is different in ventricular myocytes, atrial myocytes, and cells of the conductive system (Viragh and Challice, 1973). In the atrium and the conducting system (SA node, AV node, Purkinje fibers), which are generally composed of small-diameter myocytes, the T system is much reduced or absent. By reducing membrane area, this feature would lower total membrane capacitance and speed impulse conduction. Purkinje fibers and most mammalian atrial cells lack T tubules (Sommer and Johnson, 1970; Martinez-Palomo et al., 1970). Frog and chicken cardiac muscle lack a T system in both atria and ventricles.

B. THE MYOCYTE MEMBRANE

1. Structure

Each cell is surrounded by a plasma membrane, or sarcolemma, that acts as a selective barrier, separating ions and other substances in the cytoplasm from those outside. The sarcolemma is composed of a lipid bilayer, about 8 nm thick and associated glycolipids, proteins, and glycoproteins. Membrane proteins constitute 60–70% of the membrane mass in most cell types. The structure and properties of the plasma membrane have been the subject of several recent reviews (Robbins and Nicolson, 1975; Steck, 1974; McNutt, 1975; Wallach, 1975; Metcalfe and Warren, 1977).

To assume their lowest free-energy configuration, membrane phospholipids and glycolipids present the hydrophilic head regions of their structure to the bulk aqueous phases inside and outside the cell and sequester their nonpolar tail portions inside the hydrophobic interior of the membrane, away from the aqueous environment. This produces the bilayer configuration originally proposed by Danielli and Davson (1935) which has since been supported as the basic framework of the membrane by an overwhelming body of evidence provided by a wide range of techniques: X-ray diffraction (Blaurock, 1971), differential scanning calorimetry (Papahadjo-poulos *et al.*, 1973), electric spin resonance (Jost *et al.*, 1973) and freeze-fracture electron microscopy (Deamer and Branton, 1967). Neutral lipids such as cholesterol are intercalated into the lipid bilayer with the more hydrophilic parts of their structure facing the aqueous environment (Jain, 1975). In all cells studied, the components of the membrane are distributed asymmetrically across the two leaflets of the lipid bilayer. With chemical labeling techniques, Bretscher (1972, 1973) demonstrated that phosphatidy-lethanolamine and phosphatidylserine are located mainly in the inner membrane surfaces of the human erythrocyte, while phosphatidylcholine and sphingomyelin occur predominantly in the outer half (Zwaal *et al.*, 1973).

A continuous lipid bilayer separating the two aqueous compartments inside and outside a cell would act as a high-resistance insulator against ion movements (Parsegian, 1969). A large literature on model lipid membranes (reviewed in Ehrenstein, 1971) indicates that such layers have electrical resistances of about $10^8 \, \Omega \cdot cm^2$. This value is much greater than the resistance of biological membranes. Nonetheless, the common phospholipid component of biomembranes clearly has an important role in the maintenance of membrane potential. This is supported by experiments with phospholipase C, an enzyme that hydrolyzes the ester bond between glycerol and the phospholipid phosphate. In both heart (DeMello, 1972) and skeletal muscle (Albuquerque and Thesleff, 1967) a gradual irreversible depolarization results after exposure to phospholipase C, with a reduction of input resistance of 40–80%. Interestingly, spontaneously active cells such as those that comprise the sinoatrial node and Purkinje fibers of the heart are more resistant to the effects of phospholipase than are those that make up working ventricular muscle of the heart (DeMello, 1972).

2. Passive Electrical Parameters

The passive membrane electrical properties reported for a variety of heart tissue preparations are listed in Table I. The value of approximately 1 $\mu F/cm^2$ for membrane capacitance (C_m) is consistent with the concept that a lipid bilayer about 8 nm thick is the main site of charge separation in the heart cell membrane. However, the reported membrane resistance of about $2 \times 10^4 \, \Omega \cdot cm^2$ is four orders of magnitude less than that of an unbroken lipid layer.

Precise measurements of C_m and R_m are subject to two major sources of error. Both are dependent (in reciprocal fashion) on estimates of the area of membrane across which current flows between the cytoplasmic compartment of a tissue and the extracellular medium. Early determination of passive membrane properties based on measurements of input resistance and $\tau_m = R_m C_m$ were referred only to the outer cylindrical surface area of the

Table I

PASSIVE ELECTRICAL PROPERTIES OF HEART TISSUE

Tissue	R_m (kΩ · cm²)	R_i (Ω · cm)	C_m (μF/cm²)	$τ_m$ (msec)	$λ$ (mm)	Vel. (m/sec)
Purkinje fiber:						
mammal	19.0[d]	105[d]	1.2[d]	24.1[g]	2.2[g]	1–4[q]
	20.0[e]	181[e]	1.0[a]	19.5[e]	1.9[e]	—
Ventricular trabeculae:						
mammal	13.8[e]	470[e]	0.8[e]	4.4[e]	0.88[e]	0.75[e]
	—	—	1.25[f]	3.2[f]	—	—
Atrial trabeculae:						
dog	3.0[g]	500[a]	1.31[f]	14.6[f]	1.24[f]	—
rabbit	—	—	1.3[k]	3.0[g]	0.66[g]	0.7[j]
Crista terminalis:						
rabbit	—	500[a]	1.27[g]	2.7[g]	0.99[g]	0.5–0.8[j]
S.A. node:						
rabbit	12.0[i]	—	1.0[a]	12.0[i]	0.47[h]	0.37[i]
Frog ventricle strips:	2.6[c]	460[c]	1.2[c]	—	—	—
Cultured heart:						
chick (strand)	20.5[l]	180[l]	1.54[l]	—	—	0.002–0.3[n]
	16–136[m]	245[m]	1.46[m]	29–269[m]	0.95–3.2[m]	—
chick (aggregate)	18.2[b,o]	—	1.25[b,o]	17[p], 22.6[o,r]	—	—

[a] Assumed. [b] Measurement made at −67.0 mV. [c] Van der Kloot and Dane (1964). [d] Mobley and Page (1972). [e] Weidmann (1970). [f] Sakamoto and Goto (1970). [g] Bonke (1973a). [h] Bonke (1973b). [i] Noma and Irisawa (1976). [j] Masuda and Paes de Carvalho (1975). [k] Paes de Carvalho, Hoffman, and Paes de Carvalho (1969). [l] Lieberman et al. (1975). [m] Sachs (1976). [n] Lieberman et al. (1973). [o] Present work. [p] DeHaan and Fozzard (1975). [q] Cranefield (1975). [r] RmCm not measured from pulse rise time.

muscle bundle, and thus resulted in values of R_m of $1\text{–}2 \times 10^3$ Ω · cm² and C_m of 10–20 μF/cm², i.e., both off by a factor of about 10. The introduction of stereological techniques to provide more accurate estimates of the actual areas of membrane surrounding heart cells (Mobley and Page, 1972; McCallister and Page, 1973) brought these values into the more realistic range shown in Table I. However, in intact heart tissue cell shapes are irregular and complex, and both shape and cell size vary widely. Thus, values for these electrical parameters cannot be given with much precision.

The second large source of error in measuring R_m (though not C_m) is its voltage dependence near rest potential. It has been known for many years that as membrane voltage depolarizes from maximal diastolic potential toward threshold, membrane resistance increases (Weidmann, 1951; Noble, 1962). Where careful measurements have been made in model systems, the increase can be as much as eightfold (Sachs, 1976; see below). This fact has been virtually ignored by most previous workers, and has undoubtedly contributed to the large scatter in reported values. A third source of error arises from the use of stimuli too short to allow the membrane response to settle. It will be shown below that the settling time in heart cell membranes may be several seconds. This is long compared to nerve membrane.

Because of the low ionic permeability of lipid layers, it has often been suggested that ions must travel through channels or pores (Armstrong, 1975; French and Adelman, 1976). Many materials are known that enhance ionic permeability of model bilayers. The best studied of these are cyclic polypeptides that form conductance channels in the layer (Ehrenstein, 1971). Such molecules can produce ion-specific, voltage-dependent conductances remarkably like those observed in biological membranes (Bangham, 1968; Hall, 1975). In excitable living membranes these channels are apparently represented anatomically by globular proteins that penetrate the membrane. When cells are prepared for examination by the freeze-fracture technique (Branton, 1966) in which the plasma membranes are cleaved preferentially along the internal plane between the two lipid leaflets (Deamer and Branton, 1967) the exposed surfaces are seen to be scattered with 8–10 nm protein particles, distributed more or less randomly as singlets and small clusters. These particles, which are termed "integral membrane proteins" (IMP) (Singer and Nicolson, 1972), tend to be large globular amphipathic structures that require strong chaotropic agents (organic solvents, detergents, etc.) for solubilization (Steck and Yu, 1973). When isolated, they are often bound strongly to lipids and/or carbohydrates.

In freeze-fracture preparations, it can be seen that many of the IMP particles extend through both halves of the bilayer, thus being candidates for the putative ion channels. Other particles are restricted to only one of the two lipid leaflets, and tend to be asymmetrically distributed, with many more associated with the inner leaflet (P-face) than with the external leaflet (E-face). Raynes *et al.* (1968) found 400–700 particles/μm^2 of surface area in the E-face in the membrane of guinea pig ventricular myocytes. Similar numbers have been observed in other cell types (Hasty and Hay, 1977; Schotland *et al.*, 1977).

3. Fluidity

In view of these observations, early models of a "unit membrane" structure (Davson and Danielli, 1952; Robertson, 1959) have been replaced in recent years by the "fluid mosaic" model of Singer and Nicolson (1972). This model takes into account the evidence that the IMPs are intercalated into and may traverse the lipid bilayers, thereby permitting resistance values in the proper range. It also recognizes that the bilayer forms a dynamic fluid matrix in which the IMPs float, free to move unless stabilized by linkage to cell cytoskeletal elements at the inner membrane surface or by the thick glycoprotein coat on the outer surface.

There is good evidence for lateral mobility of membrane components.

Measurements with fluorescence optics (Edidin, 1974; Schlessinger et al., 1976) and nuclear magnetic and electron paramagnetic resonance spin-label studies indicate rapid phospholipid planar diffusion (Jost et al., 1973), but much slower flip-flop or rotation of lipid components from one side of the membrane to the other (Sherwood and Montal, 1975; Ehrenstein et al., 1975). Membrane proteins also diffuse laterally in the membrane, but less rapidly than membrane lipids (Schlessinger et al., 1976; 1977; Edidin et al., 1976). Moreover, different protein and glycoprotein components move laterally at different rates (Frye and Edidin, 1970; Edidin and Weiss, 1974; Pinto da Silva, 1972; Robbins and Nicolson, 1975) and there is a substantial class of protein or glycoprotein complexes that display only very slow lateral motion or none at all (Schlessinger et al., 1976).

Integral membrane proteins and glycoproteins are thought to form not only the ion diffusional channels, but also other "functional" units of the plasmalemma: hormone receptors (Kahn, 1976), antigenic and lectin binding sites (Phillips and Perdue, 1976; Schlessinger et al., 1976), and transmembrane transferases or pumps (Besch et al., 1976; Garrahan and Garay, 1976). According to the fluid mosaic model, functional activity of some membrane components requires lateral movement and association of two or more IMP particles. This idea was originally proposed by Singer and Nicolson (1972) and has since been amplified by others (Cuatrecasas, 1974; DeMeyts et al., 1976; Kahn, 1976). Once individual IMP particles come into association to form a functional complex, their lateral diffusion may be restricted by linkage with submembrane cytoskeletal elements (Yahara and Edelman, 1973; Wickus et al., 1975) or by interaction with proteoglycans of the outer cell coat (Hynes, 1973). Aggregations of membrane particles are often seen in cells prepared for freeze-fracture (McNutt, 1975). In some cases the function of the aggregation is recognized (Scott et al., 1973; Kahn, 1976), but often it is not certain (Tilney and Mooseker, 1976). Rash and Ellisman (1974) have found orthogonal or "square" arrays of IMP particles in the sarcolemma of rat skeletal muscle, but absent from the neuromuscular junctions. They have pointed out that this distribution correlates with the distribution of electrical excitability. Similar arrays have since been seen on heart muscle membranes (McNutt, 1975). However, since orthogonal particle groups have also been visualized on nonexcitable intestinal epithelial cell membranes (Rash et al., 1974) and on glial cells (Landis and Reese, 1974), they may have no necessary relation to excitability.

4. Cell Junctions

Functional aggregations of membrane proteins are most commonly seen in zones of cell–cell contact. Three kinds of cell surface specializations are associated with the contact regions in the intercalated disks of heart tissue:

(1) desmosomes (macula adherentes) which are generally thought to strengthen cell adhesions in the tissues (Overton, 1977); (2) fascia adherentes, regions where actin filaments associated with the myofibrils attach to the inside of the cell membrane (McNutt, 1975); and (3) nexuses ("gap junctions") that provide low-resistance diffusional channels through which ionic currents and some molecules can pass from cell to cell (Pollack, 1976). There is now an overwhelming body of evidence that heart cells of all species are electrically coupled across cell junctions (reviewed in DeHaan and Sachs, 1972; McNutt, 1975; Kensler *et al.*, 1977). When current is injected through an intracellular microelectrode into a single cell in a myocardial fiber, the degree of electrical coupling with an adjacent cell can be measured directly by comparing the change in membrane voltage produced in the two cells. The length constant λ, i.e., the distance over which the subthreshold voltage pulse decreases by a factor e, differs in various parts of the heart (Table I) but is generally reported in the range 0.5–1.5 mm (Spira, 1971; Weidmann, 1970; Bonke, 1973a,b; Seyama, 1976). Such values of λ equivalent to many cell lengths indicate that longitudinal cell–cell resistance must be very low. Measured values of longitudinal resistivity in myocardial fibers (Table I, R_i) generally fall in the range of 100–500 $\Omega \cdot$ cm. Since these measurements are usually made between intracellular electrodes spaced several cell lengths apart, they indicate that the intercellular junctional resistance does not add appreciably to the resistance of the cytoplasm. When combined with measurements of junctional area, the junctional resistance between myocardial cells has been estimated at less than 5 $\Omega \cdot$ cm^2 (Woodbury and Crill, 1961; 1970; Heppner and Plonsey, 1970; Matter, 1973). This is three to four orders of magnitude less than the specific resistance of the nonjunctional cardiac myocyte membrane, but about four orders of magnitude greater than the specific resistance of a 10-nm-thick layer of cytoplasm, assuming 100 $\Omega \cdot$ cm resistivity. This difference provides evidence that the conductance pathways between cells occupy only a small portion of the area of contact between cells.

There is substantial evidence implicating the nexus as the ultrastructural site of the low-resistance conductance path between adjacent cells in heart, liver, and a wide variety of other tissue types (Bennett, 1973; DeHaan and Sachs, 1972; McNutt and Weinstein, 1973). Viewed with the electron microscope in thin sections cut perpendicular to the membrane, the nexus is seen to be a region in which the plasma membranes of adjoining cells are brought into close apposition. The intercellular space that can be penetrated by a marker such as colloidal lanthanum is reduced to 2 nm or less (McNutt and Weinstein, 1970). In tangential sections in which the nexal structure is viewed face on, the lanthanum impregnated region appears as a network of channels, surrounding electron-lucent globular subunits. These subunits in heart cells are 7–8 nm in diameter and are generally packed in hexagonal arrays with 9–10 nm center-to-center spacing (Brightman and Reese, 1969; McNutt and Weinstein, 1973). In the center of each subunit can be seen a pit, thought to be the mouth of a channel which penetrates through each subunit to open into cytoplasmic space. When prepared by freeze-fracture, the nexal subunits appear as particles that span both the lipid bilayer of the apposed membranes and the 2-nm intercellular gap. Particles of each membrane appear to abut against each other in the intercellular space, thus providing a framework for the putative intercellular channels to connect the adjoining intercellular compartments. This structure has been modeled by several investigators (Bennett, 1973; McNutt and Weinstein, 1973; Loewenstein, 1973; Kensler et al., 1977). The penetration of the cell membranes by nexal proteins has been confirmed by X-ray diffraction studies (Makowski et al., 1977), and in a highly purified preparation of mouse hepatocyte nexal junctions, Goodenough (1976) has been able to visualize slender channels 1–2 nm in diameter by 15 nm long, spanning the full profile thickness of the nexus. From such preparations, he has also been able to isolate what appears to be a single 9000-dalton junctional protein, termed Connexin B.

The image of transnexal channels, 1–2 nm in diameter, joining apposed cells is consistent with data on the rates of intercellular diffusion of ions and molecules from cell to cell in cardiac tissue. When the cut end of a bundle of ventricular fibers is exposed to an isotopically labeled substance and the bundle is continuously washed with Tyrode's solution, the label diffuses down the length of the fibers within the myoplasm across the low-resistance intercellular junctions. With this technique, Weidmann (1966) reported that the permeability of the intercellular junctions to movement of ^{42}K from one cell to the next is about 5000 times greater than the permeability of the surface membrane to outward diffusion of the ion. By comparing the rates of intercellular diffusion of $[1 - {}^{14}C]$tetraethylammonium (TEA$^+$, MW 130) and of the fluorescent dye Procion Yellow (PY^{3-}, MW 700) with that of ^{42}K (at. wt. 39), Weingart (1974) calculated P junction/P surface membrane ratios of 21,000, 200, and 9600, respectively, for these tracers. From consideration of the van der Waal's sizes of these ions, he estimated a minimal nexal pore size of 1.0–1.4 nm. Equations describing tracer diffusion using two-compartment kinetics (MacDonald, Mann, and Sperelakis, 1974) indicate that Weidmann's original calculations (1966) may have been in error in not accounting adequately for the nexal diffusion of tracer from the cytoplasm into the intercellular space. However, the fluorescent dyes are virtually impermeant to the plasma membrane. Nonetheless, fluorescein (MW 332) injected directly through a micropipette into a Purkinje fiber cell diffuses into neighboring cells at a rate predicted precisely by its molecular weight in relation to that of K$^+$, TEA$^+$, and PY^{3-} (Pollack, 1976). Indeed, the linear relationship between molecular weight and permeability of four substances that differ greatly in there permeability across the plasmalemma argues strongly for a common mechanism in their movement from cell to cell, and is consistent with their diffusion through channels of 1.0–1.4 nm diameter. With a size-graded series of 19 fluorescent probes ranging from 251 to 4158 daltons, Simpson *et al.* (1977) have confirmed that junctional permeability between salivary gland cells is linearly (inversely) related to molecular weight, but found a sharp cut-off for molecules above 1200 daltons. From these data, they also calculated an effective channel diameter of 1.4 nm, and a lower limit of conductance of 10^{-10} mhos for a single junctional channel.

From his diffusion experiments with K^+, Weidmann (1966) calculated a resistance for the intercalated disk of about $3 \ \Omega \cdot cm^2$. Similar values were reported by Tanaka and Sasaki (1966) for mouse myocardium, and by Woodbury and Crill (1961) for rat atrium. Spira estimated a value of 1.4 $\Omega \cdot cm^2$ for dog atrial fiber junctions. Most of the authors agreed that these values were probably high, subject to reduction when better measurements of nexal area became available. On this basis, Woodbury and Crill (1970) have since revised their estimated disk resistance down to $0.3 \ \Omega \cdot cm^2$. On the basis of morphometric measurements from electron micrographs of rat ventricular muscle, Matter (1973) has estimated the number of channels in a 47 μm^2 nexus as 6.7×10^5. Assuming intercellular channels of 1.0 nm diameter, filled with aqueous solution containing 150 mM/l K^+, he calculated a specific nexal resistance of $0.05 \ \Omega \cdot cm^2$.

Some of the apparent variability in estimated nexal resistance may result from the use of different cardiac tissues. Pollack's measurements of intercellular fluorescein diffusion indicated that the nexal conductance in atrial fibers is about 5 times greater than in a Purkinje fiber, and 1000 times greater than between cells in the AV node (Pollack, 1976). This corresponds with the well-known differences in conduction-velocities in various tissues of the heart (Table I). A typical value for Purkinje fibers is 1–4 m/sec, and for AV node 0.05 m/sec (Cranefield, 1975).

It is reasonable to suppose that variation in intercellular resistance in different parts of the heart may reflect differences in number of nexal junctions per cell (DeFelice and Challice, 1969; Martinez-Palomo *et al.*, 1970; Viragh and Porte, 1973), or number of channels per nexus (Pollack, 1976). However, it is also known that within a given tissue, intercellular resistance can be modulated by several agents. Iontophoretic injection of calcium ions into a canine Purkinje fiber cell reduces transnexal electrical coupling to neighboring cells to about 20% of pre-injection values (DeMello, 1975). This suggests that nexal channel conductance may be subject to intrinsic regulation via a mechanism that controls intracellular $[Ca^{2+}]$ (Loewenstein review, 1973). It is of interest that nexal decoupling in crayfish axon septa is associated with ultrastructural alterations. Treatment with dinitrophenol (DNP) or calcium-free solutions in this preparation led to an increase in nexal resistance similar to that in the dog heart. Freeze-fracture analysis indicated that nexal subunit spacing decreased by about 30% in decoupled nexal junctions. In addition, thickness of the intercellular gap, subunit particle size, and total junctional width were all reduced in those junctions with significant decreases in ionic permeability (Peracchia and Dulhunty, 1976). Analogous changes are reported in rat stomach gap junctions after exposure to DNP or hypoxic conditions (Peracchia, 1977).

III. Impedance and Oscillation

Against this background of the molecular structure of the cardiac myocyte sarcolemma, we wish to draw the connection between the small signal impedance of such excitable membranes and the occurrence of repetitive firing of action potentials. We have seen that cardiac tissue may be treated as an electrical syncytium. With appropriate attention to isopotentiality the excitable membrane may be studied by penetrating any cell within a tissue, although this ideal condition may be approached closely only in tissue models of the heart (see Section V). The best descriptions of the currents that flow through excitable membranes derive from studies of the squid giant axon, precisely because of its simple geometry. No fragment of heart tissue has such a simple organization. Nonetheless, cardiac membrane currents have been assessed in Purkinje fiber preparations and other cardiac tissues. We begin our discussion of the electrical properties of excitable membranes by comparing the nerve axon and the Purkinje strand, which have many qualitative features in common, but differ primarily in the fact that the cardiac membrane is about one hundred times slower than the nerve membrane in its electrical response.

Hodgkin *et al.* (1949) introduced the theory of excitability by showing that the total current through membranes is composed of distinguishable specific ion currents. In the membrane of the squid giant axon, three ionic components are required to reproduce the essential features of the transmembrane potential. The conductances that control the flow of these currents are voltage sensitive (i.e., the value of the conductance depends on the electric field across the membrane), and they exhibit pronounced time delays in shifting from one value of conductance to another in response to a rapid change in voltage. This time-dependent conductance manifests itself as a reactance in the frequency domain and forms the basis for the complex impedance of excitable membranes. Cardiac membranes appear to require more than three ionic current components to account for the full range of observed transmembrane potentials. However, the conductances that control the flow of currents in heart cells are also voltage- and time-dependent and therefore display a complex impedance analogous to that found in the nerve membrane.

The basic equation that relates the current through the membrane to the voltage across it is (Hodgkin and Huxley, 1952)

$$I = C(dV/dt) + I_K + I_{Na} + I_L$$

where the capacitative, specific ionic, and general leakage currents are shown separately. We consider only the nonpropagated, or membrane potential and ignore local circuit, or cable current flow. The K current is related to voltage by

$$I_K = \bar{g}_K n^4 (V - V_K)$$

where $\bar{g}_K$ and V_K are constants and n depends on voltage and time. The variation in I_K is taken at a particular steady state. Thus

$$\delta I_K = \bar{g}_K n_V^4 \, \delta V + 4\bar{g}_K n_V^3 (V - V_K) \, \delta n_V$$

where n_V is taken at the steady-state value of n at some V. From the equations that relate n to voltage and time, n_V may be calculated and eliminated from the above equation, resulting in an expression for the ratio $\delta V/\delta I_K$. Transformation from the time to the frequency domain results in the definition of the small signal impedance (Z_K) for the K system. The equivalent circuit for the membrane conductance due to potassium is shown in Fig. 2 (cf. Mauro

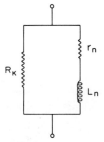

Fig. 2. Equivalent circuit for the membrane K conductance.

et al., 1970). The same circuit applies to the entire voltage range, but the values of R_K, r_n, and L_n vary for each voltage. The quantities r_n and L_n represent the time-varying conductances of the K system. They are often referred to as phenomenological circuit elements, since they are derived from a purely descriptive point of view. For example, r_n and L_n are both negative quantities for $V < V_K$.

The exact derivation of the small signal impedance is given most clearly in Mauro *et al.* (1970). A similar treatment to that above for K must be applied to the Na system and the leakage system; addition of the capacitance of the lipid bilayer yields a complete description of the impedance of the standard excitable membrane as a parallel combination of Z_K, Z_{Na}, Z_L, and $Z_C = 1/i\omega C$. The leakage current in the H-H model is time-independent and therefore gives rise to a purely resistive term. Figure 3 shows how the total

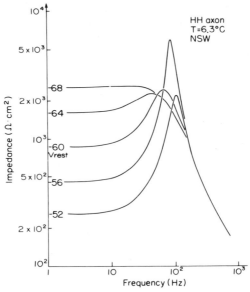

Fig. 3. The small signal impedance of the giant axon of the squid in normal sea water as calculated from the standard equations of Hodgkin and Huxley (1952) on log–log coordinates. Mean membrane potential is shown as a parameter in millivolts.

impedance (Z) of the nerve membrane depends on voltage and frequency. At −60 mV, which is the normal resting potential (V_{rest}), the impedance increases about threefold at 77 Hz from its value at dc. The maximum value of the impedance occurs at 54.35 mV at 93.3 Hz. This point may be taken as a definition of threshold, since at that potential and frequency the smallest current should drive the voltage out of the linear range and produce an action potential. This is an important concept to which we shall return, in connection with spontaneous firing in both nerve and heart cells.

In experiments with nerve or cardiac tissue preparations, it is often possible to eliminate pharmacologically one of the current components without appreciable effect on the others. For example, the application of the drug tetrodotoxin (TTX) to seawater bathing a nerve axon effectively eliminates

I_{Na} and therefore the component of the impedance due to Na (Z_{Na}). In this condition, the observed impedance of the membrane should be dramatically altered. Pharmacological agents that reduce other currents with varying degrees of specificity are also known. Tetraethylammonium (TEA⁺) blocks I_K; Verapamil or its derivative D-600 interferes with the slow inward current (I_{si}) of heart tissue (see below).

The effect of blocking I_K in the nerve axon, which may be approximated to exposing the nerve to an appropriate concentration of TEA⁺, can be computed with the standard H-H equations. Normally, $\bar{g}_K = 32.4 \, \text{mS/cm}^2$ (a Siemans is one inverse ohm, or mho). By varying $\bar{g}_K$ about its normal value, marked effects of total impedance (Z) are produced at a given potential. Figure 4a shows the effect of depolarizing the membrane from V_{rest} by 4 mV (to -56 mV) and varying $\bar{g}_K$. The normal Z relation $(\bar{g}_K = 32.4 \, \text{mS/cm}^2)$ is shown for -56 mV in Fig. 3. Reducing $\bar{g}_K$ by only about 13% to 28.2

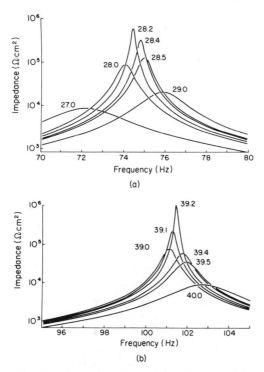

(a)

(b)

Fig. 4. The small-signal membrane impedance of the giant axon of the squid, as calculated from the standard equations of Hodgkin and Huxley (1952). The maximum conductance of the potassium system, $\bar{g}_K$, has been systematically varied from its normal value of 34.2 mS/cm², and is shown as a parameter. The impedance relations at (a) -56 mV, (b) -54 mV (unpublished calculations of L. Barfield and L. J. DeFelice).

mS/cm^2 results in a dramatic shift in peak Z from about 6 $k\Omega \cdot cm^2$ at 92 Hz to 900 $k\Omega \cdot cm^2$ between 74 and 75 Hz. The effects on Z are so great that the assumption of linearity for small signal perturbations no longer applies. This will become more obvious when we calculate the effect of a similar decrease in $\bar{g}_K$ on macroscopic behavior. The effect of increasing $\bar{g}_K$ on the total impedance at 6 mV depolarization $(-54$ mV) is shown in Fig. 4b. In this case, the impedance peaks at a resonant frequency above its normal value, between 101 and 102 Hz.

The impedances shown in Fig. 4 were calculated from the linearized Hodgkin–Huxley equations for the standard axon by varying $\bar{g}_K$ and mean membrane potential. Action potentials can be computed from the same equations. The effect of varying $\bar{g}_K$ on action potentials stimulated by a single brief depolarizing pulse is shown in Fig. 5 on a compressed time scale, to emphasize the events following the single stimulus. Lowering $\bar{g}_K$ by about 30% results in rapidly damped oscillatory behavior of the potential after the stimulus (Fig. 5b) with a period of about 15 msec. Reducing $\bar{g}_K$ to 23.0 caused the oscillating after-potential to cross threshold, and results in a second action potential (Fig. 5c). A further decrease to $\bar{g}_K = 21$ mS/cm^2 causes repetitive firing in response to the initial stimulus, with a period of about 22 msec or 46 spikes per second (Fig. 5d). Notice that the shapes of the action potentials do not change appreciably between panel a and d; only the oscillatory behavior of the membrane near rest potential changes. Reducing $\bar{g}_K$ is equivalent to decreasing the density of K channels in the membrane. One effect of this change is to reset the steady-state potential of the membrane to a level that is depolarized from normal $(-60$ mV, referred to as 0 in Fig. 5). As $\bar{g}_K$ is reduced, the impedance increases and the potential rebounds from each action potential toward a more depolarized level, which it overshoots as a result of the inertial effect of the inductance (time-varying conductance). When $\bar{g}_K$ is reduced sufficiently (Fig. 5d), the rebound after each action potential crosses threshold and the system becomes repetitive. Note that the steady potential in Fig. 5b is depolarized with respect to Fig. 5a, and that the initial action potential amplitudes increase as $\bar{g}_K$ decreases.

Figures 4 and 5 were produced from the same set of equations. It is clear that the effect of varying $\bar{g}_K$ on the small signal impedance is to increase its resonant behavior about some mean voltage, and to cause the membrane to fire repetitively in response to a single stimulus. It is difficult to derive the exact analytic relationship between the resonant behavior of membrane impedance and spontaneous firing. However, the simulation in Figs. 4 and 5 shows that the two are related and that a discussion of the mechanism of repetitive firing may be approached through the membrane's impedance.

This same line of reasoning can be applied to any excitable membrane. Figure 6 contrasts the node of Ranvier with the Purkinje heart cell mem-

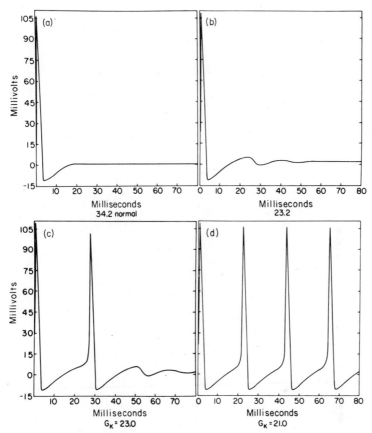

Fig. 5. A series of responses of the standard Hodgkin and Huxley (1952) equations to a brief depolarizing stimulus, using different values of $\bar{g}_K$ in units of MS/cm². The voltage response is given on a relative scale, where the resting potential is taken as zero. (Unpublished calculations of D. Strawn and L. J. DeFelice.)

brane using two sets of current-voltage relationships derived from macroscopic measurements. The impedance is shown in three-dimensional perspective with the magnitude of Z plotted vertically against frequency and mean membrane voltage. Figure 6a shows the total impedance of the frog node calculated from the complete voltage clamp equations of Hille (1971) and Dodge (1963), by linearizing the nonlinear equations about successive voltages. Similar information was expressed in a different form by Clapham and DeFelice (1976), who also calculated the effect on the impedance of omitting separately the Na and K systems. At hyperpolarized potentials (-100 mV) the impedance is approximated by an RC circuit. Upon depolarization, a resonance appears which has its maximum value near -65 mV

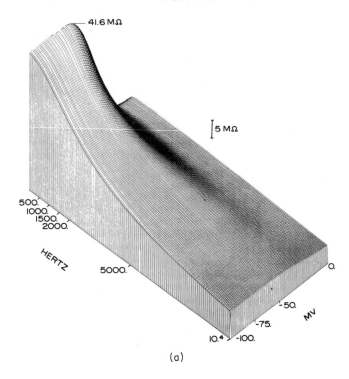

(a)

and 250 Hz. This is qualitatively the same behavior seen in the squid axon in Fig. 3. Note that the values in Fig. 6a are referred to nodal area (about 50 μm^2). Figure 6b gives a similar calculation from the Noble (1962) equations for the Purkinje heart cell fiber, except that in this case the Na conductance has been omitted. The resulting impedance, called Z_K, has a resonant value near 1 Hz of about 60 k$\Omega \cdot cm^2$ referred to the total membrane area. Resistance values given in Noble (1962) are one-tenth those used here, because Noble's calculations referred resistance and capacitance measurements to the outer cylindrical surface area of the Purkinje fiber, rather than to the total cellular membrane (Page and McCallister, 1973). This correction has been made in the calculations for all figures shown. It is important to note that the resting heart cell membrane has its maximum impedance near 1 Hz, in contrast to the squid axon and node membrane, which peak near 100 Hz. The fact that we have compared the total nerve impedance (Figs. 3 and 6a) with the heart impedance lacking its fast Na system (Fig. 6b) does not detract from the comparison since the K system dominates the nerve membrane impedance at rest (Conti and Palmieri, 1968). Similarly, the use of the original modified Hodgkin–Huxley equations for Purkinje fibers (Noble, 1962) to calculate the small signal impedance of Fig. 6b, rather than more

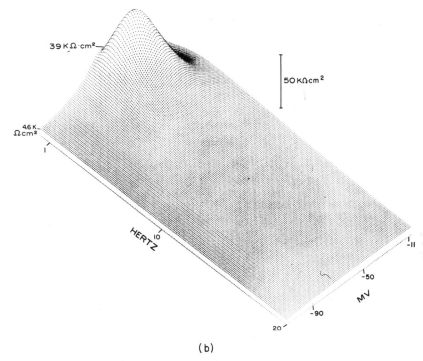

(b)

Fig. 6. The modulus of the small-signal impedance of two excitable membranes. The impedance plane is shown in three dimensional perspective, as a function of frequency and membrane voltage. The dc value is taken as 0.001 Hz. (a) Impedance of the frog node of Ranvier, calculated from the equations of Hille (1971) and Dodge (1963), referred to the area of a standard node. The resting potential is −75 mV. [Unpublished calculations of D. Clapham and L. J. DeFelice. See also Clapham and DeFelice (1976).] (b) The impedance of the Purkinje heart cell membrane calculated from the original model of Noble (1962), having omitted the fast Na system from circuit. Note the different frequency scales in (a) and (b). (Unpublished calculations of D. Clapham and L. J. DeFelice.)

recent modifications (McAllister *et al.*, 1975; Beeler and Reuter, 1977) does not materially affect the qualitative comparison being made here between nerve and heart excitable membranes, as will be seen below.

The relationships between the small signal impedance of excitable membranes and subthreshold oscillatory behavior was originally drawn by Hodgkin and Huxley in 1952. Huxley later (1959) investigated the response of the H-H equations to various values of extracellular calcium concentration. Frankenhauser and Hodgkin (1957) demonstrated that the major effect of altering calcium ion concentration is to shift the voltage dependence of the ionic conductances. At low concentrations of external Ca, subthreshold oscillations occurred for certain stimuli. These gradually increased until excitation occurred and action potentials were produced. At sufficiently low values of external calcium (approximately one-third of the normal value of 44 mM) all stimuli produced responses that ultimately grew into trains of repetitive action potentials.

We may investigate this phenomenon further experimentally by studying the spontaneous voltage fluctuations which occur across the squid giant axon membrane in low Ca^{2+} solutions. At two-thirds normal external calcium, the spontaneous voltage fluctuations that occur are similar to white noise passed through a highly tuned filter. The zero crossings are nearly regular, but the envelope of the wave is random. Figure 7 shows the voltage

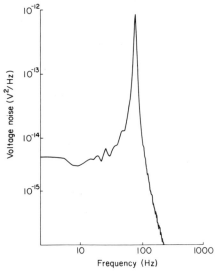

Fig. 7. The power spectral density of spontaneous voltage noise measured from the giant axon of the squid *Loligo vulgarius* under space clamped conditions in two-thirds normal external Ca^{2+}. The vertical scale is in log units of V^2/Hz versus a logarithmic frequency scale. Temperature, 8°C. Details of the preparation and set-up used for this experiment are given in Wanke *et al.* (1974). (Unpublished data of E. Wanke, L. J. DeFelice, and F. Conti.)

spectral density of a one-minute sample of voltage noise data from the squid axon in low external calcium. The maximal power occurs near 100 Hz, which coincides approximately with the impedance peak for this preparation (Wanke *et al.*, 1974).

The voltage spectral density of spontaneous fluctuations reflects primarily the small signal impedance of the membrane. Spontaneous voltage fluctuations are filtered across the time-varying voltage-dependent conductances and passive components of the membrane as discussed above. Decreasing the external Ca concentrations increases the resonant behavior of the nerve membrane. This effect is similar to that of reducing $\bar{g}_K$ (Fig. 4). Further decrease of external Ca moves the membrane from the steady state represented in Fig. 7 to a spontaneously active state similar to that shown for low $\bar{g}_K$ in Fig. 5. We will present data for the heart cell aggregate membrane that correspond to that given in Fig. 7. The main difference we will see is that the rhythmic frequency of the heart, in terms of both small signal impedance analysis and repetitive firing, is about two orders of magnitude slower than in the nerve.

IV. Cardiac Membrane Currents

A. THE MCALLISTER–NOBLE–TSIEN MODEL

We have seen that a two-current system (Na and K) in parallel with a nonspecific leakage and capacitance is capable of producing repetitive firing in the normally quiescent nerve membrane. The conditions that cause rhythmic activity, in this case, are those that sharpen and increase the subthreshold impedance about some resonant frequency. The membranes of the nodes and Purkinje fibers of the heart are spontaneously repetitive, i.e., they are already "tuned" in their natural state. It does not appear, however, that a simple two-current system can account for all of the phenomena found in cardiac membranes. A more complex model is required, and several have been proposed by cardiac electrophysiologists. Nonetheless, the concept that rhythmicity is related to the impedance of the cardiac membrane remains valid for cardiac cells.

The most complete description of the conductances in the excitable cardiac membranes (and a review of earlier models) has been provided by McAllister *et al.* (1975), based on voltage clamp analyses of the cardiac Purkinje fiber. To reconstruct the cardiac action potential these authors required nine currents (instead of only the three needed for the H-H axon model), and they have extensively reviewed the experimental evidence for each of them. In the words of McAllister *et al.*, their model (MNT model) is a "mosaic of various experimental results" that may serve as a useful framework for studying the interplay of the ionic currents. For the present section, we have repeated and extended the calculations of the MNT model to illustrate the dependence of the various currents on time and voltage, and to explore further the relation between those currents and rhythmicity.

It is worth re-emphasizing, at the outset, the multicellular nature of the cardiac preparations to be described. All of the electrical parameters of the original MNT model were referred to the outer surface area of the cylindrical Purkinje fiber. We noted above that the actual cell surface area of a fiber, which passes current and is charged and discharged, is about ten times the value of the outer area (McCallister and Page, 1973). Referred to the total membrane area, all capacitive and current values in the MNT model must be divided by ten, and all resistive values should be multiplied by ten. This assumes that all of the membrane in the cylindrical bundles is equally accessible to current flow. We have made this correction for all calculations in the present section. All values given in figures and text refer to total membrane area. We shall return to this point in connection with our own preparation of spherical aggregates of heart cells below.

The reconstructed Purkinje fiber action potential to be described is shown in Fig. 8. This action potential is calculated from the "standard" equation of McAllister, Noble, and Tsien (1975) and should be compared with their Fig. 9. Figure 8a represents the third in a sequence of repetitive action potentials produced by a single initial stimulus. The third action potential differs considerably from the first calculated in the MNT model, but only negligibly from those that follow in the series. For example, the maximal upstroke velocity of the stimulated first spike is 478 V/sec in the MNT model, but 196 V/sec for the third and all subsequent spikes. All currents calculated below are from the third computed action potential.

1. Sodium Currents

There are two time-dependent inward Na currents (Fig. 8c). The first resembles the fast inward current of the squid giant axon (I_{Na}). This current

---

Fig. 8. The Purkinje cell action potential and the currents that underlie it, calculated from the equations of McAllister *et al.* (1975). Represented is the third in a repetitive series of action potentials produced by a single brief depolarizing stimulus. All of the currents calculated in (b)–(h) are associated with this action potential. The conventions followed are inward currents are negative, outward currents are positive; inward currents depolarize the membrane, outward currents cause hyperpolarization. All currents are referred to 1 cm² of total cell surface area. Currents whose time course is identical to the action potential voltage change (I_{Na_b}, I_{Cl}, I_{K_1}) are assumed to flow through conductance channels that are both time independent and linear with voltage. These are referred to as background currents. (a) The computed third action potential. (b) Total net current (I_i) superimposed on the action potential to emphasize the time relations. Note that each current peak (inflection point) represents a local maximum in rate of voltage change ($\dot{V}$). (c) The fast Na inward current (I_{Na}) controlled by a time-dependent, linear steady-state conductance, and the slow inward current (I_{si}) controlled by a time-dependent, nonlinear steady-state conductance. (d) The outward transient Cl current (I_{qr}), controlled by a time-dependent, linear steady-state conductance. (e) The inward background Na current (I_{Na_b}) with a linear conductance. (f) The outward background K current (I_{K_1}) with a nonlinear conductance, and the outward pacemaker K current (I_{K_2}) controlled by a time-dependent, nonlinear steady-state conductance. (g) The plateau current (I_{x_1}) controlled by a time-dependent, nonlinear steady-state current. This current is outward during the action potential, but inward during the pacemaker depolarization phase. (h) The outward plateau current (I_{x_2}) controlled by a time-dependent, linear steady-state conductance, and the background Cl current, with a linear conductance. This current is outward during the action potential but inward during the pacemaker depolarization phase.

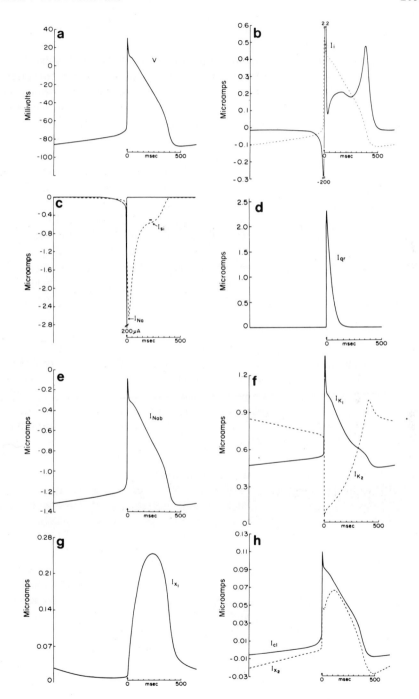

is specifically blocked by tetrodotoxin and may be described by two gating variables m and h, analogous with the Hodgkin and Huxley (1952) description of I_{Na} in nerve membrane. The value used in the MNT model for $\bar{g}_{Na}$ of the Purkinje fiber is 150 mS/cm^2. When this value is referred to actual membrane area, it becomes 15 mS/cm^2, which is considerably lower than the axon value of 120 mS/cm^2. I_{Na} carries virtually all of the current that causes the rapid upstroke of the action potential spike. It is activated and inactivated very rapidly, reaching its peak value in less than a millisecond.

The second slower inward current (I_{si}) is probably carried by both Na and Ca. It is blocked by Mn^{2+} or drugs such as Verapamil or D-600, but is unaffected by TTX (Kass and Tsien, 1975). I_{si} rises to a peak value in about 10 msec and declines in about 50 msec, with a distinct plateau phase during the decline (Fig. 8c). In early studies of cardiac electrophysiology, differential effects on the maximum rate of rise of the action potential and the height of the plateau phase were taken as evidence for two separate inward currents. For example, TTX had no effect on the plateau, but dramatically reduced maximum upstroke velocity. This literature is reviewed by Reuter (1973). Note that $\bar{g}_{si}$, i.e., the maximal conductance of the slow inward current is 0.08 mS/cm^2 referred to actual membrane area, or about 1/200 of $\bar{g}_{Na}$.

In addition to the two time-dependent Na currents, there is a time-independent Na current (I_{Na_b}), which is also an inward current (Fig. 8e). In contrast to the fast currents, I_{Na_b} is largest near rest, and its minimum value occurs at the peak of the action potential. I_{Na_b} is responsible mainly for maintaining the resting potential of the quiescent Purkinje fiber near -90 mV rather than closer to the K equilibrium potential some 20 mV more negative. Since I_{Na_b} is a linear function of potential, its time course mimics that of the action potential. The ionic basis of I_{Na_b} is not certain. In Na-free experiments, the current may be carried by other ions such as choline (McAllister *et al.*, 1975). This current plays an important role in pacemaking activity, since the decay of the pacemaker current can cause repetitive firing only against a steady inward depolarizing current (see discussion of I_{K_2} below).

2. Potassium Currents

Voltage clamp data for outward currents are more reliable than the fast inward current data because the outward currents are much slower than either I_{Na} or I_{si}. There are two large outward currents carried by K ion. These are designated as I_{K_1} and I_{K_2} (Fig. 8f). I_{K_1} is called the outward background current. Although I_{K_1} is time-independent, it is rectified, i.e., not linear with voltage. Therefore, the time course of the current during an action potential does not follow exactly the shape of the action potential itself.

I_{K_2} is a time-dependent K outward current whose magnitude varies approximately as the inverse of voltage during an action potential. I_{K_2} is called the pacemaker current because it has its largest values over the pacemaker range of voltages, i.e., during the slow depolarization from about -90 to -70 mV prior to the action potential spike. Since an outward positive current hyperpolarizes the membrane, the slow decline in I_{K_2} prior to the spike, set against the steady inward current of I_{Na_b}, results in a net slow depolarization referred to as the "pacemaker potential." The slope of this depolarization determines the length of the interspike interval, and thus sets the pace of spontaneous rhythmic firing of the membrane.

Two small outward currents have their largest values during the plateau of the action potential and are often referred to as plateau currents. These are also carried primarily by K, with small contributions from other ions. They are designated I_{x_1} and I_{x_2}. Their time course is shown in Fig. 8g and h. I_{x_1} is entirely outward, but I_{x_2} has a small inward component during the slow depolarization phase of the action potential. Note the relatively small magnitude of I_{x_1} and I_{x_2} compared to the currents previously discussed. Both currents are controlled by time-dependent conductances. I_{x_1} is rectified, but I_{x_2} has a linear dependence on voltage for fully activated currents.

3. Chloride Currents

There are two outward currents that are carried primarily by Cl ion. One is called the transient chloride current (I_{qr}) since it is strongly activated during the depolarization phase of the action potential (Fig. 8d). This time-dependent conductance gives rise to an inward flow of negative ions, i.e., an outward current. It is largely responsible for the sharp repolarization seen immediately after the inward spike of the action potential. [The inactivation formulas used in these plots are Eqs. (25) and (26) in MNT. Alternative inactivation formulas are given in their Eqs. (27) and (28).] The activation kinetics of I_{qr} are not well known and it is evident from the opposite time course and magnitudes of I_{qr} and I_{si} (Fig. 8c) that these two currents might be difficult to resolve.

Chloride ions also carry current through a time-independent pathway. Since this current (I_{Cl}) is also a linear function of potential, its time course is identical to the action potential itself. The magnitude of I_{Cl} is in the same range as I_{x_1} and I_{x_2}. It represents an inward flow of negative ions, i.e., an outward current, over most of its range although there is a slight inward current near the extreme hyperpolarization during the repetitive firing cycle.

The total ionic current (I_i) through the Purkinje cell membrane during the action potential is the sum of the currents shown in Fig. 8c–h. This is the net ionic current (Fig. 8b) that is balanced by membrane capacitive current to give zero total membrane current. In other words, the model describes a uniform membrane potential that obeys the equation

$$I_I + C(dV/dt) = 0.$$

The propagated action potential is not considered in this description. In Fig. 8b I_i is superimposed on the action potential in order to emphasize the time relations of current to voltage. Note that each peak in the current coincides with a local maximum in rate of voltage changes ($\dot{V}$) as required by the above equation. Since C_m is close to 1 $\mu F/cm^2$, the values of I_i (in $\mu A/cm^2$) are also close to the values of dV/dt (in V/sec).

4. Pacemaker Currents

We have noted that the essential feature of the pacemaker activity in the MNT model and our extension of it is the interplay between the background Inward current (I_{Na_b}) and the slowly declining value of I_{K_2}. The absolute value of each of the currents flowing during the slow depolarization of the Purkinje fiber is given in Table II (except for I_{qr}, which is negligible during this phase). During more than half of each cycle, when the membrane potential is near rest, I_{Na_b} is greater than 1 $\mu A/cm^2$ (Fig. 8c), but is more than balanced by the two large outward currents I_{K_1} and I_{K_2} (Fig. 8f). Indeed, throughout almost all of the slow depolarization represented in Table II, $I_{K_1} + I_{K_2} > I_{Na_b}$.

The maximum diastolic depolarization achieved during the MNT action potential is 87.195 mV, at which point there is no net ionic current. The first line in Table II represents the voltage at which the net current (I_i) has just begun to flow inwardly to start the slow depolarization phase. The last line is taken at the potential (about 63 mV) at which I_{Na} becomes approximately equal to the sum of all the other currents and begins to dominate I_i. Despite the fact that I_{K_2} is considered the major pacemaker current, Table I shows that minor variations in any of the other currents during the slow depolarization would have dramatic effects on its slope. For example, a decrease in I_{Na_b} of less than 2% or an increase in I_{K_1} of about 5% would abolish the pacemaker depolarization and suppress spontaneous rhythmic firing of the membrane. The net negative (i.e., depolarizing) value of I_i during the pacemaker depolarization requires I_{Cl} (Fig. 8h) in the early part of the depolarization phase (near -87 mV) and I_{Na} and I_{si} (Fig. 8c) in the later part of this phase (near -70 mV). The primary process that accounts for the slope of the pacemaker potential is indeed the decay of I_{K_2} in balance against I_{Na_b}. However, the magnitude of I_i depends critically on all the other currents, including I_{Na} and I_{si}, which are normally thought of only in association with the action potential during the activated state.

The slow decay of the I_{K_2} conductance is also primarily responsible for the

Table II

CURRENTS THAT PRODUCE SLOW DEPOLARIZATION IN THE MNT MODEL[a]

Pacemaker voltages V (mV)	Time-Dependent					Time-Independent			
	Action potential currents		Pacemaker current	Plateau currents		Background currents			Net ionic current
	Na	si	K_2	x_1	x_2	Na_b	K_1	Cl	I_i
−87.196	−0.00088	−0.00297	0.88157	0.04106	−0.02638	−1.33555	0.45990	−0.01720	−0.00045
−85.797	−0.00149	−0.00567	0.84860	0.02791	−0.01998	−1.32986	0.47205	−0.01580	−0.01524
−84.994	−0.00201	−0.00718	0.83706	0.02392	−0.01820	−1.31244	0.47858	−0.01499	−0.01526
−79.999	−0.01195	−0.02130	0.77371	0.01048	−0.00097	−1.25999	0.51279	−0.00999	−0.01622
−75.011	−0.05634	−0.04882	0.73971	0.00856	−0.00546	−1.20761	0.53762	−0.00501	−0.03735
−72.953	−0.10117	−0.06621	0.73286	0.00898	−0.00421	−1.18601	0.54567	−0.00295	−0.07304
−70.987	−0.17647	−0.08869	0.72616	0.00952	−0.00313	−1.16536	0.55239	−0.00099	−0.14457
−63.290	−1.53752	−0.18313	0.68225	0.01141	+0.00084	−1.08455	0.57183	−0.00671	−1.54558

[a] Values in $\mu A/cm^2$.

increase in dc resistance during the pacemaker depolarization. This increase in resistance of the heart cell membrane was first measured by Weidmann (1951) and played an important role in the early Noble (1962) model of the reconstructed action potential (Fig. 6b) as well as in the MNT model.

B. THE MECHANISM OF SPONTANEOUS RHYTHMICITY

In the previous sections we have viewed the rhythmicity of the heart cell membrane according to two distinct conceptual frameworks. According to the MNT model, which is the most widely accepted description of the beating heart, the slow depolarization that brings the cell from its most hyperpolarized point at the end of an action potential to threshold for the next results from the increasing magnitude of the net inward ionic current. In this view, the membrane depolarizes until the nonlinear voltage-dependent conductance (I_{Na}) is activated. In the "impedance" model, the RLC characteristics of the excitable membrane are so tuned that any small perturbation causes the voltage to oscillate at a natural resonant frequency. In this view, the pacemaker depolarization represents the depolarizing phase of one oscillatory half-cycle. If threshold is not achieved at the crest of the oscillation, it does not continue to depolarize, but merely completes the cycle. We demonstrated earlier that small changes in $\bar{g}_K$ or other parameters of the H-H equations could tune the membrane to exhibit such oscillatory behavior about a resonant frequency, characteristic for the type of membrane. We have illustrated these oscillatory voltage fluctuations in a variety of living membranes (Fig. 1). As predicted, heart cells resonate at about 1 Hz, close to their normal beat frequency. The squid giant axon has a natural oscillation frequency about two orders of magnitude greater (about 100 Hz). When the axon preparation is caused to fire spontaneously by reducing extracellular Ca^{2+}, it does so at 100–200 spikes per second (Guttman, 1969), in qualitative agreement with Fig. 7.

Recent studies (reviewed in Stevens, 1972; Verveen and DeFelice, 1974; Conti and Wanke, 1975; Neher and Stevens, 1977; DeFelice, 1977) have emphasized that biological membranes exhibit small stochastic voltage fluctuations ("noise") that result from random increases and decreases in conductance of membrane ionic channels. The concepts and mathematical techniques of noise analysis have generally been couched in terms of the fundamental electrical quantities. Thus it was inevitable that investigations of membrane noise would lead to an emphasis on membrane impedance. From our studies of the noise and impedance properties of a tissue culture model membrane preparation—the heart cell aggregate—we recognize that the idea that the slope of the pacemaker potential is determined by net ionic currents (as in the MNT model), and the concept that it represents a half-cycle of an oscillatory resonant frequency, are not alternative or conflicting mechanisms. They are simply two ways of considering the time-varying voltage-dependent properties of the various membrane conductances. In the limit of small perturbations, e.g., inherent voltage noise, the impedance formalism is most useful.

In the next section we will describe the aggregate properties in detail. We will show that the impedance of the membrane exhibits a resonant frequency near 1 Hz, and that this frequency is related to that of action potential generation. Spontaneous voltage noise from the aggregates will be related to both the impedance and macroscopic rhythmicity.

V. The Heart Cell Aggregate

A. CHARACTERISTICS AND ADVANTAGES

A major problem in any successful analysis of the electrical properties of heart tissue arises out of the multicellular nature of the preparation. We have seen that heart tissue is composed of individual cells connected by junctions of low but varying resistance, into fibers of complex geometry. Substantial voltage gradients exist within such fibers, and currents crossing the membrane of one cell may differ from those flowing in a distant cell. The magnitude and kinetics of recorded electrical events may reflect tissue geometry more than the characteristics of the cell membranes. This problem of electrical inhomogeneity of cardiac tissue has been documented and discussed at length (DeHemptinne, 1976; Eisenberg and Johnson, 1970; Johnson and Lieberman, 1971; Joyner *et al.*, 1975; Nathan and DeHaan, 1977; Ramon *et al.*, 1975).

One way to minimize the geometrical problems of heart muscle would be to study single cells isolated in tissue culture. Muscle from either adult (Pretlow *et al.*, 1972; Fabiato and Fabiato, 1972) or embryonic (DeHaan, 1967) heart can be dissociated into its component cells. Isolated adult myocytes have not remained viable in culture for more than a few hours thus far. However, embryonic heart cells survive and continue to beat rhythmically for days under conditions that are readily reproducible *in vitro* (DeHaan, 1967; Speicher and McCarl, 1974). Once such preparations recover from the initial effects of the dissociation procedures (LeDouarin *et al.*, 1974) they respond to a variety of pharmacological and other environmental perturbations in ways similar to the intact heart muscle from which they were derived (Lane *et al.*, 1977; McCall, 1976; Goshima, 1976), although under some circumstances they appear to revert back to earlier embryonic physiological characteristics (Sperelakis *et al.*, 1976). Unfortunately, although a microelectrode can be introduced into an isolated embryonic heart cell (DeHaan and Gottlieb, 1968; Sperelakis and Lehmkuhl, 1966; Hyde et al., 1969), these impalements usually last less than a minute, and efforts to control membrane voltage through the single electrode have been unsuccessful. About 50% of the single cells isolated from a 7-day-old chick embryo heart beat spontaneously (DeHaan, 1970), with pulsation rates ranging from 0.5–5 beats per second according to a skewed Gaussian distribution (median = 1.98 beats per second); only about 7% had rates greater than 3.3 beats per second. When two such independently beating cells were brought into contact, their pulsation rates synchronized to a common rhythm in 4–60 minutes, at which time newly formed primitive nexal junctions could be seen between their apposed surfaces (DeHaan and Hirakow, 1972).

Isolated cells can also be reassociated into tissuelike masses that can be formed into monolayer cell sheets (Sperelakis, 1972), cylindrical strands (Purdy et al., 1972; Sachs, 1976), or spheroidal aggregates (Sachs and DeHaan, 1973). Low-resistance nexal junctions are also quickly formed between the newly apposed cells in these groupings and their pulsation rates become entrained to a common rhythm. When such artificial tissue systems have been examined in the electron microscope, they also reveal gap junctions joining the cells (DeHaan and Sachs, 1972; Hyde et al., 1969; Purdy et al., 1972; Goshima, 1970; Shimada et al., 1974).

We have shown that the cells within a spheroidal aggregate are electrically coupled (Sachs and DeHaan, 1973). Moreover an analysis of the passive electrical properties of such spheroids (DeHaan and Fozzard, 1975) indicates that because of their close coupling and spherical geometry, such systems are virtually isopotential. All parts of the aggregate membrane experienced action potentials or small voltage perturbations simultaneously and without measurable attenuation throughout the sphere. The system behaved like a single giant spherical cell such as that modeled by Eisenberg and Engel (1970). We have recently succeeded in exploiting this unique property by completing a current-voltage analysis of these spheroids with the voltage-clamp technique (Nathan and DeHaan, 1977).

For studies of the electrical properties of the heart cell membrane, spheroidal aggregates present numerous advantages: (a) They can be prepared in a range of sizes (60–300 μm diameter) composed of a hundred cells or up to several thousand cells (Fig. 9); (b) Aggregates can be impaled with two or even three electrodes, and yield stable recordings for many minutes; (c) These spheroids beat with a more regular rhythm than do single cells, and their rate is predictable as an inverse function of size (Sachs and DeHaan, 1973).

B. Preparation of Aggregates

For all of the work to be described in this section, measurements have been made on spheroidal clusters of embryonic chick heart cells maintained in tissue culture. After incubation for seven days at 37.5°C, chicken embryos were harvested in amniotic fluid and decapitated. Hearts were dissected free, trimmed of extraneous tissue, and the ventricles were dissociated with trypsin by techniques now standard in this laboratory (DeHaan, 1970). Aggregates were prepared from an inoculum of 5×10^5 cells in 3 ml of culture medium (818A) on a gyratory shaker. Cells were allowed to aggregate during 48–72 hours gyration (62 rpm, 37.5°C) in an atmosphere containing 5% CO_2, 10% oxygen, 85% nitrogen. At the end of the gyration period, each aggregation flask contained up to 200 spheroidal clusters of heart cells (Fig. 9).

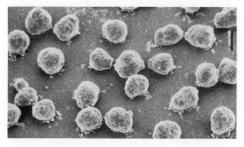

Fig. 9. Scanning electron micrograph of a field of heart cell aggregates adhering to the surface of a culture dish. Scale: 100 μm (courtesy of Dr. Claudia Baste).

For impalement with microelectrodes, aggregates were allowed to adhere to the surface of a Falcon plastic tissue culture dish maintained on a microscope warm-stage in medium containing 1.3 or 4.8 mM K^+, or higher concentrations, under conditions in which pH, temperature, gaseous atmosphere, and evaporation could be controlled, and mechanical vibration minimized. In these circumstances, aggregates continued beating rhythmically for hours or even days, unless their spontaneous activity was suppressed with tetrodotoxin or high K^+ levels.

C. Electrical Measurements from Aggregates

Figure 10 is a schematic representation of the experimental set-up. One to three microelectrodes could be used to penetrate the aggregate at one time. Two electrodes measured voltage noise, while current could be injected through the third. The output of each voltage electrode was recorded at low gain by an FET input, dc-coupled voltage follower. The dc channels were

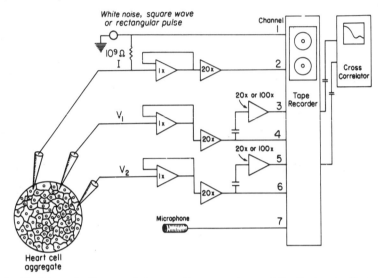

Fig. 10. Schematic diagram of the recording apparatus used for the heart cell aggregate preparation. Three low gain dc channels with FET input amplifiers (channels 2, 4, and 6), and two high-gain ac-coupled channels (channels 3 and 5) record voltage. Input current is recorded on channel 1 and physiological data through the voice channel (7). The ac-coupling time constant on the high gain channels into the FM tape recorder is 10.61 sec; that into the cross-correlator is 1.59 sec.

used to record action potentials and resting potential (E_r). Spontaneous voltage fluctuations around E_r were recorded with high-gain ac-coupled amplifiers. Since these voltage fluctuations have a prominent low frequency component, a high-pass RC filter with a 3-dB low-frequency roll off of 0.015 Hz was used. Glass microelectrodes filled with $2M$ KCl had dc resistances between 10 and 30 MΩ. The time constant of these microelectrodes was greater than 1 msec, resulting in a 3-dB high-frequency roll off of about 160 Hz. This roll off determined the upper band limit of the measured noise spectra when capacitance neutralization was not used. Both the KCl solution in the electrodes and the agar salt bridge in the medium bathing the heart cells were connected to ground through Ag–AgCl half-cells.

Typical values for the dc resistance measured between an intracellular electrode and the bathing medium in an aggregate made quiescent with TTX, ranged between 1 and 5 MΩ. The exact value depended on aggregate size, external K concentration, and E_r. Current was injected through a 10^9 Ω resistor to assure that it would be independent of changes in aggregate resistance. Rectangular pulses were applied in the nanoampere range to produce voltage deflections across the membrane no greater than 2–3 mV.

Six simultaneous records were stored on FM 7-track magnetic tape. These records included the dc transmembrane potential from each electrode, the voltage noise from each of the two voltage electrodes, the injected current, and a voice track for noting temperature, external K^+ concentration, the presence or absence of TTX, aggregate size, and other physiological data. Transmembrane potentials on the order of 100 mV required amplification of about 20-fold to reach an adequate voltage level for the tape recorder. Random voltage fluctuations on the order of 200–1000 μV required a gain of 500–2000.

The transmembrane voltage response to injected current pulses was used to define the input impedance of the aggregate. When voltage excursions from a particular resting value were no greater than 1 mV, the system behaved in an approximately linear manner.

Voltage noise was analyzed using a Hewlett–Packard 3721A cross-correlator. If A and B represent the two noise records and C the correlation function, C_{AA}, C_{AB}, and C_{BA} were measured. That is, the auto- and cross-correlation functions were measured for simultaneous records of inherent voltage noise from two cells within an aggregate. The correlation functions could be corrected for the high-pass (low frequency) roll off of 0.1 Hz of the dc input into the correlator. In order to analyze spontaneous voltage fluctuations by their power spectral densities or correlation functions, the signal must be stationary. To treat signals whose dominant frequencies are near 1 Hz requires records several hundred seconds long, during which the condition of stationarity must be satisfied. The simplest test for stationarity is to show that $C_{AB} = C_{BA}$, that is, that the cross-correlation function between two records is the same whether the signals are analyzed forward or in reverse. We have taken this approach in analyzing our data. Comparison of C_{AB} with C_{AA} and C_{BB} was used to determine the degree of intercellular communication (DeFelice and DeHaan, 1977).

1. The Aggregate Action Potential

In low-K medium (1.3 mM K^+) all aggregates beat rhythmically for as long as they were maintained in culture (usually 2–3 days). Large aggregates consistently beat more slowly than small (Sachs and DeHaan, 1973). Aggregates of about 60 μm diameter (composed of about 100 cells) averaged 2.7 beats per second; larger spheroids (mean = 170 μm, containing about 3000 cells) had pulsation rates of about 1 beat per second, while those larger than 200 μm in diameter beat more slowly still or intermittently. In the latter case, regular bursts of spontaneous rhythmic action potentials were followed by a period of quiescence lasting usually 5–20 sec, before the next spontaneous train appeared. Such a burst of action potentials is shown in Fig. 11a. Note that the first action potential of the burst is smaller than those that follow and is preceded by low-level voltage oscillations. It also has a differ-

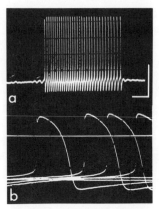

Fig. 11. A spontaneous train of rhythmic action potentials recorded from an aggregate 152 μm in diameter in standard medium containing 1.3 mM K. E_r before train, -67 mV. (a) Slow sweep record to show the entire train as well as the voltage oscillations that precede and follow it. Scale: 50 mV, 5 sec. (b) The first four action potentials shown at fast sweep speed. Zero potential is indicated by the solid line. Scale: 50 mV, 50 msec.

ent shape, since it is not preceded by a hyperpolarizing phase, and has a sharp initial spike peak followed by a notch before the plateau, that is missing from those that follow (Fig. 11b). We noted above that the maximal upstroke velocity of the first of a train of action potentials computed by the MNT model is substantially greater than those of subsequent spikes. Curiously, in experimental trains recorded from aggregates, $\dot{V}_{max}$ does not differ substantially between the first and subsequent action potentials (Fig. 11b), indicating that the model fails to describe the aggregates fully in this regard. The smaller size and shape differences are predicted by the MNT model as well as by the BVP model (Fitzhugh, 1969). Such differences are related to a well-known property in the nonlinear differential equations (Davis, 1960; Minorsky, 1947). A set of equations that have periodic solutions may approach such solutions asymptotically after adequate initial conditions are set. When such solutions are stable, the final periodic condition is known as the limit cycle (see Cole, 1968, pp. 204–228; Jack _et al._, 1975, pp. 317–330). The train shown in Fig. 11 achieves a stable limit cycle after the first action potential and revolves through 31 more cycles. It is not understood why the aggregate stops firing at this time. The last action potential does not differ perceptibly in shape from those previous to it, except that it is followed by a damped oscillation of the same form, but on a different time scale, as that computed with diminished $\bar{g}_K$ from the H-H model (Fig. 5b and c).

Increasing external K to 4.8 mM tends to suppress spontaneous activity, and to augment the size–rate relationship seen at lower K levels. In these conditions, large aggregates are often quiescent for extended periods, except

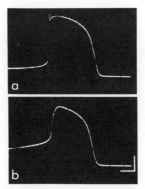

Fig. 12. Isolated action potentials recorded from two different aggregates in standard medium containing 4.8 mM K. (a) 320 μm diameter, no TTX. (b) 180 μm diameter, 3×10^{-6} M TTX. Scale: 25 mV, 50 msec for both (a) and (b).

for occasional single beats. An isolated action potential from a 320 μm aggregates is shown in Fig. 12a. Such action potentials exhibit the large initial spike characteristic of the first potential of a train. It is also notable that the duration of action potentials from large aggregates in higher levels of K is substantially longer than those from smaller spheroids at 1.3 mM K. In the presence of TTX, spontaneous action potentials and visible mechanical contractions cease in many or all aggregates (Figs. 12b and 13), depending on the size of the aggregates, the age of the hearts from which they were made, and the concentration of the drug used. Spontaneous activity was suppressed in 50% of aggregates made from 7-day-old heart at $3 \times 10^{-7}M$ TTX (DeHaan *et al.*, 1976). Furthermore, the initial spike peak disappears completely. Rise time of control action potentials was 100–200 V/sec. Those beating in TTX showed upstroke velocities of 10–20 V/sec (Fig. 14). When I_{Na} was omitted from the computed MNT Purkinje fiber action potential, the slow inward current could still support action potentials, but without the initial spike peak and with very slow rise times.

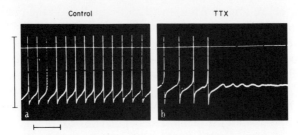

Fig. 13. Intracellular recordings from a continuous impalement of a single cell within an aggregate. (a) Control. (b) Two minutes after TTX (6×10^{-8} M). Scale: 100 mV vertical, 1 sec horizontal (from DeHaan *et al.*, 1976).

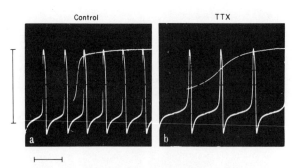

Fig. 14. Effect of TTX on action potential shape and maximum upstroke velocity. Both records are from a single long-lasting impalement of an aggregate (a) Control. (b) Five minutes after TTX (3×10^{-8} *M*). Vertical scale, 100 mV; horizontal scale, 0.5 sec (slow sweeps), 2 msec (fast sweeps) in both (a) and (b) (from DeHaan *et al.*, 1976).

The similarity between our recorded action potential shapes (Figs. 12 and 14) and the computed events (Fig. 8a of McAllister *et al.*, 1975) is striking. A major difference, however, is in action potential duration. The MNT action potential, like that recorded from experimental Purkinje fiber preparations (see Weidmann, 1956), lasts typically about 500 msec. Aggregate action potentials are generally less than half that duration and appear to be shorter in small aggregates than large. When membrane resistance collapses during the action potential, access resistance (i.e., intercellular cleft resistance) becomes a significant component of the IR drop. Access resistance in a 150 μm aggregate has been estimated at about 20 kΩ (Nathan and DeHaan, 1977). Thus, some aspects of action potential shape may result from tissue properties as well as pure membrane characteristics. This point has been discussed previously (DeHaan and Gottlieb, 1968).

2. The Resting State

In aggregates made quiescent with TTX, the oscillatory properties of the membrane are heightened and the details of voltage behavior near rest can more readily be examined. All of the data that follow in the present section are taken from TTX-suppressed preparations.

There appears to be two stable resting states in the TTX-treated aggregate: one near -70 mV and one near -50 mV. The exact value depends both on external potassium and on the aggregate size. Between 1 and 2 m*M* K, E_r remains near -50 mV (-45 to -54 mV). Between 3–6 m*M* K, E_r tends to stabilize near -70 mV (-68 to -76 mV). Above about 10 m*M* external potassium the membrane depolarizes continuously. The input dc resistance also varies with E_r. Near the -50 mV stage, the resistance is 5–6 times the value near the -70 mV state for the same size aggregate.

An aggregate in TTX at rest near -50 mV may spontaneously change its

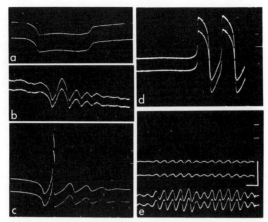

Fig. 15. Simultaneous intracellular recordings from two-electrode impalements of aggregates exposed to TTX (3×10^{-6} M), to show electrical activity in the two stable potential states, and transitions between them. All calibrations refer to scale in panel e. (a) Electrodes approximately 80 μm apart in a 140 μm aggregate. A 50-sec record showing the spontaneous transition from -49 to -76 mV (at lowest point) about three minutes after elevating K from 1.3 to 4.8 mM. Scale: 50 mV, 5 sec. (b) Transition from -52 to -68 mV after elevating K from 1.3 to 2.3 mM results in a damped oscillatory response. Scale: 20 mV, 2 sec. (c) Transition from -54 to -75 mV following shift from 1.3 to 2.3 mM K. The crest of the first oscillatory swing crosses threshold to produce an action potential. Scale: 50 mV, 1 sec. (d) Spontaneous action potentials fired from an apparently stable E_r at -49 mV in an aggregate in standard medium containing 1.3 mM K. The shape of these action potentials should be compared with those in Fig. 11b. Scale: 50 mV, 200 msec. (e) Spontaneous steady voltage oscillations recorded from the same aggregate shown in (c), several minutes later. Scale: upper pair of traces (low gain), 50 mV; lower traces (high gain) 12 mV; 2 sec. In (c)–(e) zero potential for low-gain records is marked at the right edge of the panel.

resting level to the -70 mV state. In Fig. 15a, a 50-sec record from two voltage electrodes in an aggregate illustrates the transition from one stable state to the other and back, following elevation of K from 1.3 to 4.8 mM. These tracings are from an impalement that lasted more than 40 minutes. After several more similar shifts, E_r in this aggregate stabilized at -72 mV. The extent to which the two traces are identical defines the degree of iso-potentiality of the cells (DeHaan and Fozzard, 1975). Normally, two electrodes placed in different cells within the same aggregate detect the same potential, as shown. Small constant differences can be attributed to different tip potentials of the microelectrodes.

The transition from one stable state to the other may be associated with marked oscillatory behavior. In Fig. 15b, the external potassium has been raised to 2.3 mM, causing the aggregate to stop beating and come to rest in the -50 mV state. After 2–3 minutes, the quiescent state in 2.3 mM K changes spontaneously from the -50 mV state to the -70 mV state. When this occurs, considerable ringing accompanies the change, but the macroscopic oscillations damp out to a stable E_r at -68 mV. In a different aggregate a similar sequence of events occurs (Fig. 15c), but in this case an action potential is fired from the first oscillatory rebound. Spontaneous action potentials can also arise directly from the -50 mV state (Fig. 15d).

Cardiac Purkinje fibers also exhibit two stable resting potentials, one near -50 mV and one near -90 mV (Wiggins and Cranefield, 1976). The -50 mV state is revealed by exposing the preparation to Na^+-free solution. In these fibers, as in heart cell aggregates, action potentials can be generated from either resting level; the membrane shows oscillatory responses, and input resistance is greater in the more depolarized state. A possible difference, however, is that only "slow response" action potentials with very slow rise times are generated from the -50 mV state whereas when the membrane spontaneously shifts down to -90 mV, normal action potentials with rapid upstrokes arise. In our aggregates, such normal action potentials with fast rise times are blocked by the TTX. Just as spontaneous action potentials may be generated from either stable E_r, large spontaneous oscillatory potentials that are not sufficient to cause firing may also occur from either rest level. These subthreshold oscillations wax and wane, and may grow to as large as several millivolts, without shifting from one rest level to the other. Periods of steady oscillation may last for several minutes. Figure 15e shows this behavior in the 70 state. The oscillations are shown at high amplification, ac-coupled at 0.1 Hz, and at lower amplification, dc-coupled. The onset of this activity is seen clearly in the figure. In this case, the spontaneous oscillations last only about 20 sec. These oscillations may grow to the extent that they cause an action potential to fire (Fig. 11a, no TTX); usually in the -70 mV state with TTX, the oscillations simply die out as shown in Fig. 15e, and the membrane potential returns to the quiescent state.

3. Membrane Impedance and Oscillatory Behavior

As noted earlier, we may define the impedance of the aggregate membrane by its response to injected current pulses. The voltage responses to a series of rectangular hyperpolarizing current pulses injected into an aggregate are illustrated in Fig. 16. The initial response to a short pulse (Fig. 16a) is difficult to distinguish from a typical RC exponential voltage change. Upon termination of the pulse, however, there is a perceptible overshoot of the voltage before E_r is reached. With longer pulses, the LRC character of the voltage response to both the onset and termination of current becomes progressively more apparent (Fig. 16b and c). Upon release from a 2-sec pulse, the oscillatory rebound is so large that it crosses threshold and fires an action potential (Fig. 16d). This phenomenon is usually described as the

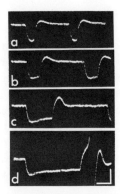

Fig. 16. Voltage responses recorded from one electrode to a series of 2 nA rectangular current pulses injected through a second electrode approximately 90 μm distant, in a 150 μm diameter aggregate. All responses are from the same impalement. $E_r = -52$ mV, K $\simeq$ 1.3 mM. Scale: 10 mV, 0.5 sec for all panels. Pulse lengths were, respectively, 300, 450, 900, and 2000 msec.

"anodal break response." In nerve axon, the length of hyperpolarizing pulse sufficient to cause an action potential when the pulse is released is the order of 10 msec. In the heart cell membrane, the pulse length must be at least 100 times longer.

Membrane time constants on the order of 10–20 msec have been reported for a variety of cardiac preparations (Table I). Most of these were from measurements of voltage responses to short rectangular current pulses. From Fig. 16c and d it is apparent that the aggregate membrane actually reaches a new stable potential in response to such a pulse only after 1–2 sec. Thus most of the measured values of τ_m in the literature are in error because the membrane was implicitly treated as an RC circuit rather than one with an inductive component. The value of C may be estimated reliably from the leading edge of the response (Clay *et al.*, 1977). However, an error of 20% or more may result if R is calculated from a short pulse. This error is more serious for depolarized states, which are more oscillatory than hyperpolarized states. The LRC circuit for the cardiac membrane may be similar to that discussed earlier for nerve preparations. An obvious difference is that the natural resonant frequency of the heart membrane is near 1 Hz, about 100 times slower than in nerve.

The 2 nA pulses, applied in Fig. 16 to a preparation in the −50 mV resting state, resulted in voltage deflections of about 12 mV at the overshoot peak and about 8 mV at the end of the 2 sec pulse (Fig. 16d). The apparent input resistance $(\Delta V/\Delta I)$ to be derived from this measurement is therefore about 4 MΩ. However, even this simple measurement is in error, since the pulses shown are much too large for linear analysis of the equivalent membrane circuit parameters. Linear responses up to several millivolts can be obtained from nerve membranes. In the heart cell membrane, nonlinearity is more apparent, and the perturbing pulse must therefore be smaller. Even pulses in the range of 0.1 to 1.0 nA show nonlinear effects. However, with the smallest of these, the values for C_m and R_m given in Table I for the aggregate were determined (Clay *et al.*, 1977). By fitting computed responses of estimated LRC circuits to applied voltage transients, we have confirmed that the aggregate membrane behaves as if it had an inductive branch in parallel with the usual RC circuit (about 25 k Henrys-cm^2). This putative inductive element results from the voltage- and time-dependent properties of the conductances that are involved in membrane behavior near rest (Clay *et al.*, 1977).

The rapid transition from the -50 mV rest level to the -70 mV state represents a large spontaneous voltage step. No current was injected to cause the change pictured in Fig. 15a. Nonetheless, the membrane behaved as if in response to a pulse of current, even to the extent of the oscillations shown in Fig. 15b and c. In this case, however, the membrane was spontaneously shifting between two resting states, which by definition exhibit zero net current.

4. Membrane Voltage Noise

The cardiac membrane can also be shown to exhibit oscillatory behavior at the microscopic level. When spontaneous activity of an aggregate was suppressed with TTX or high K^+, the pronounced oscillatory voltage swings of several millivolts (Figs. 13 and 15b) were not the most common condition observed. Instead, after a single over-damped subthreshold oscillation, the membrane quickly reached a stable rest potential (Fig. 17) where it often remained for many minutes. At the usual levels of amplification for macroscopic events ($\times$ 20), E_r showed little or no fluctuation. At a level of amplification adequate to display events at the microvolt level, however, spontaneous voltage fluctuations ranging from 50 to 500 μV were revealed. The traces shown in Fig. 16b illustrate three 6-sec records taken from the beginning, middle, and end of a 5.1-min period of stable potential ($E_r = 54$ mV) uninterrupted by voltage fluctuations greater than 1 mV peak to peak. At this gain, although much of the activity is represented by random fluctuations (noise), a distinct periodicity of approximately 1.2 sec can still be observed.

Similar noise can be recorded from membranes at either the -50 or -70 mV rest level. The tracings shown in Fig. 18 illustrate this behavior, and emphasize the remarkable degree of voltage homogeneity within an aggregate, as evidenced by the similarity of microscopic events recorded simultaneously from widely separated cells.

To distinguish random from periodic events, an autocorrelation analysis

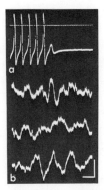

Fig. 17. Intracellular recording from a 154 μm diameter aggregate, 1.3 mM K, 3 × 10^{-6} M TTX. (a) End of a spontaneous train of action potentials, followed by an extended period (5.1 min) of "stable" E_r at −54 mV. Scale: 40 mV, 1 sec. (b) Three 6-sec high-gain traces of voltage noise recorded at the beginning, middle, and end of the 5 min period of stable E_r, following the action potential train shown in (a). Scale: 0.5 mV, 1 sec.

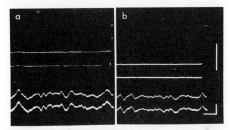

Fig. 18. Voltage noise recorded in an aggregate in the −50 mV state and from the same aggregate in the −70 mV state, after raising K from 1.3 to 2.3 mM. In both cases the upper (low-gain) traces record E_r while the lower (high-gain) traces show voltage noise simultaneously from two cells about 118 μm apart in a 150 μm aggregate. (a) E_r = −51 mV. (b) E_r = −73 mV. Scale: low gain, 50 mV; high gain, 1 mV; 1 sec.

of 56.8 sec of the voltage record shown in Fig. 17 was done. The autocorrelation function (Fig. 19) has a period of about 1.5 sec, which may be taken as the "average" oscillatory period of the microscopic fluctuations. The voltage noise is stationary within the time analyzed. From similar records taken from aggregates exposed to higher levels of K at the −70 mV rest level— that is, those conditions that tend to suppress macroscopic electrical activity—we have noted that the oscillatory component of the microscopic noise may be much reduced or disappear altogether on the scale shown (DeFelice and DeHaan, 1977).

D. THE MICROSCOPIC BASIS OF RHYTHMICITY

The "stable" resting membrane exhibits a range of microscopic voltage events. The smallest of these are the elemental random voltage fluctuations

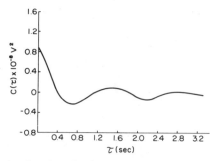

Fig. 19. Autocorrelation function of a 56.8 sec sample of the voltage noise shown in Fig. 17b. The membrane noise was so much greater than background microelectrode noise that the auto- and cross-correlation functions were the same within error of the measurement.

pictured (Figs. 17 and 18) that range in magnitude from a few microvolts to a fraction of a millivolt. There is a large body of literature (reviewed in DeFelice, 1978) indicating that these fluctuations of potential result from the random opening and closing of individual current channels in the membrane.

Occasionally, in aggregates exhibiting a very stable rest potential, unitary events which we have termed "microspikes" may be seen (Fig. 20a). These

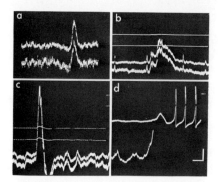

Fig. 20. Microscopic electrical events recorded from aggregates showing no macroscopic activity. 3×10^{-6} M TTX. All calibrations refer to scales shown in panel (d). (a) A microspike, recorded simultaneously from two cells in the same aggregate as that recorded in Fig. 18, but after elevating K to 6.3 mM. Scale: 100 μV, 0.1 sec. (b) Same impalement as (a) at slow sweep speed, to show repetitive microspikes occurring at about 3 sec intervals, and in a burst during one interspike interval. $E_r = -62$ mV. Scale: upper traces, 25 mV; lower traces 0.5 mV; 1 sec. (c) "Giant" microspike from the -70 mV stable state. 2.3 mM K, $E_r = -73$ mV. (d) One-electrode record from a 150 μm aggregate to show the transition from microscopic oscillatory voltage noise (lower trace) to a macroscopic oscillatory swing (upper trace) which leads to a train of overshooting action potentials. The lower trace ceases at the point where E_r had depolarized about 1.5 mV. At this point, the high-gain amplifiers saturate. $E_r = -66$ mV, 2.3 mM K. Scale: upper trace, 25 mV; lower trace, 0.75 mV; 1 sec.

are brief spontaneous voltage swings 200–300 μV in size, lasting about 100 msec. With single-electrode recordings, such events might be discounted as spurious "popcorn" noise generated by a clogged electrode. The fact that they are seen with two widely spaced electrodes in an aggregate simultaneously and with identical form attests to their origin in the membrane, and again illustrates the isopotential nature of the spheroidal aggregate. Microspikes may occur repetitively at 2–3 sec intervals, or in bursts that tend to sum to a barely visible deflection at the macroscopic level (Fig. 20b). Similar events have recently been observed in a mammalian cardiac preparation (G. H. Pollack, personal communication). Larger microspikes, 2–3 mV in size, may erupt suddenly out of random voltage noise (Fig. 20c). However, the most common behavior of aggregates in low levels of K (1.3 mM) is that a period of quiescence is broken by one or a train of action potentials. In the macroscopic (upper) trace in Fig. 20d the stable rest potential spontaneously depolarizes by about 15 mV, but fails to activate an action potential. A second depolarizing swing apparently crosses threshold to produce the first of a train of spikes. The microscopic (lower) trace reveals that even during the stable state, voltage oscillations of 0.5 mV or larger were buried in the stochastic fluctuations. (A similar gradual growth of oscillatory deflections leading to an action potential train was pictured in Fig. 11.) From the crest of one such oscillation, a large microspike-like event takes off. This apparently starts the membrane into the subthreshold macroscopic oscillatory swing which then rebounds to initiate an action potential.

We may speculate about how these microscopic events may be related to the macroscopic rhythmic properties of the cardiac membrane. Presumably, the channels whose random openings and closings produce the intrinsic voltage noise of the membrane are the same as those that carry the macroscopic conductances of the action potential, I_{Na}, I_{si}, I_{K_1}, etc. (Fig. 8). We have already emphasized that the phenomenological inductance that results in voltage oscillation around a resonant frequency is, in fact, a manifestation of voltage- and time-dependent conductances. Presumably, these also are carried through one or more of the same conductance channels. Recent estimates indicate that these conductance channels may be present in excitable membranes at densities of several hundred/μm^2 (see DeFelice, 1978, for review). It is not difficult to imagine three kinds of behavior of the conductance channels.

(1) *Completely random opening and closing,* leading to stochastic voltage noise. Filtered across the impedance of the membrane, such conductance fluctuations would set the voltage oscillating at the resonant frequency of the membrane, and would result in records such as those illustrated in Fig. 17.

(2) *Partial cooperativity,* which could cause patches of channels to open and close simultaneously. If we assume an average channel conductance of

about 10 pS (where a pS = $10^{-12}\Omega^{-1}$; DeFelice, 1978) and a dc resistance of the membrane of 3 MΩ, the sudden initiation of a cooperative state in a few hundred channels would result in a microspike of the size shown in Fig. 20a.

(3) *Tuning* the membrane to increase its impedance and sharpen its resonance. The heart cell membrane has a maximum impedance near 1 Hz upon depolarization in the range -80 to -50 mV, as mentioned earlier. An increase in a steady inward current would drive the membrane to a more depolarized potential and increase the magnitude of its oscillatory voltage swings. In the -50 mV state, even small depolarizing perturbations lead to macroscopic rhythmic activity. We may describe this as bringing the membrane potential close to "threshold," or as tuning it to an increased impedance. This was the case illustrated in Fig. 5 where rhythmic action potentials were induced by reducing $\bar{g}_K$. Similar rhythmic behavior has been elicited in quiescent frog atrial fibers by applying small steady depolarizing (inward) currents (Brown and Noble, 1969).

The correlation function of stationary voltage noise from a quiescent aggregate has the form of a damped sine wave, indicating that the fundamental oscillator is not periodic but random. The microscopic oscillations have average properties that define a fundamental oscillatory period. When the impedance of the membrane in this state is measured experimentally by injecting very small current pulses, the LRC voltage response exhibits a resonant frequency that approaches that extracted from the noise by correlation analysis, near 1 Hz. We are presently fitting the voltage responses produced by very small current pulses to those computed from specific equivalent circuit models (Clay et al., 1977). The evidence indicates that an equivalent RC circuit with at least one parallel inductor is required to yield a satisfactory fit to the pulse data. This is the phenomenological inductance manifested by the voltage- and time-dependent conductances that are active in the cardiac membrane at rest. The role that the inductive element plays in the endogenous behavior of the aggregate has been illustrated in the oscillatory rebound following spontaneous cessation of firing (Fig. 17a) and that resulting from the rapid shift from one stable state to the other (Fig. 15d). The similarity between these "responses" to intrinsic voltage perturbations, and the responses to injected current pulses (Fig. 16) is inescapable. In both cases, an LRC response is required to describe the behavior.

We have interpreted the spontaneous voltage fluctuations of the membrane as resulting from random opening and closing of ionic channels, filtered through the phenomenological impedance associated with the time-varying, voltage-dependent conductance pathways. The time dependence of the conductances is such that the equivalent LRC circuit is highly resonant, having a natural frequency near the normal frequency of macroscopic rhythmicity. We have shown that the resonant frequency derived from correlation analysis of voltage noise is near that measured from the small signal impedance. The cardiac aggregate membrane is thus originally in a tuned state similar to a nerve membrane whose impedance is artificially increased to augment its resonant behavior (Fig. 5). Support for this view may be derived from the similarity between the resonant period extracted from noise correlation functions, that obtained from pulses responses, and the rhythmic period (interbeat interval) of macroscopic beating. The fact that rhythmic frequency is not identical with the LRC resonant frequency is due to the strong nonlinearity of the excitable aggregate membrane. The 10–20 mV excursion of the slow pacemaker potential is not within the linear voltage range. Thus, any single resonant frequency determined at a particular potential within that range would not be applicable to the macroscopic firing frequency. The exact analytic relationship between the linearized LRC resonance and the nonlinear macroscopic frequency is not well understood. We are currently attempting to clarify this relationship.

The fact that we have chosen to describe the oscillatory behavior of the aggregate preparation in terms similar to those that yield rhythmic activity in the H-H model does not indicate that we view the aggregate membrane as a simple extension of the classical nerve membrane. But the extent to which it is similar as regards oscillatory phenomena is worth stressing. Our main message is that the two views of rhythmicity, as arising from time-varying conductances (MNT model) on the one hand, and from voltage oscillations, on the other, are merely different manifestations of the same phenomenon. One view translates readily into the other. We continue to use the impedance framework because it simplifies the interpretation of voltage noise data.

VI. Summary and Conclusions

Based upon the evidence examined here, we wish to propose the following model for the excitable cardiac membrane and its rhythmic electrical behavior, with the understanding that some of the ideas we put forth are working hypotheses or mere speculations.

The most convincing view of the structure of the cardiac membrane (like all excitable membranes) is that it is a high-resistance, fluid bilayer of phospholipids, thickly coated on the outside with glycoprotein, and studded with proteinaceous intramembrane particles. Some of these particles float in one lipid lamina or the other, but many pierce the entire sarcolemma and can thus form conductance channels that permit ion flow across the membrane. Some of these channels are penetrated by simple diffusive "holes," but many exhibit differential permeability to specific ion species, and have the following properties. (1) A given channel opens and closes on a random basis, but the average number of channels that are open at any instant depends upon the electric potential across the membrane and can be influenced by temperature, ion concentrations, and a variety of pharmacological agents. (2) The time it takes to open or close an individual channel is short (microseconds). However, the time required to force a large number of one type of channel to change its average open time ranges from milliseconds to seconds depending on the channel type. (3) There are nine different types of conductance channels, according to the MNT model, distinguishable by their ion specificity, time- and voltage-dependence, and pharmacological responsiveness.

We have attempted to marshal evidence that the resistive properties of the sarcolemma, the capacity of the lipid bilayer to separate ionic charges, and the time- and voltage-dependent nature of the conductances, all combine to cause the membrane to respond to an electrical perturbation like a tuned LRC circuit. We have compared simple models of oscillation in excitable nerve membranes with the more complicated oscillatory behavior seen in cardiac preparations, and have discussed the mechanism of rhythmicity in light of two separate theoretical frameworks. In one, repetitive firing was described in terms of time-varying conductances. In particular, the slow pacemaker depolarization was attributed to an inward background sodium current against which outward time-varying potassium currents are balanced to yield a net slow depolarization. We saw that most of the nine currents of the MNT model play important roles in this process. According to the second conceptual framework, pacemaker activity arises from the intrinsic oscillatory behavior of the cardiac membrane that derives from its LRC properties. We have shown that these oscillations resemble those that can be induced in the nerve membrane by "tuning," except on a much different time scale.

We have described a tissue culture cardiac membrane model—the heart-cell aggregate—which exhibits the oscillatory properties listed above. When the spontaneous macroscopic electrical activity of such an aggregate is suppressed, a variety of microscopic, subthreshold events are revealed, the most common being spontaneous random noise in the range of 50–500 μV, that arises from the random opening and closing of individual current channels in the membrane. Presumably these are the same channels that carry the macroscopic currents of the pacemaker potential.

We have argued that these random microcurrents, filtered across the impedance of the membrane, produce voltage perturbations that set the membrane oscillating at the resonant frequency of its LRC circuit. The oscillatory frequency produced in response to intrinsic membrane noise is related to that obtained from small-signal impedance analysis by injecting small current pulses into the aggregate, and is similar (except for the nonlinearity of the system) to the frequency of rhythmic firing. We conclude that the two apparently contrasting views of rhythmicity are not really different; that the time-varying conductances translate into an equivalent resonant small-signal impedance. The small-signal impedance is a convenient bridge between the current and voltage fluctuations in excitable membranes. The study of fluctuations in heart cell membranes offers an understanding of oscillatory properties and excitability at the macroscopic level.

ACKNOWLEDGMENT

The authors acknowledge the support of the Georgia Heart Association, and of the National Institutes of Health Grants HL-17827, HL-16567, and LM-02502.

References

Albuquerque, E. X., and Thesleff, S. (1967). *J. Physiol.* **190**, 123–137.
Armstrong, C. M. (1975). *Biophys. J.* **15**, 932–933.
Arvanitaki, J. (1939). *Arch. Int. Physiol.* **49**, 209–256.
Bangham, A. D. (1968). *Prog. Biophys. Mol. Biol.* **18**, 31–95.
Beeler, G. W., and Reuter, H. (1977). *J. Physiol.* **268**, 177–210.
Bennett, M. V. L. (1973). *Fed. Proc., Fed. Am. Soc. Exp. Biol.* **32**, 65–75.
Bernstein, J. (1902). *Pfluegers Arch. Gesamte Physiol. Menschen Tiere.* **92**, 521–562.
Bernstein, J. (1912). "Elektrobiologie." Braunschweig.
Besch, H. R., Jones, L. R., and Watanabe, A. M. (1976). *Circ. Res.* **39**, 586–595.
Blaurock, A. E. (1971). *J. Mol. Biol.* **56**, 35–52.
Bonke, F. I. (1973a). *Pfluegers Arch.* **339**, 1–15.
Bonke, F. I. (1973b). *Pfluegers Arch.* **339**, 17–23.
Bozler, E. (1943). *Am. J. Physiol.* **138**, 273–282.
Branton, D. (1966). *Proc. Natl. Acad. Sci. U.S.A.* **55**, 1048–1056.
Bretscher, M. S. (1972). *J. Mol. Biol.* **71**, 523–528.
Bretscher, M. S. (1973). *Science* **181**, 622–629.
Brightman, M. W., and Reese, T. S. (1969). *J. Cell Biol.* **40**, 648–678.
Brown, H. F., and Noble, S. J. (1969). *J. Physiol.* **204**, 717–736.
Chandler, W. K., Fitzhugh, R., and Cole, K. S. (1962). *Biophys. J.* **2**, 105–127.
Clapham, D. E., and DeFelice, L. J. (1976). *Pfluegers Arch.* **366**, 273–276.
Clay, J. R., DeFelice, L. J., and DeHaan, R. L. (1977). *Physiologist* **20**, 18.
Cole, K. S. (1928). *J. Gen. Physiol.* **12**, 29–36.
Cole, K. S. (1968). "Membranes, Ions and Impulses." Univ. California Press, Berkeley, California.
Cole, K. S., and Baker, R. F. (1941). *J. Gen. Physiol.* **24**, 535–549.
Cole, K. S., and Marmont, G. (1942). *Fed. Proc., Fed. Am. Soc. Exp. Biol.* **1**, 15–16.
Conti, F., and Palmieri, G. (1968). *Biophysik* **5**, 71–77.
Conti, F., and Wanke, E. (1975). *Rev. Biophys.* **8**, 451–506.
Cranefield, P. R. (1975). "The Conduction of the Cardiac Impulse." Futura, New York.
Cuatrecasas, P. (1974). *Annu. Rev. Biochem.* **43**, 169–214.
Danielli, J. F., and Davson, H. (1935). *J. Cell Physiol.* **5**, 495–508.

Davis, H. T. (1960). U.S. Atomic Energy Commission. Reissued by Dover Publ., New York (1962).
Davson, H., and Danielli, J. F. (1952). "The Permeability of Natural Membranes," 2nd ed. Cambridge Univ. Press, London and New York.
Deamer, D. W., and Branton, D. (1967). *Science* **158**, 655–657.
DeFelice, L. J. (1977). *Intl. Rev. Neurobiol.* **20**, 169–208.
DeFelice, L. J. (1978). (tr.) "Oliver Heaviside": a lecture by B. van der Pol. *In* "Electronics and Power" (in press).
DeFelice, L. J., and Challice, C. E. (1969). *Circ. Res.* **24**, 457–474.
DeFelice, L. J., and DeHaan, R. L. (1977). *Proc. IEEE (Biol. Signals)* **65**, 796–799.
DeHaan, R. L. (1967). *Dev. Biol.* **16**, 216–249.
DeHaan, R. L. (1970). *Dev. Biol.* **23**, 226–240.
DeHaan, R. L., and Fozzard, H. A. (1975). *J. Gen. Physiol.* **65**, 207–222.
DeHaan, R. L., and Gottlieb, S. H. (1968). *J. Gen. Physiol.* **52**, 643–665.
DeHaan, R. L., and Hirakow, R. (1972). *Exp. Cell Res.* **70**, 214–220.
DeHaan, R. L., and Sachs, H. G. (1972). *Curr. Top. Dev. Biol.* **7**, 193–228.
DeHaan, R. L., McDonald, T. F., and Sachs, H. G. (1976). *In* "Developmental and Physiological Correlates of Cardiac Muscle" (M. Lieberman and T. Sano, eds.), pp. 155–168. Raven, New York.
DeHemptinne, A. (1976). *Pfluegers Arch.* **363**, 87–95.
DeMello, W. C. (1972). *In* "Electrical Phenomena in the Heart" (W. C. DeMello, ed.). pp. 89–110. Academic Press, New York.
DeMello, W. C. (1975). *J. Physiol.* **250**, 231–245.
DeMeyts, P., Bianco, A. R., and Roth, J. (1976). *J. Biol. Chem.* **251**, 1877–1888.
Dodge, F. (1963). "A study of ionic permeability changes underlying excitation in myelinated nerve fibers of the frog." Ph.D. Thesis, The Rockefeller Institute, New York.
Edidin, M. (1974). *Annu. Rev. Biophys. Bioeng.* **3**, 179–201.
Edidin, M., and Weiss, A. (1974). *In* "Control of Proliferation in Animal Cells" (B. Clarkson and R. Baserga, eds.), pp. 213–220. Cold Spring Harbor Symp., Cold Spring Harbor, New York.
Edidin, M., Zagyansky, Y., and Lardner, T. J. (1976). *Science* **191**, 466–468.
Ehrenstein, G. (1971). *In* "Biophysics Physiology of Excitable Membranes" (W. J. Adelman, ed.), pp. 463–476. Van Nostrand-Reinhold, New York.
Ehrenstein, G., Gilbert, D. L., and Lipicky, R. J. (1975). *Biophys. J.* **15**, 847–849.
Eisenberg, R., and Engel, E. (1970). *J. Gen. Physiol.* **55**, 736–757.
Eisenberg, R., and Johnson, G. (1970). *Prog. Biophys. Mol. Biol.* **20**, 1–65.
Fabiato, A., and Fabiato, F. (1972). *Circ. Res.* **31**, 293–307.
Fänge, R., Persson, H., and Thesleff, S. (1956). *Acta Physiol. Scand.* **38**, 173–183.
Fitzhugh, R. (1969). *In* "Biological Engineering" (H. Schwan, ed.), pp. 1–115. McGraw-Hill, New York.
Forssman, W. G., and Girardier, L. (1970). *J. Cell Biol.* **44**, 1–19.
Frankenhaeuser, B., and Hodgkin, A. L. (1957). *J. Physiol.* **137**, 218–244.
French, K. J., and Adelman, W. J. (1976). *Curr. Top. Membr. Transp.* **8**, 161–207.
Fricke, H. (1925). *Phys. Rev.* **26**, 678–781.
Frye, L. D., and Edidin, M. (1970). *J. Cell Sci.* **7**, 319–335.
Garrahan, P. J., and Garay, R. P. (1976). *Curr. Top. Membr. Transp.* **8**, 29–97.
Goodenough, D. A. (1976). *J. Cell Biol.* **68**, 220–223.
Goshima, K. (1970). *Exp. Cell Res.* **63**, 124–130.
Goshima, K. (1976). *J. Mol. Cell. Cardiol.* **8**, 713–725.
Guttman, R. (1969). *Biophys. J.* **9**, 269–277.

Guttman, R., and Barnhill, R. (1970). *J. Gen. Physiol.* **55**, 114–118.

Hall, J. E. (1975). *Biophys. J.* **15**, 934–940.

Hasty, D. L., and Hay, E. D. (1977). *J. Cell Biol.* **72**, 667–686.

Heaviside, O. (1893). "Electromagnetic Theory, I–III." Bern Brothers, London (see I, p. 291 and II, p. 445).

Heppner, D. B., and Plonsey, R. (1970). *Biophys. J.* **10**, 1057–1075.

Hermann, L. (1899). *Pfluegers Arch. Gesamte Physiol. Menschen Tiere.* **75**, 574–590.

Hille, B. (1971). *In* "Biophysics and Physiology of Excitable Membranes" (W. J. Adelman, ed.), pp. 230–246. Van Nostrand-Reinhold, New York.

Hirakow, R. (1970). *Am. J. Cardiol.* **25**, 195–203.

Hodgkin, A. L., and Huxley, A. F. (1952). *J. Physiol.* **117**, 500–544.

Hodgkin, A. L., and Rushton, W. A. H. (1946). *Proc. R. Soc. London* **133**, 444–479.

Hodgkin, A. L., Huxley, A. F., and Katz, B. (1949). *Arch. Sci. Physiol.* **3**, 129–150.

Huxley, A. F. (1959). *Ann. N.Y. Acad. Sci.* **81**, 221–246.

Hyde, A., Blondel, B., Matter, A., Cheneval, J. P., Filloux, G., and Girardier, L. (1969). *Prog. Brain Res.* **31**, 282–311.

Hynes, R. O. (1975). *Proc. Natl. Acad. Sci. U.S.A.* **70**, 3170–3174.

Jack, J. J. B., Noble, D., and Tsien, R. W. (1975). "Electrical Current Flow in Excitable Cells." Oxford Univ. Press (Clarendon), London and New York.

Jain, M. K. (1975). *Curr. Top. Membr. Transp.* **6**, 1–57.

Johnson, E. A., and Lieberman, M. (1971). *Annu. Rev. Physiol.* **33**, 479–532.

Jost, P. C., Griffith, O. H., Capaldi, R. A., and Vanderkooi, G. (1973). *Proc. Natl. Acad. Sci. U.S.A.* **70**, 480–484.

Joyner, R. W., Ramon, F., and Moore, J. W. (1975). *Circ. Res.* **35**, 654–661.

Kahn, C. R. (1976). *J. Cell Biol.* **70**, 261–286.

Kass, R. S., and Tsien, R. W. (1975). *J. Gen. Physiol.* **67**, 599–617.

Kelvin, Lord William Thompson (1872). "Papers on Electrostatics and Magnetism." Macmillan, New York.

Kensler, R. W., Brink, P., and Dewey, M. M. (1977). *J. Cell Biol.* **75**, 768–782.

Landis, D. M. D., and Reese, T. S. (1974). *J. Cell Biol.* **60**, 316–321.

Lane, M. A., Sastre, A., and Salpeter, M. M. (1977). *Dev. Biol.* **57**, 254–269.

LeDouarin, G., Renaud, J. F., Renaud, D., and Coraboeuf, E. (1974). *J. Mol. Cell. Cardiol.* **6**, 523–529.

Lieberman, M., Kootsey, J. M., Johnson, E. A., and Sawanobori, T. (1973). *Biophys. J.* **13**, 37–55.

Ling, G. N., and Gerard, R. W. (1949). *J. Cell. Comp. Physiol.* **34**, 383–396.

Loewenstein, W. R. (1973). *Fed. Proc., Fed. Am. Soc. Exp. Biol.* **32**, 60–64.

MacDonald, R. L., Mann, J. T., Jr., and Sperelakis, N. (1974). *J. Theor. Biol.* **45**, 107–130.

Makowski, L., Caspar, D. L. D., Phillips, W. C., and Goodenough, D. A. (1977). *J. Cell. Biol.* **74**, 629–645.

Martinez-Palomo, A., Alanis, J., and Benitez, D. (1970). *J. Cell Biol.* **47**, 1–17.

Masuda, M. O., and Paes de Carvalho, A. (1975). *Circ. Res.* **37**, 414–421.

Matter, A. (1973). *J. Cell Biol.* **56**, 690–696.

Mauro, A., Conti, F., Dodge, F., and Schor, R. (1970). *J. Gen. Physiol.* **55**, 497–523.

McAllister, R. E., Noble, D., and Tsien, R. W. (1975). *J. Physiol.* **251**, 1–59.

McCall, D. (1976). *Circ. Res.* **39**, 730–735.

McCallister, L. P., and Page, E. (1973). *J. Ultrastruct. Res.* **42**, 136–155.

McNutt, N. S. (1975). *Circ. Res.* **37**, 1–13.

McNutt, N. S., and Fawcett, D. W. (1969). *J. Cell Biol.* **42**, 46–67.

McNutt, N. S., and Weinstein, R. S. (1970). *J. Cell Biol.* **47**, 666–668.

McNutt, N. S., and Weinstein, R. S. (1973). *Prog. Biophys. Mol. Biol.* **26**, 45–101.

Metcalfe, J. C., and Warren, G. B. (1977). *In* "International Cell Biology, 1976–1977" (B. R. Brinkley and K. R. Porter, eds.), pp. 15–23. Rockefeller University Press, New York.

Minorsky, N. (1947). "Introduction to Non-linear Mechanics." J. W. Edwards, Ann Arbor, Michigan.

Mobley, B. A., and Page, E. (1972). *J. Physiol.* **220**, 547–563.

Nathan, R. D., and DeHaan, R. L. (1977). *J. Gen. Physiol.* (submitted).

Neher, E., and Stevens, C. F. (1977). *Annu. Rev. Biophys. Bioeng.* **6**, 345–381.

Noble, D. (1962). *J. Physiol.* **160**, 317–352.

Noma, A., and Irisawa, H. (1976). *Pfluegers Arch.* **364**, 45–52.

Overton, J. (1977). *Dev. Biol.* **55**, 103–116.

Paes de Carvalho, A., Hoffman, B. F., and Paes de Carvalho, M. (1969). *J. Gen. Physiol.* **54**, 607–635.

Palti, Y., and Adelman, W. J. (1969). *J. Membr. Biol.* **1**, 431–458.

Papahadjapoulous, D., Poste, G., and Schaeffer, B. E. (1973). *Biochim. Biophys. Acta.* **323**, 23–42.

Parsegian, A. (1969). *Nature (London)* **221**, 844–846.

Peracchia, C. (1977). *J. Cell Biol.* **72**, 628–641.

Perrachia, C., and Dulhunty, A. F. (1976). *J. Cell Biol.* **70**, 419–439.

Phillips, E. R., and Perdue, J. F. (1976). *J. Supramol. Struct.* **4**, 27–44.

Pinto da Silva, P. (1972). *J. Cell Biol.* **53**, 777–787.

Pollack, G. H. (1976). *J. Physiol.* **255**, 275–298.

Pretlow, T. G., Glick, M. R., and Reddy, W. J. (1972). *Am. J. Pathol.* **67**, 215–226.

Purdy, J., Lieberman, M., Roggeveen, A. E., and Kirk, R. G. (1972). *J. Cell Biol.* **55**, 563–578.

Ramon, F., Anderson, N., Joyner, R. W., and Moore, J. W. (1975). *Biophys. J.* **15**, 55–69.

Rash, J. E., and Ellisman, M. H. (1974). *J. Cell Biol.* **63**, 567–586.

Rash, J. E., Staehelin, L. A., and Ellisman, M. H. (1974). *Exp. Cell Res.* **86**, 187–190.

Rayleigh, J. W. S. (1894). "Theory of Sound, I and II," 2nd ed. Dover Reprint, New York (see I, p. 81).

Raynes, O. G., Simpson, F. O., and Bertrand, W. S. (1968). *J. Cell Sci.* **3**, 467–474.

Reuter, H. (1973). *Prog. Biophys. Mol. Biol.* **26**, 1–43.

Robbins, J. C., and Nicolson, G. L. (1975). *In* " Biology of Tumors " (F. E. Becker, ed.), pp. 1–54. Plenum, New York.

Robertson, J. D. (1959). "The Ultrastructure of Cell Membranes and their Derivatives," Biochem. Soc. Symp. Cambridge Univ. Press, London.

Sachs, F. (1976). *J. Membr. Biol.* **28**, 373–399.

Sachs, H. G., and DeHaan, R. L. (1973). *Dev. Biol.* **30**, 223–240.

Sakamoto, Y., and Goto, M. (1970). *Jpn. J. Physiol.* **20**, 30–41.

Schlessinger, J., Koppel, D. E., Axelrod, D., Jacobson, K., Webb, W. W., and Elson, E. L. (1976). *Proc. Natl. Acad. Sci. U.S.A.* **73**, 2409–2413.

Schlessinger, J., Axelrod, D., Koppel, D. E., Webb, W. W., and Elson, E. L. (1977). *Science* **195**, 307–308.

Schotland, D. L., Bonilla, E., and Van Meter, M. (1977). *Science* **196**, 1005–1007.

Scott, R. E., Furcht, L. T., and Kersey, J. B. (1973). *Proc. Natl. Acad. Sci. U.S.A.* **73**, 3631–3635.

Seyama, I. (1976). *J. Physiol.* **255**, 379–397.

Sherwood, D., and Montal, M. (1975). *Biophys. J.* **15**, 417–434.

Shimada, Y., Moscona, A. A., and Fischman, D. A. (1974). *Dev. Biol.* **36**, 428–446.

Simpson, I., Rose, B., and Loewenstein, W. R. (1977). *Science* **195**, 294–296.

Singer, S. J., and Nicolson, G. L. (1972). *Science* **175**, 720–731.

Sommer, J., and Johnson, E. A. (1970). *Am. J. Cardiol.* **25**, 184–194.

Speicher, D. W., and McCarl, R. L. (1974). *In Vitro* **10**, 30–41.
Sperelakis, N. (1972). *In* "Electrical Phenomena in the Heart" (W. C. DeMello, ed.), pp. 1–61. Academic Press, New York.
Sperelakis, N., and Lehmkuhl, D. (1966). *J. Gen. Physiol.* **49**, 867–895.
Sperelakis, N., Shigenobu, K., and McLean, M. J. (1976). *In* "Developmental and Physiological Correlates of Cardiac Muscle" (M. Lieberman and T. Sano, eds.), pp. 209–234. Raven, New York.
Spira, A. W. (1971). *J. Ultrastruct. Res.* **34**, 409–425.
Steck, T. L. (1974). *J. Cell Biol.* **62**, 1–19.
Steck, T. L., and Yu, J. (1973). *J. Supramolec.* **1**, 220.
Stevens, C. F. (1972). *Biophys. J.* **12**, 1028–1047.
Tanaka, I., and Sasaki, Y. (1966). *J. Gen. Physiol.* **49**, 1089–1110.
Tilney, L. G., and Mooseker, M. S. (1976). *J. Cell Biol.* **71**, 402–416.
Trautwein, W., and Kassebaum, D. G. (1961). *J. Gen. Physiol.* **45**, 317–330.
Van der Kloot, W. G., and Dane, B. (1964). *Science* **146**, 74–76.
van der Pol, B. (1926). *Philos. Mag.* **2**, 978–992.
van der Pol, B., and van der Mark, J. (1928). *Philos. Mag.* **6**, 763–775.
Verveen, A. A., and DeFelice, L. J. (1974). *Prog. Biophys. Mol. Biol.* **28**, 189–265.
Viragh, S., and Challice, C. E. (1973). *J. Ultrastruct. Res.* **42**, 1–24.
Viragh, S., and Porte, A. (1973). *Z. Zellforsch. Mikrosk. Anat.* **145**, 191–211.
Wallach, D. (1975). "Membrane Molecular Biology of Neoplastic Cells." Elsevier, New York.
Wanke, E., DeFelice, L. J., and Conti, F. (1974). *Pfluegers Arch.* **347**, 63–74.
Weidmann, S. (1951). *Am. J. Physiol.* **115**, 227–236.
Weidmann, S. (1966). *J. Physiol.* **187**, 323–342.
Weidmann, S. (1970). *J. Physiol.* **210**, 1041–1054.
Weidmann, S. (1971). *In* "Research in Physiology" (F. F. Kao, K. Koizumi, and M. Vassale, eds.), pp. 3–25. Aulo Gaggi, Bologna.
Weingart, R. (1974). *J. Physiol.* **240**, 741–762.
Wickus, G., Gruenstein, E., Robbins, P. W., and Rich, A. (1975). *Proc. Natl. Acad. Sci. U.S.A.* **72**, 746–749.
Wiggins, J. R., and Cranefield, P. F. (1976). *Circ. Res.* **39**, 466–474.
Woodbury, J. W., and Crill, W. E. (1961). *In* "Nervous Inhibition" (E. Florey, ed.), pp. 124–135. Pergamon, New York.
Woodbury, J. W., and Crill, W. E. (1970). *Biophys. J.* **10**, 1076–1083.
Woodbury, A. L., Woodbury, J. W., and Hecht, H. H. (1970). *Circulation* **1**, 264–266.
Yahara, I., and Edelman, G. M. (1973). *Nature (London)* **246**, 152–154.
Zwaal, R. F. A., Roelofsen, B., and Colley, C. M. (1973). *Biochim. Biophys. Acta* **300**, 159–182.

Selected Topics from the Theory of Nonlinear Physico-Chemical Phenomena

P. Ortoleva

Department of Chemistry,
Indiana University, Bloomington, Indiana

I. Introduction

One of the most significant advances of the macroscopic theory of continuum systems in recent years is the analysis of a variety of biological, geological, and other systems in terms of nonlinear physico-chemical phenomena. It is the purpose of this article to describe some of these developments and to set forth the framework in which several other phenomena may be studied in the future. I shall refer to existing reviews (Hess and Boiteaux, 1971; Chance et al., 1973; Nicolis and Portnow, 1973; Noyes and Field, 1974; Ross, 1976; Hanusse et al., 1978) when appropriate to avoid duplication and shall point out certain features not emphasized or covered by them. In many cases I shall draw on unpublished results or on work in progress so that the coverage is up to date. In addition, an attempt is made to point out possible directions of future development and in some cases to map out the methods that will most likely be useful in solving and formulating the mathematical problems associated with the physical phenomena.

The wealth of phenomena found in nonequilibrium physico-chemical systems seems almost endless. These systems are usually describable by nonlinear partial differential equations and in a certain sense one can say "anything can happen," as a well-known statistical mechanician once told me. Although this is probably somewhat alarming to those wishing to make simple statements delineating the kinds of phenomena that can occur, we should be thankful for this " principle " since its application by nature has apparently led to much of the beauty and variation of the universe.

Before proceeding with the main body of this article I would like to stress that I have not tried to write an " unbiased " review. The reader is referred to reviews (Hedges and Myers, 1926; Glansdorff and Prigogine, 1971; Hess and Boiteaux, 1971; Chance et al., 1973; Nicolis and Portnow, 1973; Noyes and Field, 1974; Ross, 1976; Hanusse et al., 1978) where other authors have made varying degrees of effort to be objective and complete in their coverage. It has been my hope that the present article would cover some topics not already covered in the reviews and to choose areas that I feel will lead to interesting new results in the near future. To the degree that the examples have been chosen too heavily from my own present interests, I apologize to my fellow workers in the field.

A. MACROSCOPIC THEORY

We shall discuss a variety of physico-chemical systems within the framework of the continuity equations describing the temporal evolution of macroscopic variables such as concentrations, temperature, and local center of mass velocity (DeGroot and Mazur, 1962; Fitts, 1962; Glansdorff and Prigogine, 1971). The appropriate equations corresponding to these variables are, respectively, the equations of conservation of mass, energy, and fluid momentum. In addition, if the chemical species involved are charged, then we must also find the electric and magnetic fields. In this section we briefly discuss the general features of the macroscopic theory and the relationship between boundary condition and the maintenance of a system far from equilibrium.

We concentrate on processes that are slow compared to the time required for the passage of light through the system and to the dielectric relaxation times. Hence the electric field $\mathbf{E}$ satisfies Poisson's equation. To provide a full macroscopic description of a system we need the form of the continuity equations, the dependence of the dielectric constant on composition, temperature and electric field, and finally the boundary and initial conditions for the descriptive variables.

We shall focus our attention mainly on systems that are isothermal, isobaric and at rest, and neglect magnetic effects. [Magnetic phenomena have been studied extensively in the magnetohydrodynamic (Chandrasekar, 1964) and plasma physics (Davidson, 1973) literature.] For this case the system is completely described by the concentration $C(\mathbf{r}, t)$ and the electric field $\mathbf{E}(\mathbf{r}, t)$ at all points $\mathbf{r}$ and times t. To solve a specific problem we need to know these quantities at $t = 0$ and the boundary conditions.

B. BOUNDARY CONDITIONS AND CONSTRAINTS

The phenomena of interest in this article arise in systems maintained far from equilibrium (Glansdorff and Prigogine, 1971). This condition may be obtained by the application of either homogeneous or heterogeneous constraints. Heterogeneous constraints are those conditions applied to the system at boundary surfaces. Mathematically, they enter the macroscopic description through the boundary conditions. Homogeneous constraints are the ideal limit of heterogeneous constraints when certain species are so mobile that they are maintained constant throughout the system despite their consumption in chemical reactions within the system.

Let $\mathbf{J}_i$ be the flux (moles/area-time) of species i in the system. Then, if G_i is the rate of passage of i through the boundary surface S out of the system, we have

$$-\mathbf{n} \cdot \mathbf{J} = G. \tag{1.1}$$

In Eq. (1.1) we have denoted the unit normal vector to S pointing into the system by $\mathbf{n}$ and have introduced the vector notation $C = \{C_1, C_2, ..., C_N\}$, $\mathbf{J} = \{\mathbf{J}_1, \mathbf{J}_2, ..., \mathbf{J}_N\}$, etc. convenient for the N species problem. For systems in which the boundary surface is mobile, the "free boundary problem," an additional term must be added to (1.1) as we shall discuss in Section VI.

If the system is bounded by a catalytic surface, then G_i has a contribution due to the rate of consumption of i in the heterogeneous reaction. Of course if i is produced, this contribution to G_i is negative. Also, as mentioned above, G_i may represent the rate of passage of i out of the system. For example, if the system is surrounded by a thin membrane and kept in contact with a well-stirred reservoir at constant composition C^0, then for simple passive permeation we have

$$G_i = h_i(C_i - C_i^0) \tag{1.2}$$

where h_i is a permeability factor. For this case C_i^0 is a typical constraint parameter since its value controls the degree to which the system is maintained out of equilibrium. If Fick's law without cross-coupled transport is assumed valid for charge neutral species at low concentrations, we have

$$\mathbf{J}_i = -D_i \nabla C. \tag{1.3}$$

Hence combining (1.1) with (1.3) results in

$$\mathbf{n} \cdot D_i \nabla C_i = h_i(C - C^0). \tag{1.4}$$

Since $h_i \to \infty$ as the membrane becomes thinner, we see that this reduces to the case $C_i = C_i^0$ where the concentrations are fixed at the boundaries. To rigorously arrive at the connection between homogeneous constraints, $C_j = \text{constant}$ within the system for some j, we must invoke the time scale of the fastest reaction in which species j participates. Letting this time be τ_j the limit of homogeneous constraints arises when the diffusion length l_j $[\equiv (D_j \tau_j)^{1/2}]$ is much longer than the maximum system dimension l,

$$l/l_j \ll 1, \text{ homogeneous constraint.} \tag{1.5}$$

This is clearly a rather limiting restriction and probably not valid for most cases.

One notable exception is the situation of the maintenance of nonequilibrium conditions by the imposition of light. If the Beer's law length (the distance that the light intensity decreases by a factor e^{-1} due to absorption) is much greater than the dimensions of the system, we may consider the system to be under homogeneous photochemically driven constraint (Nitzan and Ross, 1973; Nitzan, et al., 1974a; Creel and Ross, 1976; Del-leDonne and Ortoleva, 1977).

Transient situations provide another approximation to homogeneous constraints. If within the system certain parts of the overall chemical kinetic network reach a steady state of production of certain species used as reactants for other parts of the scheme, then this provides a homogeneous source of these species and hence a homogeneous constraint may be established. Another approximation to this occurs if certain reactant species are in large excess and are only slowly consumed. These methods for keeping the system out of equilibrium are the basis for many known nonequilibrium chemical systems—the Belousov–Zhabotinsky–Zaikin (BZZ) reaction, oscillatory glycolysis, and the Bray reaction. (See appropriate chapter in this volume.) Homogeneous constraints have received a great deal of theoretical analysis both because of its experimental applicability and its mathematical simplicity.

In this article we shall discuss a variety of nonequilibrium phenomena arising in reaction–transport systems. To start our discussion (Section II) we consider some simple model systems that lend themselves to analytically tractible solutions. In doing so we introduce many of the concepts that we shall use throughout this article. In later sections we discuss a variety of phenomena and the theories that have been used to describe them.

II. Nonequilibrium Phenomena in Soluble Model Systems

Let us now study several simple soluble model systems that introduce some important concepts.

A. Branching of Homogeneous Steady States

First consider examples of the simplest situation, steady states. We assume mass action kinetics to hold in all examples and that the systems are under homogeneous constraint. Finally, we take all rate coefficients to be unity unless stated otherwise.

1. Exchange of Stability

Consider the reaction scheme

$$A + X \rightarrow 2X$$
$$B + 2X \rightarrow X + P_1$$
$$X \rightarrow P_2. \tag{2.1}$$

The rate equation for X is

$$dX/dt = B(\lambda - X)X$$
$$\lambda \equiv (A - 1)/B. \tag{2.2}$$

Clearly this system has steady states $X^* = 0, \lambda$. These are shown in Fig. 1 along with their stability as determined from linear stability analysis. Note

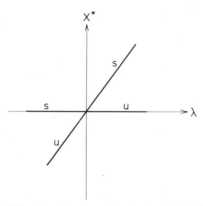

Fig. 1. Exchange of stability of steady states of mechanism (2.1). Labels s and u denote stable and unstable to small perturbations.

that the behavior of this system changes qualitatively if we add a small term ε to (2.2); the steady states are solutions of $X^2 - \lambda X - \varepsilon/B = 0$, i.e., $X_{\pm}^* = \frac{1}{2}[\lambda \pm (\lambda^2 + 4\varepsilon/B)^{1/2}]$. For $\varepsilon > 0$ the system only has one state, X_+^*, which is stable for all λ. The state X_-^* is negative for all λ and hence is unphysical.

As λ passes through zero the two branches of steady states intersect and an exchange of stability takes place. The point $\lambda = 0$ is the bifurcation point for this system—a new branch of states, $X^* = \lambda$, is said to bifurcate from the branch of states $X^* = 0$. This phenomenon is studied in more generality in Appendix A in the framework of bifurcation theory to illustrate the latter as a powerful tool in analyzing these problems. The bifurcation theory has been applied to a variety of reaction diffusion problems including static dissipative structures (Achmuty and Nicolis, 1975), chemical waves (Kopell and Howard, 1973; Ortoleva and Ross, 1974), and the problem of secondary branching, i.e., the bifurcation of secondary branches from new branches (Keener, 1976).

2. Exchange of Stability without Branch Intersection: Continua of States

Consider the three species system X, Y, Z evolving according to a mechanism of competitive preying by X and Y on Z:

$$R \longrightarrow Z$$

$$X + Z \xrightarrow{\alpha} 2X$$

$$Y + Z \xrightarrow{\beta} 2Y$$

$$X \longrightarrow P_1$$

$$Y \longrightarrow P_2. \tag{2.3}$$

The rate equations take the form

$$dX/dt = (\alpha Z - 1)X$$
$$dY/dt = (\beta Z - 1)Y$$
$$dZ/dt = R - (\alpha X + \beta Y)Z \qquad (2.4)$$

where we have taken all rate coefficients equal to 1 (except α and β) for the second and third steps in (2.4). There are two steady states of the system,

$$(X_1^*, Y_1^*, Z_1^*) = (R, 0, \alpha^{-1})$$
$$(X_2^*, Y_2^*, Z_2^*) = (0, R, \beta^{-1}), \qquad (2.5)$$

except for the special case $\alpha = \beta$ where we have a continuum of states connecting the branches,

$$(X_3^*, Y_3^*, Z_3^*) = (\mu R, (1 - \mu)R, \alpha^{-1}), \qquad \alpha = \beta, 0 \leq \mu \leq 1. \qquad (2.6)$$

Note that at $\alpha = \beta$ the stability of states 1 and 2 is exchanged without the intersection of these branches. We also see the possibility of a continuum of states. This model has a clear ecological interpretation in terms of mutants X and Y whose competition advantage for the prey Z exchanges at $\alpha = \beta$. At this point the system becomes degenerate because X and Y become indistinguishable, accounting for the continuum of states. The bifurcation diagram for this system is shown in Fig. 2.

3. Bifurcation of a Pair of Branches: Critical Index $\frac{1}{2}$

The mechanism

$$A + X \rightarrow 2X$$
$$X \rightarrow P_1$$
$$3X \rightarrow P_2 + 2X \qquad (2.7)$$

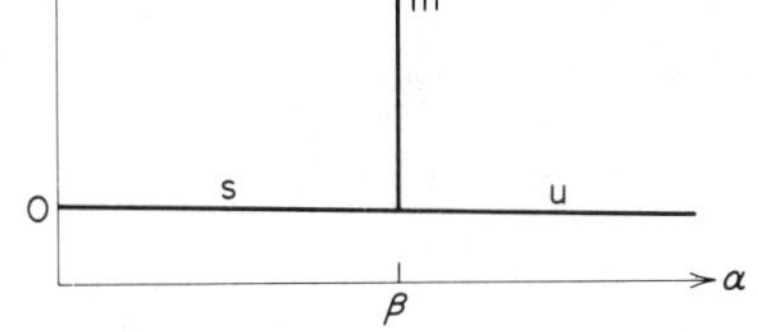

Fig. 2. Exchange of stability without branch crossing in mechanism (2.3). Note the continuum of marginally stable states (m) at $\alpha = \beta$.

leads to the kinetic equation

$$dX/dt = \lambda X - X^3$$
$$\lambda = A - 1. \tag{2.8}$$

It is clear that the pair of steady states

$$X_{\pm}^{*}(\lambda) = \pm\sqrt{\lambda} \tag{2.9a}$$

bifurcate from the branch

$$X_0^{*} = 0 \tag{2.9b}$$

at $\lambda = 0$ as shown in Fig. 3. The branch X_{-}^{*} is negative and is unphysical since concentrations are positive. A slight modification of (2.7) leads to a cubic with three positive roots as in the Edelstein mechanism (Glansdorff and Prigogine, 1971) and below. Note that here as in mechanism (2.1) the intersection of branches is accompanied by an exchange of stability.

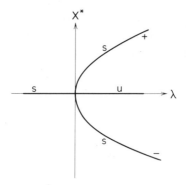

Fig. 3. Bifurcation of steady states for the mechanism (2.7) with critical index $\frac{1}{2}$.

4. *Finite Amplitude Instability*

The kinetic equation

$$dX/dt = -(X^2 - X + \tfrac{1}{4}\lambda)X \tag{2.10}$$

may be derived from the mechanism

$$A + X \rightarrow P_1$$
$$B + 2X \rightarrow 3X$$
$$3X \rightarrow 2X + P_2 \tag{2.11}$$

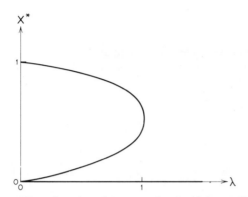

Fig. 4. Bifurcation of steady states at $\lambda = 1$ at finite amplitude.

where $B = 1$, $\lambda = 4A$. The system has steady states

$$X_0^* = 0, \qquad X_\pm^* = \tfrac{1}{2}[1 \pm (1 - \lambda)^{1/2}]. \tag{2.12}$$

In Fig. 4 we see that as λ passes through the bifurcation point at $\lambda = 1$ the new states arise at finite distances away from the state $X_0^* = 0$. Thus the transition between the stable branches $X_+ \to X_0$ that occurs if we are on branch X_+^* and increase λ beyond 1 will be a "hard transition." This is unlike the "soft" transitions of mechanisms (2.1) and (2.7) where the system variable (X concentration) changes continuously as λ passes through the bifurcation point. The hard and soft transitions are analogous to first- and second-order equilibrium phase transitions (Nitzan *et al.*, 1974). Note that typically a hard transition is characterized by a region of multiple states where the system can reside in more than one state. However this multiplicity is not necessary as we have seen in the example of Section II,A,2 where a hard transition occurs without an overlapping region (and accompanying hysteresis).

For cases with regions of multiple stable steady states, the system has a finite amplitude instability as transitions between stable states can occur when perturbations of sufficient magnitude, suprathreshold perturbations (Nitzam, Ortoleva, and Ross, 1974b), are applied.

B. HOMOGENEOUS OSCILLATIONS

We now consider several soluble oscillatory systems introduced long ago in the study of nonlinear oscillations (Caesari, 1971). The model demonstrates several important properties of nonlinear oscillations. To my knowledge there has been no soluble model oscillation based on chemical kinetics, although the present model can be viewed as an approximate description of a model in terms of deviations of chemical concentrations from their steady-state values.

Let X and Y be two variables evolving according to (Caesari, 1971; Kopell and Howard, 1973; Ortoleva and Ross, 1974; Dreitlein and Smoes, 1974).

$$\frac{d}{dt}\begin{bmatrix} X \\ Y \end{bmatrix} = \begin{bmatrix} B & -A \\ A & B \end{bmatrix}\begin{bmatrix} X \\ Y \end{bmatrix}$$

$$R^2 = X^2 + Y^2, \qquad A = A(R), \qquad B = B(R). \tag{2.13}$$

Transformation of variables according to the polar variables R and θ,

$$X = R \cos \theta \qquad Y = R \sin \theta, \tag{2.14}$$

yields

$$dR/dt = RB(R) \tag{2.15}$$

$$d\theta/dt = A(R). \tag{2.16}$$

From (2.15) we see that the simplification in this model is that R decouples from θ, the latter of which can be obtained from $R(t)$ through (2.16). Since R obeys a first-order ordinary differential equation, it can only asymptotically $(t \to \infty)$ obtain either steady states R_0,

$$R_0 B(R_0) = 0 \tag{2.17}$$

or unbounded growth for which $|R| \to \infty$. Let us assume that the system is globally stable so that the last possibility is eliminated although we note that such "runaway states" are not without physical interest as has been shown recently (DelleDonne and Ortoleva, 1977).

The steady states R_0 correspond to oscillations about the point $X = Y = 0$ which, if $|B(R)| > 0$ as $R \to 0$, is a steady state of (2.13). In the subsections to follow we examine various cases of $A(R)$ and $B(R)$ to illustrate important oscillatory phenomena in nonequilibrium systems. Note that the oscillation at amplitude R_0 is stable if

$$[(d/dR)RB(R)]_{R_0} < 0, \text{ stable cycle.} \tag{2.18}$$

1. Bifurcation of a Cycle from a Steady State

Taking $B = \lambda - R^2$, we find the family of oscillatory states

$$R_0 = \lambda^{1/2}. \tag{2.19}$$

The oscillation bifurcates from the steady state $X = Y = 0$ for $\lambda > 0$. The steady state at the origin exchanges stability with the cycle at the bifurcation point. This type of bifurcation of a cycle with a critical index of $\frac{1}{2}$ was first studied by Hopf (1942) for a general system. [See the Appendix of DelleDonne and Ortoleva (1977) for a simple introduction to this theory.] It is interesting that the lowest order bifurcation analysis gives the exact result for this system. The bifurcation diagram is analogous to that of Fig. 3.

2. Threshold Oscillation

In the previous example the cycle surrounded an unstable steady state at the origin and hence the latter was not a physically obtainable state. However, for the example $B = 1 - \lambda - (R_1^2 - R^2)^2$, we obtain two distinct oscillatory states:

$$R_{0,\pm} = [R_1^2 \pm (1 - \lambda)^{1/2}]^{1/2}. \tag{2.20}$$

Note that in analogy to the steady-state example of Section II,A,4 a pair of oscillatory states bifurcates at finite amplitude at $\lambda = 1$ from the stable steady state $X = Y = 0$. Stability analysis shows that $R_{0,+}$ is stable while $R_{0,-}$ is not. This system demonstrates the common property of hard transition systems that in a given range of parameter values (λ, R_1) the system may reside in either of two (or more) stable states. This system demonstrates "threshold" or "finite amplitude instability" for although the states $R_{0,0} = 0$ and $R_{0,+}$ are stable to small perturbations they are not stable to perturbations of sufficient magnitude.

3. Frequency and Continua of Steady States

The frequency of the various models one constructs depends on a parameter λ through the relation

$$\omega(\lambda) = A(R_0(\lambda)) \tag{2.21}$$

as can be seen by integrating (2.16), noting that R_0 is constant in time. Since A is an arbitrary function of R, it is clear that a great variety of frequency variations with λ are possible.

In the above examples we have implicitly assumed that $\omega \neq 0$ in stating that the solutions of $B(R) = 0$ correspond to oscillations. If, however, there is an R_L such that

$$A(R_L) = B(R_L) = 0, \tag{2.22}$$

then there is a family of steady states at R_L (i.e., oscillations with zero frequency). If we vary λ through the value $\lambda_L(R(\lambda_L) = R_L)$, then the frequency vanishes and the system may reside in any of the continuum of marginally stable states $R = R_L$, $0 \leq \theta \leq 2\pi$.

C. Propagating Structures

Models with known spatio-temporal solutions have been studied and some of them are presented here.

1. Propagating Front

A front is a propagating transition between two states of a system. A solution of this type has been found for the following equation (Montrol, 1972; Nitzan, Ortoleva, and Ross, 1974b) for a one-dimensional system

$$\partial X/\partial t = D(\partial^2 X/\partial r^2) - q(X - X_1)(X - X_2)(X - X_3), \qquad (2.23)$$

where X_1, X_2, and X_3 are homogeneous steady states for the infinite system. The solution given involves a propagating transition between the two stable states X_1 and X_3 for $X_3 > X_2 > X_1$. Letting ϕ be the spatial coordinate in a reference frame moving with the front, one finds the wave profile taking the form

$$X = X_1 + (X_3 - X_1)(1 + e^{\pm \beta \phi})^{-1} \qquad (2.24)$$

where $\phi = r - vt$ is the wave coordinate and β is the inverse length of the transition layer

$$\beta = (q/2D)^{1/2}(X_3 - X_1) \qquad (2.25)$$

and the velocity of propagation is given by

$$v = (\tfrac{1}{2}qD)^{1/2}(X_1 + X_3 - 2X_2). \qquad (2.26)$$

Wave solutions of other equations closely related to the reaction diffusion equations have been also found (Montrol, 1972). Much work has been done on variants of (2.23) with different nonlinear terms and are reviewed in Aronson and Weinberger (1975) and Murray (1977).

2. Periodic Wave Trains

Periodic wave trains for the model of Section II,B have been given for various choices of the functions $A(R)$ and $B(R)$ (Kopell and Howard, 1973; Ortoleva and Ross, 1974; Dreitlein and Smoes, 1974). As was shown for the homogeneous oscillations in Section II,B, it was found that the wave solutions can demonstrate a number of nonlinear wave phenomena (as stressed in the Appendix of Ortoleva and Ross, 1974).

For the problem of periodic wave trains we seek solutions of the form

$$X = X(\rho), \qquad Y = Y(\rho) \qquad (2.27)$$

where

$$\rho = kr - \omega t = k\phi.$$

Here k is the wave vector (i.e., 2π divided by the wavelength) and ω is the wave frequency showing dispersion because of diffusion

$$\omega = \omega(k^2). \qquad (2.28)$$

If we let $D_x = D_y = D$ for the diffusion coefficients of X and Y of the model of Section II,B and introduce the polar transformation of (2.14), we obtain

$$k^2 D[R'' - R\theta'^2] + \omega R' + RB(R) = 0 \qquad (2.29)$$

$$k^2 D[\theta'' + 2R'\theta'/R] + \omega\theta' + A(R) = 0. \qquad (2.30)$$

From this we see that there are wave solutions with $R = R_0(k^2)$ such that

$$B(R_0) - k^2 D\theta'^2 = 0 \qquad (2.31)$$

$$k^2 D\theta'' + \omega\theta' + A(R_0) = 0. \qquad (2.32)$$

Periodic solutions for X and Y are obtained with $\theta'' = 0$. Hence

$$B(R_0)\bar{\theta}k^2 D[A(R_0)/\omega]^2 = 0. \qquad (2.33)$$

Integration of (2.32) with $\theta(\rho = 0) = \bar{\theta}$, $\theta'' = 0$, yields

$$\theta = \bar{\theta} - [A(R)/\omega]\rho. \qquad (2.34)$$

Since X and Y are 2π periodic functions of ρ and since by definition of the polar coordinate transformation the solutions must be 2π periodic in θ, we have

$$\omega(k^2) = -A(R_0(k^2)) \qquad (2.35)$$

giving the dispersion relation for the waves.

Using this model one can take various cases as in Section II,B and obtain examples of bifurcation of families of waves as a function of k^2 and of threshold wave phenomena. For these developments the parameter k^2 plays the role of λ in Section II,B. Details are found in Kopell and Howard (1973), Ortoleva and Ross (1974), and Dreitlein and Smoes (1974). These models verify the bifurcation (Kopell and Howard, 1973; Ortoleva and Ross, 1974) and limit-cycle (Kopell and Howard, 1973; Ortoleva and Ross, 1973a, 1974; Ortoleva, 1976) perturbation theories of chemical waves [see Hanusse *et al.* (1978) for a review of the theory of chemical waves].

D. STATIC PATTERNS

Here we consider three cases where time-independent spatial structures may be found in soluble model systems.

1. Coexistence Structure

For the special case $X_1 + X_3 = 2X_2$ of the multiple steady-state system studied in Section II,C,1 the velocity v is seen to vanish and the system has a static interface between the two states X_3 and X_1. This state, as all non-equilibrium structures in an infinite system, is marginally stable to a perturbation that uniformly translates the structure. The translation operator is d/dr and indeed this translation type perturbation is proportional to dX^*/dr where X^* is the static structure given by (2.24) for this special case.

2. Periodic Structures

For the case $A = 0$ of the model of Section II,C,2 we have static periodic structures. Thus the system

$$\frac{\partial}{\partial t} \begin{bmatrix} X \\ Y \end{bmatrix} = D \frac{\partial^2}{\partial r^2} \begin{bmatrix} X \\ Y \end{bmatrix} + B(R) \begin{bmatrix} X \\ Y \end{bmatrix}, \qquad R^2 = X^2 + Y^2 \qquad (2.36)$$

has static periodic structures in the form

$$\begin{bmatrix} X \\ Y \end{bmatrix}^* = R_0 \begin{bmatrix} \cos kr \\ \sin kr \end{bmatrix} \qquad (2.37)$$

where R_0 is determined by the condition

$$B(R_0) - k^2 D = 0. \qquad (2.38)$$

The stability and other properties of these solutions have not been studied. It is clear that by making appropriate choices for $B(R)$ one may construct a model with a cutoff wave vector k_c beyond which the system has no solution. For example, in the case $B = [Q^2 R^2 - M^2]/[R^2 + M^2]$, $k_c^2 = Q^2/D$. This system has a stable homogeneous steady state at $X = Y = 0$ and a continuum of homogeneous states at $R_0 = |M/Q|$. This continuum has a patterned extension that exists for k in the interval $0 < k < k_c$. The properties of this and other cases warrants further investigation.

3. Symmetry Breaking Instability in a Model Reaction

Using the mechanism

$$B + X \rightarrow Y + D$$
$$2X + Y \rightarrow 3X \qquad (2.39)$$

it has recently been found possible to determine many of the properties of the static dissipative structures analytically (Lefever *et al.*, 1977). We do not review the details here, but in short bifurcations of symmetry breaking states can be found. As the system length is changed the number of undulations that are possible takes on discrete (i.e., quantized) values.

E. RUNAWAY STATES

In a far from equilibrium system open to energy exchange with an external bath one might imagine that under certain conditions the temperature of the system would increase without bound (unless at elevated temperatures new kinetic processes become important). This phenomena has been termed the "runaway state" (DelleDonne and Ortoleva, 1977).

Consider a closed illuminated system with two species X and Y involved in the reactions

$$X + Y \xrightarrow{\alpha} 2Y \qquad (2.40a)$$

$$2X + Y \xrightarrow{\beta} 3X \qquad (2.40b)$$

$$X \xrightarrow{\mu} Y \qquad (2.40c)$$

$$Y \xrightarrow{\eta} X. \qquad (2.40d)$$

The last two steps are photochemically induced isomerizations and the first two are collision induced. The first reaction is taken to proceed with negligible activation energy barrier whereas the second involves an activation energy E_a such that

$$\beta = \beta_0 e^{-E_a/k_B T} \qquad (2.41)$$

where β_0 is a constant, k_B is Boltzmann's constant, and T is the absolute temperature. The state X is taken to have the highest internal energy, $E_x > E_y$. The only exchange of energy between the system and the environment is via (2.40c,d).

It should be noted that as stated the model (2.40) violates microscopic reversibility. This inconsistancy may be removed by reinterpreting the mechanism with intermediate photochemical steps (Noyes, DelleDonne, and Ortoleva, 1978). Furthermore, it should be pointed out that (2.40) represents a reduction of the model of Prigogine and Lefever (1968) (see DelleDonne and Ortoleva, 1977).

Letting x be the mole fractions of X and taking the system to be closed (so that the total number of particles X or Y is constant), we have, assuming simple mass action kinetics,

$$dx/dt = \eta - [\alpha + \mu + \eta]x + [\alpha + \beta(T)]x^2 - \beta(T)x^3. \qquad (2.42)$$

Assume that the time scale of the reactions is long relative to that of attaining a Maxwellian velocity distribution, so that the average kinetic energy is given by $C_v' T$ where C_v' is a kinetic specific heat. With this one finds that for a gas the temperature obeys the equation

$$dT/dt = h[\eta - (\mu + \eta)x - (dx/dt)] \qquad (2.43)$$

where $h = (E_x - E_y)/C_v'$.

The steady states of the system may be obtained from (2.41)–(2.43). Although the steady-state equations are transcendental, it is interesting that for this system they can be solved: one finds

$$x^* = \eta/(\mu + \eta) \qquad (2.44)$$

$$T^* = -E_a/\ln[\alpha(\mu + \eta)/\beta_0 \eta]. \qquad (2.45)$$

Note that for physical steady states T^* must be positive. This can only be obtained if

$$\alpha(\mu + \eta) > \beta_0 \eta. \tag{2.46}$$

It has been shown that in the domain of parameters (μ, η) where this is not satisfied the system temperature increases linearly in time and without bound.

Physically, the temperature will not increase without bound for even if the vessel in which the experiment was carried out was not destroyed, the integrity of the molecules X and Y would not be maintained. Thus new processes and chemical species would eventually arise and the runaway would be terminated. However the important point is that for some far from equilibrium systems runaway states may indeed be possible wherein the temperature increases over many order of magnitude.

F. Remarks

An examination of soluble model systems serves as an introduction to a variety of nonequilibrium phenomena and, furthermore, allows for the detailed checking of various general approximation schemes such as bifurcation theory (Stakgold, 1971; Sattinger, 1973; Kopell and Howard, 1973; Ortoleva and Ross, 1974; Achmuty and Nicolis, 1975; Keener, 1976) and limit-cycle perturbation techniques (Kopell and Howard, 1973; Ortoleva and Ross, 1973a, 1974; Ortoleva, 1976) [see also the review in Hanusse et al. (1978)].

In addition to the phenomena introduced by these soluble model systems, there is the possibility of homogeneous (Rossler, 1976) or inhomogeneous (DelleDonne and Ortoleva, 1978) chaotic evolution of a reacting diffusing system. Also these systems may propogate single pulses of activity as has been shown experimentally (see A. T. Winfree, this volume) and theoretically via an analytical solution of a piecewise-linear kinetic system (Rinzel and Keller, 1973).

Let us now turn to the study of a number of physical systems and to general formulations and approximate descriptions of these systems.

III. Symmetry Breaking Transitions to Static Patterns

One of the most interesting possibilities that arises far from equilibrium is that of the instability of a homogeneous state to patterned perturbations and the subsequent development and sustaining of a patterned distribution of concentration of a lower symmetry. This idea for reaction diffusion systems was first introduced by Turing (1952) in his prophetic invocation of this phenomena as a mechanism for biomorphogenesis. The idea has been extensively worked on by other authors (Prigogine and Lefever, 1968; Nitzan, Ortoleva, and Ross, 1974a; Auchmuty and Nicolis, 1975; Keener, 1976; Herskowitz–Kaufman, 1978). In this section we discuss several symmetry breaking instabilities that do not appear to have received a great deal of attention.

A. SPONTANEOUS PATTERN FORMATION IN PRECIPITATING SYSTEMS

1. The Patterning Phenomena

In the classic experiment of Liesegang (Liesegang, 1896; Hedges and Myers, 1926) a salt such as $Pb(NO_3)_2$ was mixed with a gel and allowed to set in a test tube. Upon placing a solution of KI over the gel solution, precipitation of PbI_2 occurs in the gel. The striking feature of this system is that the precipitation does not necessarily occur with monotonic decrease in intensity down the tube. It may vary in an undulatory way, with many bands of precipitate alternating with clear areas, under certain ranges of concentrations. The historical explanation of the phenomena was given by Ostwald (1925) and Prager (1956). Their theory depends on the presence of cross gradients of reactants and is based on the periodic attainment of the critical concentrations of Pb^{2+} and I^- for nucleation of precipitate with subsequent elimination of these ions in the vicinity of each successively forming band as the more mobile ion (I^- in the PbI_2 example) moves down the tube.

Flicker and Ross (FR) (1974) proposed describing this phenomena as a chemical instability leading to symmetry breaking in the Liesegang experiment. To test this idea I suggested that a homogeneous PbI_2 sol be prepared and allowed to age. If a symmetry breaking was indeed present, then the homogeneous experiment would be a clear cut demonstration of this fact since, unlike the classic Liesegang experiment, cross gradients were not present. In the experiments carried out by FR they found that yellow patterns of PbI_2 precipitate did indeed develop (Flicker and Ross, 1974).

In order to explore this phenomena further, some new experiments have been devised (Feinn et al., 1978). In one experiment a solution of PbI_2 agar and water is heated until all solid is dissolved. Since PbI_2 is about three times more soluble at 100°C than it is at room temperature, this allows for the creation of a supersaturated PbI_2 solution that is free of irrelevant ions such as NO_3^- and K^+ as were present in the earlier experiment. It was found that after the precipitation occurred in a Petri dish one obtains speckled, mosaic, or wavy patterns at different initial concentrations of PbI_2 and conditions of cooling. The result of one experiment is shown in Fig. 5. Note the presence of spirallike phenomena that arise spontaneously under certain conditions. In a second experiment (Feinn et al., 1978) three gel layers were

252

P. Ortoleva

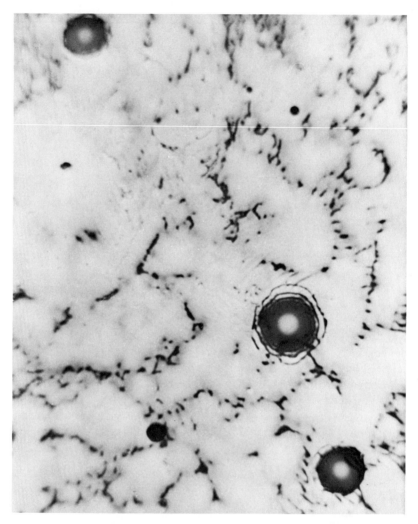

Fig. 5. Photograph of supercooling PbI_2 experiment for an agar solution [see Feinn et al. (1978) for experimental details].

prepared separately and stacked. They contained in sequence $Pb(NO_3)_2$, pure gel, and KI. When Pb^{2+} and I^- diffused into the "reaction layer" precipitation may be made to occur in a single band totally within the center gel layer by appropriate choice of layer thickness and concentrations. Since the gel was already set before the experiment started, convection could not be a factor in pattern formation. Furthermore, this second experiment provides much better controls since the supercooling experiment involves possible evaporation and temperature gradients as the poured gel cools.

The results of the second experiment is first the development of a uniform haze in a narrow band in the reactant layer. Within times of the order of 1–4 hours, depending on concentrations in the reactant layers, the uniform haze evolves to speckled or mosaic patterns of PbI_2 precipitate. The final result of one experiment is shown in Fig. 6.

2. Instability of the Particle Size Distribution Function

In order to explain this phenomena we must invoke a different mechanism than the Ostwald–Prager theory (Ostwald, 1925; Prager, 1956) since the

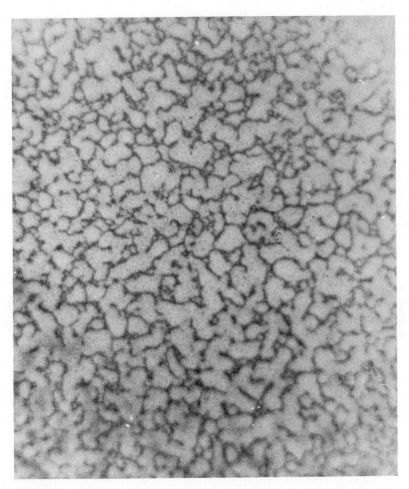

Fig. 6. Three-layer PbI_2 experiment of Section III,A showing precipitation patterning.

latter requires the presence of cross gradients, clearly absent in the spontaneous pattern formation experiments described above.

Both the three layer and the supercooled PbI$_2$ systems develop a uniform yellow haze indicating particles of the order of microns, the wavelength of visible light, are initially essentially uniformly distributed throughout the system. Because of surface tension (Defay and Prigogine, 1956; Overbeek, 1971) a larger particle can exist in equilibrium with a PbI$_2$ solution of lower concentration than one of smaller radius. Thus in any small volume centered about a point **r**, particles significantly smaller than the mean particle radius $R(\mathbf{r}, t)$ are rapidly consumed since the particle growth is a rapid process. However, the competition of neighboring volume elements is on a much slower time scale since the rate of exchange of material is the diffusion time associated with the length scale of the gradients involved. It is clear that if the average particle size at a point **r** is bigger than that in its surroundings, then the particles at **r** will grow at the expense of those in the vicinity of **r**. This is a runaway effect since the more adjacent regions become separated in particle size the greater the difference in the equilibrium salt concentration becomes. The result of this situation is the creation of a dot of precipitate particles as found in Petri dish experiments of relatively low concentration. The mosaic pattern may be viewed as more complex initial conditions on $R(\mathbf{r}, t)$. Furthermore, as initial concentration increases it becomes less frequent to find regions where $R(\mathbf{r}, t)$ is sufficiently small after the initial burst of nucleation that the equilibrium concentration associated with them is appreciably sensitive to radius. (Note that as R increases the equilibrium concentration reaches its usual thermodynamic value obtained for a planar interface.)

To formulate these ideas mathematically one may study the local particle size distribution $f(\tilde{R}, \mathbf{r}, t)$ giving the density of particles of radius $\tilde{R}$ at point **r** and time t. For the purpose of illustrating the theory here we consider a simple approximation that appears to explain many features of interest. We assume that f is sufficiently narrowly peaked about $\tilde{R} = R$ as per our earlier discussion vis-a-vis particle competition in a small volume element. Thus we need only formulate an equation for $R(\mathbf{r}, t)$.

The rate of change of R is approximated by a simple rate law

$$\partial R/\partial t = q(C - C^{\text{eq}})/\rho_s \tag{3.1}$$

where q is a positive quantity (which may depend on R and C in general) and ρ_s is the solid density. We have assumed a simple growth rate (moles/unit area-time) proportional to the difference between the salt concentration C at $(\mathbf{r}, t)$ and C^{eq}, the equilibrium concentration for particles of radius R. We assume the particles to be immobilized so there is no flux term in (3.1). Let E represent the rate of increase in C due either to evaporation in the supercooling experiment or influx from the reactant layers in the three-layer experiment. Then we have a continuity equation for C,

$$\partial C/\partial t = D\nabla^2 C - 4\pi R^2 nq(C - C^{eq}) + E \tag{3.2}$$

where D is the diffusion coefficient of PbI_2 and n is the number of precipitate particles per unit volume. The factor $4\pi R^2$ arises because the consumption of salt is proportional to the surface area of the local particles.

Equations (3.1) and (3.2) provide a mathematical formulation of our picture. It is found that the state of initially homogeneous evolution where R and C are independent of $\mathbf{r}$ at $t = 0$ becomes unstable to small spatial deviation in R or C, even under the assumption that n is constant.

Since q is typically very large, one can look at the limit of $q \to \infty$. In this limit one finds that $C = C^{eq}(R)$ and that R evolves according to

$$f(R)\frac{\partial R}{\partial t} = D\left\{\left(\frac{\partial C^{eq}}{\partial R}\right)\nabla^2 R + \left(\frac{\partial^2 C^{eq}}{\partial R^2}\right)|\nabla R|^2\right\} + E$$

$$f(R) = 4\pi n\rho R^2 + \frac{\partial C^{eq}}{\partial R}.$$

The factor f is found to be positive for physically realizable initial sols. Since $\partial C^{eq}/\partial R$ is negative for particle radii greater than the critical nucleation size, it is seen that the effective diffusion coefficient $f^{-1}D(\partial C^{eq}/\partial R)$ is negative and hence patterns are amplified. Finally we note that the theory presented here (based on the physical chemistry of the present problem) has a formal similarity to the theory of Glass (1973) on the random generation of pattern in the presence of expanding inhibitory fields.

B. PATTERN FORMATION IN SINGLE-CELL SYSTEMS

The most familiar electrochemical propagating phenomena in electrophysiology is that of axonal transmission. However, this is apparently not the only spatially patterned nonequilibrium structure to be found in cellular electrophysiological systems. For example, it has been shown that well in advance (and apparently independent of) mitotic processes there exists a transition from a state of spherically symmetric membrane potential to a polarized state in the spherical egg of Fucus (Jaffe *et al.*, 1974). This phenomenon has been termed self-electrophoresis and is believed to be the essential step in the development of the egg that leads to dramatically different daughter cells after the first mitosis. It appears that the Fucus system is an example of the chemical instability based biomorphogenesis theory of Turing (1952).

1. Two-Box Model

The idea of symmetry breaking instabilities leading to asymmetric mitosis was first considered using a two-box model (Ortoleva and Ross, 1973b,c).

Such two-box models had been used to demonstrate sustained stable patterns in reacting diffusing systems (Glansdorff and Prigogine, 1971). By representing the cell by two homogeneous compartments a tractable theory resulted. It was shown that the symmetric state (concentration in both boxes equal) could become unstable to a small asymmetric concentration perturbation if certain nonlinear membrane transport and reactions in the compartments took place. The theory was limited to neutral species and so self-electrophoresis, involving electrical fields and currents, was not demonstrated. It was shown, however, that the symmetric state of the cell dynamics could become unstable and that spontaneous asymmetric mitosis could result at the system, and not necessarily only at the genetic, level. Furthermore, this developed and sustained asymmetry could exist in a homogeneous environment, free of imposed gradients. This was an application of Turings general hypothesis (Turing, 1952) that certain biomophogenic events could be attributed to a pattern forming instability in the equations of reaction and transport.

2. A Theory of Self-Electrophoresis

Writing down equations for membrane and cytoplasmic transport and reaction, it has been shown (Ortoleva, 1978a) that the spherically symmetric state of membrane potential and cell composition may become unstable to small asymmetric perturbations leading to a polarized self-electrophoretic state of the cell with net transcellular ionic currents. It was further shown that the axis of the polarization was unstable to rotations in agreement with early behavior of the system. The later freezing in of the axis of the polarization was conjectured to be attributable to condensation phenomena associated with or driven by the symmetry breaking self-electrophoretic instability. Since the types of cell dynamical features that lead to such symmetry breaking instabilities are membrane transport nonlinearities, product activated catalytic reactions, cross catalysis, etc., and since these phenomena are common in cellular systems, it appears that self-electrophoretic and other symmetry breaking or wave and oscillatory electrophysiological phenomena should be found in a great variety of systems. Slow (minute) timescale cellular oscillations are in fact common in cellular systems from heart, nerve, algal, and many other systems [see Hess and Boiteaux (1971) and Chance et al. (1973) for a review]. It is probable that many new developments should be forthcoming in this area in the near future.

C. LOCAL STEADY STATES IN SYSTEMS UNDER
   HOMOGENEOUS CONSTRAINT

Single propagating disturbances as for pulses along the axon or in Winfree's modification of the BZZ system (see Winfree, this volume) are now well-established phenomena. It has not been shown, however, that there can exist inhomogeneous steady states that are localized in space in an infinite system under homogeneous constraint. In the mechanism of spontaneous pattern formation considered in Section III,A it can be shown that if initially the particle size R is at a value R_0 greater than the value R_h in the rest of the system, then eventually all the particles (except those in the central region) will be eliminated and those that started out at R_0 will grow indefinitely causing an intense localized dot. [This picture actually breaks down as the particles become a size of the order of the interparticle distance. This leads to a multicrystalline mass and, in fact, has been observed in gel precipitation experiments in our laboratory (Feinn *et al.*, 1978).] The question arises as to whether local states can be found in chemically reacting diffusing systems.

A suggestive model example of this comes from the following system. A species X has a diffusion coefficient $D(X)$ that is proportional to X. Defining length so that this proportionality factor is 1, consider the following system:

$$\partial X/\partial t = (\partial/\partial r)X(\partial/\partial r)X + F(X).$$

Taking $F(X) = 2X^2(1 + 4 \ln X)$, one may verify that

$$X^* = e^{-r^2}$$

is a steady-state solution for the system. It is an isolated disturbance in a system that has homogeneous steady states at $X_h^* = 0$, $e^{-1/4}$. (Note that this model is globally unstable since as $X \to \infty$, $F \to \infty$ although the state $X^* = 0$ is stable.) The question arises as to whether this phenomenon can arise in more physically interesting systems. It would be interesting to find stable localized structures in unbounded systems under homogeneous constraint.

The analogy of these localized self-sustaining regions of organized behavior to life is an attractive one to make. Indeed such an analogy of non-equilibrium structures to life is not new, but has been proposed by others in the field.

D. Pattern Formation, Scaling, and Spatial Dimensionality

Before concluding our discussion on spatial patterning, we will briefly discuss two general concepts regarding these phenomena.

1. Intrinsic and Extrinsic Length Scales

Consider a large system under homogeneous constraint. We assume that it has a homogeneous steady-state solution with descriptive variable ψ at the value ψ^*. Then perturbations from this state may be written as a linear combination of the eigen perturbations of the form $e^{i\mathbf{k}\cdot\mathbf{r}}$. We assume that there is one real branch of the stability eigenvalue spectrum $Z_u(k)$ that indicates instability by being positive for some range of wavevectors. We distinguish the following two cases (Nitzan, Ortoleva, and Ross, 1974a):

   a. *Intrinsic Scaling:* Z_u is negative for k near 0, but is positive for an interval around some value k_c of k. Thus the pattern that arises has a wavelength of order of $2\pi/k_c$, the intrinsic length of the chemical kinetics and transport.

   b. *Extrinsic Scaling:* Z_u is zero at $k = 0$ and as some constraint λ passes through a critical value λ_c. Z_u is positive for an interval $0 < k < k_m(\lambda)$ that increases for λ beyond λ_c. Thus the first mode that becomes unstable as λ increases beyond λ_c is that with a wavelength of order of the system size and hence the pattern length is not intrinsic to the reaction transport phenomonological relations, but is extrinsic to it, being fixed by the size of the particular system the reaction is being carried out in.

2. *Space Dimensionality and Critical Bifurcation Exponent*

   A discussion of spatial patterning would hardly be complete without mentioning the relationship between the spatial dimensionality and the analytic behavior of bifurcating spatial patterns (Segel, 1965; Ponmarenkov, 1968). In one dimension only a single mode becomes unstable as a constraint λ passes through a critical value λ_c. However, in two and higher dimensions it is possible that because of rotational or other symmetries a number of modes may simultaneously become unstable at λ_c. It is this difference between one and higher dimensions and the fact that the critical modes may couple to each other in lower order nonlinear terms than the single mode case that leads to marked differences in the dependence of the amplitude A of the pattern on $(\lambda - \lambda_c)$. Thus near λ_c we have $A \sim (\lambda - \lambda_c)^v$ where the bifurcation exponent v depends strongly on the dimensionality of the system. This point has been pursued in the Be'nard problem (Segel, 1965; Ponmarenkov, 1968), but apparently has not been discussed in detail for chemical instabilities.

IV. Chemical Waves

   The theory of chemical waves has recently been reviewed (Hanusse *et al.*, 1978) and hence a summary of these results is not included here. Rather, we shall now discuss several theoretical aspects of chemical waves that have not received much attention until recently. First we discuss the electrochemical nature of chemical waves, then we pose the problem of a scattering theory for nonlinear reaction–diffusion waves, and finally study inhomogeneous evolution in systems with limit cycle kinetics.

A. ELECTROCHEMICAL WAVES AND WAVE–FIELD INTERACTIONS

   In the most well-studied example (see Field and Winfree, this volume) of a system that can support chemical waves, most of the chemical species of interest are ions, yet little attention has been paid to the electrochemical nature of wave propagation in an electrochemical system. In a recent study (Schmidt and Ortoleva, 1977) a new equation for plane waves valid for ionic media was derived. Furthermore, it is important to point out that a great number of instabilities in electrochemical systems are known to exist (de Levie, 1970), but these will not be reviewed here.

We now consider the derivation of a wave equation for ionic media. In one dimension (r) the equation of continuity takes the form

$$\partial C/\partial t = -\partial J/\partial r + \mathscr{R}[C] \qquad (4.1a)$$

where $J = \{J_1, J_2, \cdots J_n\}$ is the flux. Assuming linear relations between the flux and the concentration gradient and electric field E, we have

$$J_i = -[D_i \partial C_i/\partial r - z_i u_i C_i E] \qquad (4.2)$$

where D_i, z_i, and u_i are, respectively, the diffusion coefficient, valence, and mobility of species i (DeGroot and Mazur, 1962; Fitts, 1962). This is valid for sufficiently dilute solutions. To complete the theory, we have Poisson's equation relating E to the charge density $\sum_i z_i \mathscr{F} C_i \equiv \mathscr{F} zC$,

$$dE/dr = (4\pi/\varepsilon)\mathscr{F} zC, \qquad (4.1b)$$

where $\mathscr{F}$ is Faraday's constant and ε is the dielectric constant of the system (assumed independent of C and E). Introducing a coordinate $\rho(= r - vt)$ moving along with a constant velocity (v) wave, we obtain coupled wave equations for E and C that can be combined (by integrating Poisson's equation with respect to ρ) into an integro-differential equation for C only.

By scaling the equation with typical mobilities and diffusion constants, reaction time scales and concentrations a smallness parameter α becomes manifest in this integro-differential equation:

$$\alpha = (l_D/l)^2 \qquad (4.3)$$

where l is a typical reaction diffusion length (i.e., minimum length associated with the wave profile) and l_D is the Debye length. Since l_D is typically on the order of 10–1000 Å and l is of order of about a tenth of a millimeter (say for quasi-discontinuous waves), it is clear that $\alpha \ll 1$. Thus using α as an expansion parameter a new wave equation was derived. To lowest order the condition of charge neutrality resulted. In the next higher order the new wave equation was obtained:

$$(d/d\rho)[D_i(dC_i/d\rho) - \mathscr{E} u_i C_i] + v(dC_i/d\rho) + R_i[C] = 0 \qquad (4.4a)$$

$$\mathscr{E} = \sum_{i=1}^{N} z_i D_i dC_i/d\rho \left/ \sum_{j=1}^{N} z_j^2 u_j C_j \right. . \qquad (4.4b)$$

The novel feature of this equation is that added to the transport term $D(d^2/d\rho^2)$ of the neutral species plane-wave equation is a term due to a Planck liquid junction type potential $\mathscr{E}$. The presence of the liquid junction term is in fact what coordinates the various ionic motions so that local charge neutrality is assured. With this it becomes clear that a compositional wave in an ionic system is associated with a propagating Planck liquid junction potential.

These electrochemical effects are not important in the presence of a large background electrolyte that does not take part in the chemistry of the wave propagation [because of the large denominator in (4.4b)]. Thus for the BZZ system under conditions studied to date, this effect may not be important except for the determination of the H^+ profile. Several possible changes in the composition of the BZZ reagent compatible with wave propagation, but which lowered or immobilized the background electrolyte, were suggested in Schmidt and Ortoleva (1977). It seems likely that interesting new propagating electrochemical phenomena will be discovered along these lines.

Another important feature of electrochemical systems is the possibility of studying the interaction of chemical waves with applied electrical fields. Indeed it is possible to stop, reverse, or destroy chemical waves by imposed electric fields (Schmidt and Ortoleva, 1977; 1978) and interesting effects even arise in the presence of a significant background electrolyte where the Planck term is not present.

Preliminary results (Heiney and Ortoleva, 1977) have been obtained using electric fields across the classic Liesegang band phenomena (Liesegang, 1896; Ostwald, 1925; Prager, 1956). A silver nitrate–potassium dichromate system was set up with the potassium dichromate in the gel. Voltages were applied making the silver nitrate solution above the gel positive. It was found that as the voltage increased the band spacing became strikingly regular as can be seen in Fig. 7. The spacing between bands also decreased and it was found that a critical voltage V_c exists such that for voltages beyond V_c the precipitation occurs as a single front of continuous precipitation propagating down the tube.

B. Scattering of Chemical Waves

Although scattering theory is well developed in field theories such as quantum mechanics, optics, and acoustics, this has not been the case for nonlinear chemical waves (Ortoleva, 1978c). Consider, for example, the following scattering problem. Suppose that localized in the vicinity of a point $\mathbf{r}_0$ there is a disturbance of the chemical kinetics due to say a local hot spot or catalyst. The reaction diffusion equation for a dilute solution of neutral species then reads

$$\partial C/\partial t = D\nabla^2 C + \mathscr{R}(C) + \mathscr{S}(\mathbf{r}, C) \tag{4.5}$$

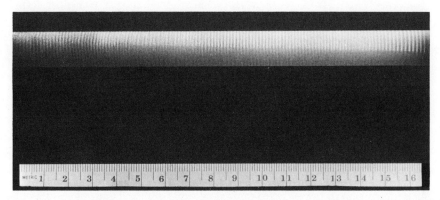

Fig. 7. Silver dichromate regularized patterns arising in a Liesegang-type experiment with an imposed voltage as described at the end of Section IV,A.

where the scattering source function $\mathscr{S}(\mathbf{r}, C)$ vanishes for large values of $|\mathbf{r} - r_0|$. Let us assume that in the absence of $\mathscr{S}$ the system propagates planar chemical waves and in particular a pulse. We induce such a planar pulse far from r_0 so that it propagates towards r_0. The question arises as to what will be the resultant wave form long after the plane wave would have passed through r_0 in the absence of the scattering center $\mathscr{S}$. Depending on the form and magnitude of $\mathscr{S}$, one might expect that either essentially no effect would occur, that a circularly radiating wave could be induced with r_0 as the center, that a local oscillation could be induced at r_0, or any number of more complex phenomena such as the induction of a pair of counter rotating spiral waves as the wave passes around the center.

One method of studying scattering phenomena is the introduction of a smallness parameter into $\mathscr{S}$, i.e.,

$$\mathscr{S} = \sigma S, \tag{4.6}$$

and developing a theory in the limit of weak centers, $\sigma \to 0$, via appropriate expansion techniques. A second approach to the problem is in terms of matching solutions valid near the center and far from the center. Finally there is the nonlinear analog to the WKB theory where the characteristic length of the wave is much smaller than the length scale on which $\mathscr{S}$ varies. These are developed in Ortoleva (1978c).

For one-dimensional systems the scattering problem stated above is somewhat simpler in the sense that if we limit ourselves to the question of what will be the behavior far from the center, there are only four possibilities: annihilation, reflection, transmission, or reflection and transmission. However even here the situation is rather complicated since, for example, it is possible that we could send a single pulse at the scattering center, have excitation of a local oscillation at the center, and subsequent emission of a periodic wave train in the forward or reverse directions. Finally, a wave could be captured or pinned in the vicinity of the scatterer.

A second type of scattering situation is that which occurs when a wave collides with an impenetrable object. For example, it may be possible that if a plane wave impinges upon a wall of finite dimensions in the transverse direction, then a spiral wave can be the scattered wave.

Finally, if there are objects embedded in the medium with catalytic properties, then it is possible that, for example, planar chemical waves can be reflected from a catalytic wall. Clearly, nonlinear chemical waves have a rich scattering theory, although one might add, a difficult one.

C. Inhomogeneous Evolution in Systems with Limit-Cycle Kinetics

In this section we consider the inhomogeneous evolution of a system with chemical kinetics $\mathscr{R}[C]$ that supports a limit cycle in the absence of diffusion. When the frequency ω_c of the unperturbed limit cycle reflects a time scale short relative to other times in the phenomena of interest, it is reasonable to consider a multiple time scale development. In a sense the resulting theory should be in the spirit of the WKB approximation. In the latter the scale separation occurs between the wave length and the characteristic length over which the medium changes. This idea has been used to study acoustic instabilities in a reacting system where the acoustic period is much shorter than the chemical relaxation time (Toong, 1972; Gilbert et al., 1973; Nitzan, Ortoleva, and Ross, 1974a).

We consider the reaction diffusion equation as introduced earlier,

$$\partial C/\partial t = \mathscr{D}\nabla^2 C + \mathscr{R}[C] + \mathscr{G}[\mathbf{r}, C] \tag{4.7}$$

where we envision $\mathscr{G}$ to be an inhomogeneous perturbation on the kinetics due to a catalyst or hot spot. We assume the unperturbed system has a homogeneous cycle solution $\psi_c(\omega_c t)$

$$d\psi_c/dt = \mathscr{R}[\psi_c] \tag{4.8}$$

with frequency ω_c. Note that ψ_c is a 2π periodic function of its argument. We consider both autonomous cases, $\mathscr{G} = 0$, and cases with heterogeneous kinetics, $\mathscr{G} \neq 0$.

1. Introducing the Scaling Parameter

We introduce a scaling of terms via characteristic values as follows:

$$\mathscr{D} = \bar{D}D$$

$$\mathscr{R} = \omega_c R$$

$$\mathscr{G} = vG \tag{4.9}$$

where v is a characteristic inverse time for $\mathcal{G}$. The time scale v^{-1} introduces a length scale l given by

$$l^2 = \bar{D}v^{-1}. \tag{4.10}$$

We consider the case for which l is much longer than the cycle diffusion length l_c defined by

$$l_c^2 = \bar{D}\omega_c^{-1}. \tag{4.11}$$

The heterogeneous term $\mathcal{G}$ will only introduce the length scale l into the problem when $\mathcal{G}$ itself contains no shorter length scale variations; we limit our considerations here to this case. With this we define a dimensionless spatial coordinate ρ,

$$\rho = l^{-1}\mathbf{r}, \tag{4.12}$$

and assume that the explicit spatial variations in $\mathcal{G}$ may be written as a function of ρ, i.e., $\mathcal{G} = vG[\rho, C]$. With this we divide (4.7) by ω_c and obtain

$$\partial C/\partial \tau = \varepsilon D\nabla^2 C + R[C] + \varepsilon G[\rho, C] \tag{4.13}$$

where

$$\tau = \omega_c t \qquad \varepsilon = v/\omega_c \tag{4.14}$$

and we write ∇^2 for the Laplacian in the scaled spatial variable ρ.

For the case of autonomous phenomena, l in the above development corresponds to the characteristic length of the concentration profile.

2. Multiple Time Scales

The scaled formulation (4.13) is convenient for introducing a multiple time theory. One time (on the order of the cycle period) is τ, which we denote as τ_0. A slower time τ_1 is associated with the dimensionless time vt which is related to τ via $\tau_1 = \varepsilon\tau$. Further, we note that even for the simple case of homogeneous evolution ($\nabla^2 C = 0$) with G independent of ρ we must be able to account for the shift of frequency $\omega_c \to \omega(\varepsilon)$ that we expect in general will occur. To include this possibility we must, furthermore, introduce an infinite sequence of time scales such that

$$\tau_n = \varepsilon^n\tau. \tag{4.15}$$

This will allow for the recovery of the renormalized frequency $\omega(\varepsilon)$ in an expansion of the form

$$\omega = \sum_{n=0}^{\infty} \omega_n \varepsilon^n \tag{4.16}$$

where $\omega_0 = \omega_c$, as we shall see later in our development.

With this we write C in the form

$$C = C(\rho, \tau_0, \tau_1, \tau_2, \ldots)$$

and seek an expansion in the form

$$C = \sum_{n=0}^{\infty} C_n \varepsilon^n \tag{4.17}$$

where $C_n = C_n(\mathbf{\rho}, \tau_0, \tau_1, \ldots)$. In terms of the times τ_n, (4.13) becomes

$$\sum_{n=0}^{\infty} \varepsilon^n \frac{\partial C}{\partial \tau_n} = \varepsilon D \nabla^2 C + R[C] + \varepsilon G[\mathbf{\rho}, C]. \tag{4.18}$$

3. Derivation of an Augmented Phase Diffusion Equation

To order $\varepsilon^0 = 1$ we obtain

$$\partial C_0 / \partial \tau_0 = R[C_0]. \tag{4.19}$$

Since we seek solutions that locally lie close to the limit cycle at all points, we have

$$C_0 = \psi_c(\tau_0 + \alpha) \tag{4.20}$$

where the local phase α depends on $\mathbf{\rho}$ and the slow times

$$\alpha = \alpha(\mathbf{\rho}, \tau_1, \tau_2, \ldots). \tag{4.21}$$

To order ε we have [with $\Omega = (\partial R / \partial \psi)_{\psi = \psi_c}$]

$$\left(\frac{\partial}{\partial \tau_0} - \Omega \right) C_1 = - \left(\frac{\partial C_0}{\partial \tau_0} \right) \frac{\partial \alpha}{\partial \tau_1}$$

$$+ D \left\{ \frac{\partial C_0}{\partial \tau_0} \nabla^2 \alpha + \frac{\partial^2 C_0}{\partial \tau_0^2} |\nabla \alpha|^2 \right\} + G[\mathbf{\rho}, C_0]. \tag{4.22}$$

Equation (4.22) is a differential equation for C_1 with respect to τ_0 parametrized by $\mathbf{\rho}$ and the slow times. Noting that the quantity $(\partial C_0 / \partial \tau_0)$ satisfies the homogeneous equation

$$(\partial / \partial \tau_0 - \Omega)(\partial C_0 / \partial \tau_0) = 0, \tag{4.23}$$

we see that we cannot solve for C_1 unless the right-hand side of (4.22) is orthogonal to the solution of the adjoint equation corresponding to $(\partial C_0 / \partial \tau_0)$

$$(\partial / \partial \tau_0 - \Omega^T) f = 0. \tag{4.24}$$

The calculations are similar to those given earlier (Ortoleva and Ross, 1973; 1974; Ortoleva, 1976), and the solubility condition for (4.22) in light of the null space of $(\partial / \partial \tau_0 - \Omega)$ as seen in (4.23) leads to

$$\partial \alpha / \partial \tau_1 = D_P \nabla^2 \alpha + \Delta |\nabla \alpha|^2 + g(\mathbf{\rho}) \tag{4.25}$$

where

$$\begin{Bmatrix} D_p \\ \Delta \\ g(\mathbf{\rho}) \end{Bmatrix} = \frac{1}{2\pi} \int_0^{2\pi} d\tau_0 f \begin{Bmatrix} D(\partial\psi_0/\partial\tau_0) \\ D\Omega(\partial C_0/\partial\tau_0) \\ G[\mathbf{\rho}, C_0] \end{Bmatrix}. \tag{4.26}$$

The notation fA for any vector A implies a summation over indices where f is given by (4.24). Equation (4.25) is a diffusionlike equation for the phase α with nonlinear transport term $\Delta|\nabla\alpha|^2$ and source term $g(\mathbf{\rho})$ due to the perturbation $\mathscr{G}$. This "augmented phase diffusion" equation is a generalization of a linear theory that did not include the nonlinear transport term (Ortoleva and Ross, 1973a; Ortoleva, 1976). In another treatment the equivalent to (4.25) for the special case of temporarily periodic solutions was obtained (Ortoleva and Ross, 1974). Finally we note that the result (4.25) similar to that obtained by Kuramoto and Yamada (1976) for the case of no inhomogeneity, $\mathscr{G} = 0$, and two chemical species. The treatment of the latter authors was not a consistent theory in that, for example, it neglected corrections in C of order ε and higher.

4. Special Cases of the Augmented Phase Diffusion Equation

Let us now consider some simple cases in order to clarify the meaning of the various terms.

a. Homogeneous Renormalization. Suppose we uniformly raise the temperature of a homogeneous oscillation. Thus G would be independent of space. Then what is the effect of the perturbation on the homogeneous evolution? For this simple case (4.25) becomes

$$\partial\alpha/\partial\tau_1 = g \tag{4.27}$$

and hence

$$\alpha = g\tau_1 + \alpha_1(\tau_2, \tau_3, \ldots). \tag{4.28}$$

Neglecting the phase shift α_1 (which is a constant to order ε), we obtain in lowest order

$$C \underset{\varepsilon\to 0}{\simeq} \psi_c((1 + \varepsilon g)\omega_c t) \tag{4.29}$$

and we see that the system oscillates with the renormalized frequency $\omega(\varepsilon)$ such that

$$\omega \underset{\varepsilon\to 0}{\simeq} \omega_c(1 + g\varepsilon). \tag{4.30}$$

b. Periodic Autonomous Phenomena. Since gradients of frequency eventually lead to infinite gradients, the most general periodic solution to the reaction diffusion equation (4.7) takes the form

$$C = \chi[\mathbf{\rho}, \tau + p(\mathbf{\rho})] \tag{4.31}$$

where χ is periodic in $\tau + p(\rho)$ with a constant frequency ω. The spatially dependent distribution of phase of oscillation is denoted $p(\rho)$. To lowest order $\chi \sim \psi_c(\tau_0 + \alpha)$ and hence for periodic solutions we must have

$$\alpha = \sum_{n=1}^{\infty} \beta_n \tau_n + p. \tag{4.32}$$

From (4.25) we obtain

$$\beta_1 = D_p \nabla^2 p + \Delta |\nabla p|^2. \tag{4.33}$$

This is the equation derived earlier by directly seeking periodic solutions (Ortoleva and Ross, 1974). For plane waves of wavelength l we have $p = \mathbf{n} \cdot \rho$ where $\mathbf{n}$ is a unit vector along the direction of propagation. Hence

$$\beta_1 = \Delta, \text{ plane waves.} \tag{4.34}$$

Similarly, for one- or two-dimensional centers or spiral waves with essentially planar form far from the core we see that $\beta_1 = \Delta$ also.

5. Reduction of the Phase Equation to Linear Form

Interestingly enough we can transform the nonlinear phase equation into a linear equation. Letting

$$\alpha = (D_p/\Delta) \ln A,$$

$$\rho = (D_p/\Delta)\mathbf{y}, \qquad \alpha(\mathbf{y}) = (\Delta/D_P)g[(D_P/\Delta)\mathbf{y}], \tag{4.35}$$

we find

$$\partial A/\partial \tau_1 = [\nabla_y^2 + \alpha(\mathbf{y})]A. \tag{4.36}$$

This equation warrants further investigation and applications. It is rather surprising that the nonlinear equation (4.25) can so generally be transformed to linear form.

6. Breakdown of the Theory

This theory breaks down under situations where new length scales enter the problem and shock structures develop. An example arises in applying the theory to find periodic centers. The theory yields divergent phase at the core of a circular wave for some systems and always does so for spiral waves (Ortoleva and Ross, 1974). The theory is believed to be valid far from the core of these phenomena, however.

A second source of breakdown is the presence of shorter time scales in the chemical kinetics or diffusion coefficients. For example, in the case of relaxation oscillations one has a periodic sequence of very short time scale changes separating regions of smooth variation on the time scale of the period of the cycle. Such problems have been treated using matched asymptotic techniques and catastrophe theory [see Hanusse *et al.* (1978) and Feinn and Ortoleva (1977) for a review and references].

7. *Application of the Theory*

The augmented phase diffusion theory may be used to study the transient evolution toward periodic states. It thus may be applied to the study of the stability of chemical waves (Ortoleva, 1976). A variety of other applications suggest themselves, including the ability of a limit-cycle system to preserve phase coherence over long distances in the presence of random fluctuations [Ortoleva, 1972 (unpublished)]† and the response of a limit-cycle system in the presence of gradients in conditions such as temperature or catalyst concentrations. Note that the phase diffusion coefficient D_p is not necessarily positive. Thus for the case $D_p < 0$ inhomogeneities in the phase will be amplified, leaving these systems inherently unstable to pattern formation. This is an alternate explanation of the tendency in the melonic acid wave medium for homogeneous oscillation to become unstable to pattern formation.

V. Membrane and Surface Localized Structures

Interfaces between a bulk phase and a membrane, dust particle, and catalytic particles or wires provide special localized sites of reaction. It appears that all the phenomena that occur in bulk media may also occur in association with these localized sites. For example, it is found in Liesegang experiments with chromate ions imposed on a silver nitrate gel solution that the bands of precipitate may be localized to the walls of the tube and not in the gel itself. In fact, the pattern may take the form of a corkscrew curling down the tube. Clearly, under situations where this may occur, the conditions at the glass–gel interface provide a favorable localized site for nucleation of pricipitate particles. In this section we discuss some theoretical results that have been obtained on pattern formation, propagating waves, and other nonequilibrium phenomena that occur in association with the interaction of localized sites and bulk media.

A. Bulk and Surface Kinetics

We consider an isothermal medium at rest, the "bulk," in contact with an object (such as a wall, membrane or wire) through an interface defined by

---

† Using a Langevin approach one may calculate the phase correlation function. The random source term is identified with G in (4.7). In the limit when the noise is small in mean square amplitude one obtains simple, physically transparent results. It is an instructive exercise and may be of interest in certain applications such as chemical and ecological oscillatory systems.

the surface S. In the bulk we have the continuity equation for the chemical concentrations which we assume to be described by Fick diffusion (with constant diffusion coefficient matrix D) according to

$$\partial C/\partial t = D\nabla^2 C + \mathscr{R} \tag{5.1}$$

where $\mathscr{R}$ is the rate of bulk phase reaction.

We distinguish two processes that can occur at the surface. If σ_i is the surface area density (moles/area) of species i that is immobilized to the surface, then we have (neglecting surface diffusion for simplicity)

$$\partial \sigma/\partial t = g(\sigma, C) \tag{5.2}$$

where g is the rate of incorporation of material into the surface (moles/area-time) and depends on the bulk concentrations at S and on σ (and for electrochemical systems on the voltage difference between the surface and the bulk across the interface). A second kind of surface reaction does not lead to incorporation in the surface and has a rate denoted by $h(\sigma, C)$ (in moles consumed per unit time per unit area). For example, g may represent the rate of surface precipitation in the Liesegang problem mentioned above and h may represent the rate of surface catalyzed reaction. Finally, g may have two distinct types of contributions—g_1 representing reactions among interfacially bound species that do not consume bulk phase molecules and g_2 that do consume the latter. Let the total rate of loss of bulk phase species per unit area be denoted G; we have

$$G = g_2 + h. \tag{5.3}$$

Since the amount of bulk material consumed must be just balanced by the flux $\mathbf{n} \cdot D\nabla C$ from the bulk ($\mathbf{n}$ is the unit normal to S at the point in question directed into the bulk), we obtain at a point $\mathbf{r}$ on S

$$\mathbf{n} \cdot D\nabla C = G(\sigma, C), \qquad \mathbf{r} \in S. \tag{5.4}$$

With this the local site problem consists of solving the continuity equation (5.1) in the bulk consistent with the boundary condition (5.4) and the surface kinetics (5.2). It is clear that the three sources of nonlinearity ($\mathscr{R}$, g, and G) yield the possibility for a richness of phenomena in this system.

It is convenient to incorporate the boundary condition (5.4) directly into the continuity equation (Ortoleva and Ross, 1972; Ortoleva, 1976). The method involves replacing (5.4) by the boundary condition $\mathbf{n} \cdot D\nabla C = 0$ for an impenetrable wall and adding a delta function localized source term to the continuity equation (5.1). Let S^+ be a surface parallel to S a distance 0^+ in the bulk and let $\delta^+(\mathbf{r})$ be the one-dimensional delta function localized to S^+ so that for any path integral $\int A(\mathbf{r})\delta^+(\mathbf{r})\,ds$ passing through S^+ at $\mathbf{r}^+$

we obtain the value $A(\mathbf{r}^+)$. Then, if we replace (5.1) by

$$\partial C/\partial t = D\nabla^2 C + \mathscr{R} + G\delta^+(\mathbf{r}), \qquad (5.5)$$

it is easy to show that (5.4) is satisfied by integrating (5.5) along a small path in the direction $\mathbf{n}$ passing through S^+.

B. LINEAR BULK KINETICS

We now show that under certain circumstances the present problem reduces to the solution of integral equations involving only σ and the surface values of the bulk concentrations. We assume that the absence of surface localized processes, $G = 0$, there is a steady state $C_b(\mathbf{r})$,

$$D\nabla^2 C_b + \mathscr{R} = 0. \qquad (5.6)$$

Furthermore, if we introduce the relative concentration $C - C_b$, which we denote C for convenience, we have, neglecting nonlinearities in the bulk,

$$\partial C/\partial t = LC + G\delta^+(\mathbf{r}) \qquad (5.7)$$

$$L(\mathbf{r}) = D\nabla^2 + \Omega(\mathbf{r}) \qquad (5.8)$$

$$\Omega = (\partial \mathscr{R}/\partial C)_{C = C_b(\mathbf{r})} \qquad (5.9)$$

where Ω is the matrix of the linearized bulk kinetics.

We introduce a bulk propagator $\Xi(\mathbf{r}, \mathbf{r}'; t)$ such that

$$\partial \Xi/\partial t = L(\mathbf{r})\Xi \qquad (5.10)$$

$$\Xi(\mathbf{r}, \mathbf{r}'; 0) = I\delta(\mathbf{r} - \mathbf{r}') \qquad (5.11a)$$

$$\mathbf{n} \cdot D\nabla\Xi = 0, \qquad \mathbf{r} \in S, \mathbf{r}' \in V \qquad (5.11b)$$

where $\delta(\mathbf{r} - \mathbf{r}')$ is the three-dimensional Dirac delta function, I is the identity matrix, and V is the volume of the bulk. With this, the solution to (5.7) may be written

$$C(\mathbf{r}, t) = \int d^3r' \Xi(\mathbf{r}, \mathbf{r}'; t)C(\mathbf{r}', 0)$$

$$+ \int_0^t dt' \int_S d^2r' \Xi(\mathbf{r}, \mathbf{r}'; t - t')G[\sigma(\mathbf{r}', t'), C(\mathbf{r}', t')]. \qquad (5.12)$$

The first term in (5.12) is the initial value contribution, whereas the second term is the accumulated effect of the surface reaction propagated into the system volume V from S. This is a great simplification of the problem since if we locate $\mathbf{r}$ on the surface, then (5.12) represents a closed equation for the bulk concentrations at the surface coupled to σ. A complete solution for the system is obtained by simultaneously solving (5.12) for the bulk concentrations at the surface consistent with (5.2), the latter of which may be converted into an integral equation by integration with respect to t. This reduces the problem to the solution of coupled nonlinear integral equations for surface quantities only.

1. Steady States

A reduction of (5.12) comes about for steady states. Let $\sigma^*(\mathbf{r})$, $C^*(\mathbf{r})$ be a steady-state solution of (5.12). Let P be the projection operator (in the N-dimensional concentration space) onto the subspace of species that have no bulk reaction [or, more generally, onto the null space of L, which for simplicity we assume to correspond to homogeneous ($\mathbf{r}$ independent) eigenfunctions]. Then $\Xi \to V^{-1}P$ as $t \to \infty$.

For simplicity we consider the case where such species are not present. Then since $\Xi \to 0$ as $t \to \infty$ for a stable bulk, we have by evaluating (5.12) in the limit as $t \to \infty$

$$C^*(\mathbf{r}) = \int_S d^2r' Z(\mathbf{r}, \mathbf{r}') G[\sigma^*(\mathbf{r}'), C^*(\mathbf{r}')] \tag{5.13}$$

where

$$Z(\mathbf{r}, \mathbf{r}') \equiv \int_0^\infty dt \Xi(\mathbf{r}, \mathbf{r}'; t). \tag{5.14}$$

To solve (5.13) we need a second equation for σ^*. This comes straightforwardly from (5.2) and reads

$$g(\sigma^*, C^*) = 0. \tag{5.15}$$

The solution of (5.15) allows elimination of σ^* in favor of C^* to derive a closed equation for C^* from (5.13). Since (5.13) and (5.15) are nonlinear, it is clear that the system can have multiple solutions in certain cases (Shymko and Glass, 1974; Bimpong-Bota et al., 1974; Bimpong-Bota et al., 1977).

2. Stability of the Steady States

The stability of the steady states to infinitesimal perturbations may be studied by linearizing (5.12) about the steady-state solution

$$C = C^* + \delta C$$

$$\sigma = \sigma^* + \delta\sigma. \tag{5.16}$$

Introducing the rectangular stability matrices as follows,

$$M = (\partial G/\partial C)^*$$

$$N = (\partial G/\partial \sigma)^*$$

$$\mu = (\partial g/\partial C)^*$$

$$v = (\partial g/\partial \sigma)^*, \tag{5.17}$$

we have

$$\delta C(\mathbf{r}, t) = \int d^3 r' \Xi(r, \mathbf{r}', t) \delta C(r', 0)$$

$$+ \int_0^t dt' \int_S d^2 r' \Xi(\mathbf{r}, \mathbf{r}'; t - t') \{ M \delta C(r', t') + N \delta \sigma(\mathbf{r}', t') \} \quad (5.18)$$

$$\partial \delta \sigma / \partial t = \mu \delta C + v \delta \sigma. \quad (5.19)$$

It is easy to see that $\delta \sigma(\mathbf{r}, t) = e^{vt} \delta \sigma(\mathbf{r}, 0) + \int_0^t dt' e^{v(t-t')} \mu \delta C(\mathbf{r}, t')$ and we obtain a closed equation for δC from (5.18) by eliminating $\delta \sigma$. For general surface geometry the problem is rather formidable. The theory can be carried out in great detail for regular arrays of local sites (Bimpong-Bota *et al.*, 1977) and for the planar case which we now consider.

C. PLANAR SURFACES

We now consider the simple case of a planar interface. Let the system be bounded by a plane at $x = 0$ with a bulk in $x > 0$. For the case of a homogeneous bulk steady state $C_b(\mathbf{r}) = C_b$ the propagator Ξ takes the form

$$\Xi(\mathbf{r}, \mathbf{r}'; t) = \Xi_0(\mathbf{r} - \mathbf{r}', t) + \Xi_0(\mathbf{r} - \bar{\mathbf{r}}', t) \quad (5.20)$$

where Ξ_0 is the propagator for an infinite medium and $\bar{\mathbf{r}}'$ is the image point to $\mathbf{r}'$, i.e., if $\mathbf{r}' = (x', y', z')$, then $\bar{\mathbf{r}}' = (-x', y', z')$. Since at the surface $x' = 0$, $\bar{\mathbf{r}}' = \mathbf{r}'$ and for points $(0, y, z)$ (5.20) simplifies as follows. Introducing a two-dimensional vector $\mathbf{r}_S = (0, y, z)$, we obtain

$$C(\mathbf{r}_S, t) = \int_0^\infty dx' \int d^2 r'_S [\Xi_0(\mathbf{r}_S - \mathbf{r}', t) + \Xi_0(\mathbf{r}_S - \bar{\mathbf{r}}', t)] C(\mathbf{r}', 0)$$

$$+ 2 \int_0^t dt' \int d^2 r'_S \Xi_0(\mathbf{r}_S - \mathbf{r}'_S, t - t') G[\sigma(\mathbf{r}'_S, t'), C(\mathbf{r}'_S, t')]. \quad (5.21)$$

This equation provides a convenient framework for studying questions of stability in the planar surface problem by linearizing this equation around a given reference state. Instability may arise due either to perturbations uniform along the wall or to those corresponding to spatial patterning, either static or propagating, parallel to the wall (Bimpong-Bota *et al.*, 1974). These disturbances decay as one goes away from the surface.

1. Spatio-temporally Periodic Patterns

To illustrate the application of the nonlinear methods developed for pure bulk phase systems we consider the phenomena of surface localized periodic patterns. For the case of waves or patterns periodic in the y direction and constant in the z direction we seek solutions

$$C = \chi(\phi, x) \quad (5.22)$$

$$\phi = ky - \omega t \quad (5.23)$$

where χ is 2π periodic in ϕ and where k and ω are the wave vector and frequency, respectively. Assuming the bulk kinetics in the absence of local reactions yields a homogeneous state, the stability matrix Ω is independent of $\mathbf{r}$ and χ obeys the equation

$$D\{(\partial^2/\partial x^2) + k^2(\partial^2/\partial \phi^2)\}\chi + [\Omega + \omega(\partial/\partial\phi)]\chi = -2G[\chi]\delta(x)$$

$$-\infty < x < \infty \qquad (5.24)$$

where for simplicity we have neglected the dependence of G on σ. We have used the convenient devise of extending the domain of the system into the unphysical region $x < 0$ with the proviso that we must seek solutions symmetric in x. (Integrating this equation over a small interval about $x = 0$ shows that the boundary condition $D\partial\chi/\partial x = G$ is satisfied, provided χ is symmetric in x.) Taking Fourier series transforms with respect to ϕ and Fourier integral transforms with respect to x, we may show that for the surface value of the pattern $\chi(\phi) \equiv \chi(\phi, 0)$ we have, including the factor of 2 in G for convenience,

$$\chi(\phi) = \int_{-\pi}^{\pi} d\phi' S_{k^2,\omega}(\phi - \phi')G[\chi(\phi')] \qquad (5.25)$$

$$S_{k^2,\omega}(\phi) = \frac{1}{(2\pi)^2} \int_{-\infty}^{\infty} dq \sum_{n=-\infty}^{\infty} e^{in\phi}[((nk)^2 + q^2)D - in\omega - \Omega]^{-1}. \qquad (5.26)$$

It is useful to write (5.25) as an operator equation

$$\chi = \mathscr{S}_{k^2,\omega}G[\chi] \qquad (5.27)$$

with the ϕ convolution being accounted for in the operator $\mathscr{S}_{k^2,\omega}$. Equation (5.27) presents itself as a convenient starting point for the study of surface waves and static patterns ($\omega = 0$).

a. Bifurcation Analysis of Surface Structures. Consider a bifurcation analysis of the onset of patterning as a small deviation of the concentrations from a static state uniform along the surface. In analogy to the bulk formalism (Kopell and Howard, 1973; Ortoleva and Ross, 1974; Achmuty and Nicolis, 1975; Keener, 1976) (see also Appendix A) we consider a family of solutions that arise in the vicinity of a wave vector k_0 with corresponding frequency ω_0 and expand all quantities in terms of an amplitude parameter A [Ortoleva (1974, unpublished)],

$$\begin{Bmatrix} \chi \\ \omega \\ k^2 \end{Bmatrix} = \sum_{n=0}^{\infty} \begin{Bmatrix} \chi_n \\ \omega_n \\ k_n^2 \end{Bmatrix} A^n. \qquad (5.28)$$

In carrying out the expansion care must be taken not to neglect the dependence of $S_{k^2,\omega}$ on A via k^2 and ω. With this procedure it may be shown that if there exists a complex conjugate pair of eigenvalues (of the stability problem) with imaginary part $\pm i\omega_0$ at k_0 and with real part passing through the origin linearly with $k^2 - k_0^2$ at k_0^2, then a family of wave solutions typically bifurcates at k_0^2 with an amplitude proportional to $|k^2 - k_0^2|^{1/2}$. The bifurcation may occur for $k^2 \gtrless k_0^2$, depending on the particular system (i.e., Ω, D, and G). The result is very similar to the results of the Hopf bifurcation theorem [see Hopf (1942), Stakgold (1971), Sattinger (1973), and Appendix A] as applied to chemical waves in a bulk medium [see Hanusse *et al.* (1978) for a review].

b. Limit-Cycle Perturbation Scheme. It has been shown that there can exist uniform oscillations along a wall (Bimpong-Bota *et al.*, 1974). Using techniques closely related to those in Section IV,C, Kopell and Howard (1973), and Ortoleva and Ross (1974), one may show [Ortoleva (1974, unpublished)] that there must be a family of long wavelength extensions of this solution for $k^2 > 0$ at least for k^2 near the origin. In deriving this result one must account for the dispersion (i.e., k^2 dependence of ω). The theory proceeds by expanding $S_{k^2, \omega}$, ω, and χ in powers of k^2 and putting the result in (5.27). The coefficients of ω are fixed by the condition of solubility of the theory to various orders. The only kinetic assumption is that the uniform oscillation be a limit-cycle stable modulo a uniform shift of phase.

c. Soluble Models. Using local kinetics as in Section II,B for the variables X and Y and letting X and Y decompose and diffuse with the same rate in the bulk one may construct model systems with periodic wave train solutions (Bimpong-Bota *et al.*, 1974) which verify the bifurcation and limit-cycle calculations outlined in the two preceding paragraphs. One may also construct models like those in Section II,A to find multiple steady state and other features of bifurcation-diagrams discussed in Section II,A (Bimpong-Bota *et al.*, 1974).

d. Multiple-Site Systems. When there are multiple localized sites of reaction a variety of interesting phenomena can occur (Bimpong-Bota *et al.*, 1977). Oscillations and multiple steady states may exist due to the interaction of two or more sites that cannot occur for one site (Shymko and Glass, 1974). For infinite arrays of sites the transition to new far from equilibrium states as a function of density of sites has the features of cooperative and critical behavior (Bimpong-Bota *et al.*, 1977). For the case of a dispersion of catalyst particles, the system may be described in terms of effective or renormalized transport and reaction in a course grained description (Ortoleva, 1978b).

e. Future Developments. Interesting problems to be solved in this area by applying or extending the results cited here include phenomena at electrode surfaces, periodic precipitation on surfaces, BZZ or other chemical waves on surfaces due to localized ferroin or other reagents on a surface, and membrane waves associated with a membrane containing an enzyme and placed in a bath of reactants. Systems with localized kinetics are of great importance in biological and chemical engineering systems.

VI. The Morphological Stability of Growing Bodies

Crystal growth under conditions far from equilibrium leads to stellar, dendritic, hollow, or feathery crystals. In contrast to this, near-equilibrium growth yields solid prismatic crystals. Tumor or colony growth (such as in corals) seem to be closely related to this type of phenomena in that the colony often grows in a dendritic or undulated pattern rather than with a smoother surface. We now discuss these types of situations in terms of the morphological stability of a growing body.

For growing bodies there are two opposing tendencies regarding the stability of simple geometries to the outcropping of dendritic protrusions. The gradient of growth material away from the object makes it advantageous for the growth of a protrusion relative to growth at a larger radius of curvature. Furthermore, in systems with a high heat of fusion a protrusion can eliminate heat more rapidly than a flatter surface. However surface tension tends to maximize the radius of curvature, hence opposing the tendency towards outcropping. Morphological instability in physico-chemical systems comes about when the balance between these two tendencies is turned in favor of outcropping. An analogous tendency towards outcropping to take advantage of nutrient gradients occurs in biological systems. The analog to surface tension is less straightforward and involves factors such as sharing of metabolites to advantage and the resistance to loss of essential chemical species or external attack when the surface area is minimized.

A. KINEMATICS OF GROWING BODIES

We consider first the growth of a pure amorphous solid from a solution of concentration $C(\mathbf{r}, t)$ (Mullins and Sakerka, 1967; Langer and Turski, 1976; Langer, 1976; Chadam et al., 1978). Consider a body defined by a surface S with unit normal $\mathbf{n}(\mathbf{r})$ at points $\mathbf{r}$ on S directed out of the body. Because of the incorporation of material from the environment, a surface area element δA moves a distance $v\delta t$ in the direction $\mathbf{n}$ where v is the local velocity of advancement of the surface. Associated with this advance, $\rho_s v\delta A\delta t$ moles of material are incorporated where ρ_s is the density of the growing body material. The material that was incorporated came from both diffusion to the surface, $-\mathbf{n} \cdot \mathbf{J}\delta A\delta t$ (where $\mathbf{J}$ is the flux of material), plus the material already present in the swept out volume $v\delta t\delta AC$ (where C is the concentration of the material in the bulk near the surface). Equating these to the quantity of moles incorporated we obtain

$$\rho_s v = Cv - \mathbf{n} \cdot \mathbf{J}. \tag{6.1}$$

This kinematic relation between the advancement velocity v and the flux of material from the solution can be further developed into a boundary condition for the reaction transport equations in the bulk. Let G be the rate of incorporation of material (moles per unit area per unit time). Because of surface tension effects, G can depend on the radius of curvature as well as the solution concentration at the surface. Since $G\delta A\delta t$ is the amount of moles incorporated in time δt on an area δA and since this is also given by $\rho_s v\delta A\delta t$, we see that

$$v = G/\rho_s \tag{6.2}$$

and hence we obtain the boundary condition

$$(1 - C/\rho_s)G = -\mathbf{n} \cdot \mathbf{J}. \tag{6.3}$$

This boundary condition differs from that found in the previous section for fixed boundaries. The new term is due to the fact that the motion of the surface sweeps out material in addition to that which comes to the surface by the flux $\mathbf{J}$. The term C/ρ_s can be ignored without much loss of accuracy for growth from a vapor or from a dilute solution, but cannot be ignored for growth from a concentrated solution or from a pure melt of the growth material.

Next we must introduce an equation for the surface defining the growing body. A surface can always be written as the set of points $\mathbf{r}$ such that

$$S(\mathbf{r}, t) = 0. \tag{6.4}$$

By convention we take S to be $\gtrless 0$ outside/inside the growing body. In a time δt a point $\mathbf{r}$ on S moves to a point $\mathbf{r} + v\mathbf{n}\delta t$. Since at time $t + \delta t$ S also satisfies (6.4) we have, after expanding in δt and setting the coefficient of δt to zero since this must hold for arbitrary but small δt,

$$\partial S/\partial t + v\mathbf{n} \cdot \nabla S = 0. \tag{6.5}$$

Noting that $\mathbf{n} = \nabla S/|\nabla S|$ and using (6.2), we obtain

$$\partial S/\partial t + G|\nabla S| = 0. \tag{6.6}$$

Since G depends on radius of curvature, it is a functional of S and hence (6.6) is a nonlinear equation in S (from the $|\nabla S|$ term as well as from G).

B. Diffusion Limited Growth Theory

In many cases of far from equilibrium growth the incorporation rate G is much more rapid than the rate of transport of material to the surface of the growing body. Since this case has been the most widely studied (Mullins and Sekerka, 1967; Langer and Turski, 1976; and Langer, 1976), it is derived here as a special case of the more general theory presented above (Chadam *et al.*, 1978).

Since there is a time scale in G that is fast relative to the other processes in the system, we introduce a smallness parameter ε to emphasize this fact:

$$G = \varepsilon^{-1}\bar{G}. \qquad (6.7)$$

The parameter ε represents the ratio of time scale associated with the attainment of equilibrium near the surface to that of crystal growth. For example, if $G = q(C - C^{eq})$, where C^{eq} is the (radius of curvature dependent) equilibrium concentration, then ε^{-1} is τq and $\bar{G} = \tau^{-1}(C - C^{eq})$ where τ is a typical time scale on which the object grows. Using (6.7), the boundary condition becomes

$$-\varepsilon \mathbf{n} \cdot \mathbf{J} = (1 - C/\rho_s)\bar{G}. \qquad (6.8)$$

If we expand C and S in ε,

$$C = \sum_{m=0}^{\infty} C_{(m)} \varepsilon^m$$

$$S = \sum_{m=0}^{\infty} S_{(m)} \varepsilon^m \qquad (6.9)$$

and similarly for $\mathbf{J}$ and $\bar{G}$ we obtain to lowest order

$$\bar{G}_{(0)} = \bar{G}(C_{(0)}, S_{(0)}) = 0 \qquad (6.10)$$

which states that at each point $\mathbf{r}$ on the solid surface C is at equilibrium with the solid at the local radius of curvature,

$$C_{(0)} = C^{eq}(S_{(0)}). \qquad (6.11)$$

To the next order we obtain

$$-\mathbf{n} \cdot \mathbf{J}_{(0)} = (1 - C_{(0)}/\rho_s)G_{(1)} \qquad (6.12)$$

and to lowest order from the surface equation we have, using (6.10),

$$\rho_s(\partial S_{(0)}/\partial t) = |\nabla S_{(0)}|\mathbf{n} \cdot \mathbf{J}_{(0)}/(1 - C_{(0)}/\rho). \qquad (6.13)$$

Finally, recalling that $\mathbf{n} = \nabla S/|\nabla S|$, we obtain

$$\frac{\partial S_{(0)}}{\partial t} = \frac{\nabla S_{(0)} \cdot \mathbf{J}_{(0)}}{(\rho_s - C_{(0)})}. \qquad (6.14)$$

This equation coupled with the continuity equation

$$\partial C_{(0)}/\partial t = -\nabla \cdot \mathbf{J}_{(0)} \qquad (6.15)$$

and the boundary condition (4.11) provides a complete theory for the study of morphological stability for the case of transport limited growth. Thus we have arrived at the starting point for the classical morphological stability theory of Mullins and Sykerka (1967).

C. Discussion

Morphological stability is an interesting free boundary problem, the latter term referring to the fact that the determination of the location of the boundary is part of the problem to be determined. It seems of great importance in its biological applications. This problem has been considered in materials science where important problems including the shape of precipitate particles and the roughness of the surface of solid deposited from vapors and melts have been studied (Mullins and Sekerka, 1967; Langer and Turski, 1976; Langer, 1976).

It is important to note that crystal anisotropy often plays a very dominant role in the determination of shape and that this effect is not covered by the simple theory presented here. Furthermore, special sites of rapid growth such as skrew dislocations may lead to the growth of whiskerlike projections out of an otherwise smooth surface (Ruth and Hirth, 1964). It is possible that the growth rate G can be made anisotropic and heterogeneous accounting for special localized sites, but this will not be developed here. Finally, the theory presented here does not account for the growth of solid mixtures. This can be handled in a manner similar to that presented above, although the results are not presented here for brevity and can be found in Walkind and Segal (1970) and Hasse *et al.* (1978).

It is clear that this area presents a wealth of possibilities for interesting and important nonequilibrium structures in physico-chemical and biological systems.

VII. Mechano-Chemical Phenomena

There are several phenomena involving center of mass motion of the system that is sustained because of the far from equilibrium conditions of the chemical reactions. When chemical reactions are taking place that involve heats of reactions or changes in numbers of moles, it was shown that acoustic modes in gases could become unstable and that such a system would spontaneously emit sound or amplify it as an acoustic pulse passed through such a reacting medium (Toong, 1972; Gilbert *et al.*, 1973). Another type of mechano-chemical coupling apparently occurs in peristaltic motion. Here slow waves of contraction sweep along various organs like the stomach and the large intestine. These motions are believed to be autonomous and not directly coordinated by the central nervous system. Clearly if a composition wave passes through the tissue comprising such an organ and if the associated concentration changes affect the contractile mechanism, one could readily see how such a peristaltic motion could take place.

A. Destabilization of a Damped Oscillator by a Chemical Reaction: Model of Insect Flight

For the purpose of illustration we consider a simple model of a mechanochemical instability that may, in fact, be of interest in understanding the mechanism of flight of certain insects (Pringle, 1968; Wigglesworth, 1972). It is known that the wing beat rate of some insects is much greater than the maximum rate at which individual nerve signals can be transmitted. For these insects the wing thorax system acts like an oscillator (often a nonlinear click oscillator) with frequency typically close to that of the wing beat frequency of flight. However the question arises as to how can the energy losses due to internal friction and losses to the demands of flight be overcome? In what follows we consider a simple model of a mechanical oscillator that is coupled to a contractile system in such a way that just enough energy is added in each cycle to overcome energy losses. In this model the neural impulses would, for example, serve to change the resonance frequency through the tension of the indirect flight muscles and nonequilibrium conditions, but not trigger individual wing beats.

Consider a harmonic oscillator characterized by a coordinate X with linear damping coefficient γ, frequency ω, and rest position $\bar{X}(Y)$ that depends on the composition of a chemical species Y (specifying the state of the contractile fibrils in the muscle):

$$(d^2 X/dt^2) + \gamma(dX/dt) + \omega^2(X - \bar{X}(Y)) = 0. \qquad (7.1)$$

The chemical species Y (located within the spring) undergoes a reaction whose rate depends on the extension of the spring and hence on X,

$$dY/dt = F(X, Y). \qquad (7.2)$$

At rest (steady state) the configuration is X^*, Y^* where

$$X^* = \bar{X}(Y^*) \qquad (7.3)$$

$$F(\bar{X}(Y^*), Y^*) = 0. \qquad (7.4)$$

The question arises under what conditions, if any, will the system spontaneously lapse into auto-oscillation?

Carrying out ordinary linear stability analysis, we can find the stability eigenvalues that answer the above question. The answer is particularly simple in the case where ω is large relative to all the other inverse times in the system—i.e., γ and those in F. First we expand the stability eigenvalues z in powers of ω^{-1},

$$z = \omega\zeta_{-1} + \zeta_0 + \omega^{-1}\zeta_1 + \cdots$$

with coefficients ζ_m. Keeping the first two terms in the oscillatory roots $z_\pm$, we have

$$z_\pm \approx \pm i\omega + \tfrac{1}{2}[(\partial F/\partial X)^*(\partial \bar{X}/\partial Y)^* - \gamma].$$

For instability, Re $z \pm > 0$, we must have

$$(\partial F/\partial X)^*(\partial \bar{X}/\partial Y)^* > \gamma,$$

i.e., the product of the change in rate of reaction with distension times the change in the rest position with the active species concentration (Y) must exceed the damping factor γ. Thus with each cycle the distension changes the concentration of Y which in turn shifts the rest position just enough to give the oscillator a kick to overcome damping.

This simple result is analogous to the results found for the coupling of reaction to acoustic modes (Toong, 1972; Gilbert *et al.*, 1973). There X is replaced by pressure or density and γ by viscosity. The quantity $\partial \bar{X}/\partial \bar{Y}$ is analogous to the heat or volume change of reaction.

The connection of the present model to the problem of insect flight obviously requires a more detailed and realistic analysis of the biological system. However the experiments on flight muscle clearly point out this tendency toward auto-oscillation (Pringle, 1968; Wigglesworth, 1972). Finally, mechano-chemical instabilities appear to play an important role in fibril auto-oscillation in sperm and other systems (Brakaw and Benedict, 1968).

B. Remarks

It appears that the variety of nonlinear methods have not been applied to problems in mechano-chemical instabilities. Furthermore, biological mechano-chemical (i.e., peristaltic) waves present themselves as interesting and important problems for future research in applied nonlinear analysis.

VIII. Concluding Remarks

The spate of activity over the past decade in the field of physico-chemical instabilities has served to greatly increase our understanding of these phenomena. Perhaps one of the greatest recent contributions has been to put a variety of engineering, biological, geological, and other widely varying problems within a unified framework.

It is important to note that the developments within the macroscopic or phenomenological theory has been paralleled by advances in the fundamental studies in nonequilibrium statistical mechanics of instabilities. Preliminary results along these lines have demonstrated the close analogy of nonequilibrium transitions to various types of equilibrium phase changes (McQuarrie, 1967; Nitzan *et al.*, 1974; Nitzan, Ortoleva, and Ross, 1974b; Keizer, 1975; Malek-Mansour and Nicolis, 1975; Gardiner *et al.*, 1976; Ortoleva and Yip, 1976; Van Kampen, 1976; Oppenheim *et al.*, 1977).

The great wealth of phenomena in chemically reacting systems can be traced for the most part to the almost limitless nonlinear analytical forms that chemical reaction rates may take on and still be consistent with basic physical principles of conservation and symmetry. Thus it should not be a surprise if in the next decade many examples of oscillatory, propagating, and other media are discovered. In fact, the surprise would be if this were not the case.

In light of this I believe that theory in this field will play a very important role in terms of its ability to predict new phenomena and to classify the multiplicity of possibilities according to rather general properties of the phenomenological relations for reaction and transport. An attempt along these lines has been done for chemical waves using catastrophe theory (Feinn and Ortoleva, 1977).

An area of great importance is the application of the theory and concepts of the study of nonlinear physico-chemical phenomena to explain processes in biological, geological, ecological, and other systems. In order to make useful contributions along this line, close collaboration between experimental and theoretical efforts must be made. Not only should the experimental data guide the theorist, but theoretical results should be used as a guide in the design of experiments so that central conceptual points may be resolved. In surveying possible areas of application and from work in progress in our and in other's laboratories, I believe that major advances along these lines are forthcoming in the next few years. And furthermore, there is much more room for other efforts along these lines.

In conclusion, it is a reasonable approximation of reality to assert that "Oh, nonlinear partial differential equations—anything can happen " and I would furthermore add that it is beautiful that it indeed does.

Appendix

A. BIFURCATION ANALYSIS

To introduce the ideas of bifurcation theory we consider a most simple example, that of homogeneous steady-state systems. Let λ be a homogeneous constraint. We wish to study the bifurcation behavior of the steady-state solutions of the equation

$$dC/dt = \mathscr{R}(C, \lambda) \tag{A.1}$$

as a function of λ. We assume there is a known branch of solutions $C^*(\lambda)$ and rewrite the equation in terms of the deviation from this state. Letting

$$C = C^* + c,$$

we obtain for the steady-state value of c

$$L(\lambda)c + \mathscr{N}(c, \lambda) = 0 \tag{A.2}$$

where

$$L(\lambda) \equiv (\partial \mathscr{R}/\partial C)_{C^*(\lambda)} \tag{A.3}$$

$$\mathscr{N} \equiv \mathscr{R} - Lc. \tag{A.4}$$

Note that $\mathscr{N}$ is of order c^2 or higher and that for the present case of a single variable that if $L(\lambda)$ passes through zero at a critical value λ_c, then the state C^* is marginally stable there. We assume that C^* becomes unstable beyond λ_c and let

$$z' \equiv (dL/d\lambda)_{\lambda_c} > 0. \tag{A.5}$$

We expect that in the vicinity of λ_c we might have the onset of a new branch of steady states. This is indeed the case under the assumptions given thus far. We introduce a parameter A measuring the amplitude of the new bifurcating branch and expand c and λ in A:

$$c = \sum_{n=1}^{\infty} c_n A^n \tag{A.6}$$

$$\lambda = \lambda_c + \sum_{n=1}^{\infty} \lambda_n A^n. \tag{A.7}$$

Substitution into (A.2) yields to lowest order

$$L(\lambda_c)c_1 = 0 \tag{A.8}$$

which is automatically satisfied since $L(\lambda_c) = 0$ when the steady state C^* is marginally stable at λ_c. To second order we have

$$\lambda_1 z' c_1 + b c_1^2 = 0, \qquad b \equiv \tfrac{1}{2}(\partial^2 \mathscr{N}/\partial c^2)_{c=0, \lambda=\lambda_c}. \tag{A.9}$$

If $b \neq 0$, we have a new branch of states (i.e., $c_1 \neq 0$)

$$c_1 = -\lambda_1 z'/b \tag{A.10}$$

since $z' \neq 0$ by assumption. Thus the onset of a new branch (solution) requires both a loss of stability of one branch and a condition on the nonlinearity ($b \neq 0$) in the vicinity of the marginal stability point. To complete the theory we must make a more specific assumption about the relation between λ and A. A convenient choice is to let $\lambda_n = 1$ for the first value of n for which λ_n may be chosen to be nonzero. Since this is the case for λ_1, we put $\lambda_1 = 1$ and obtain

$$c \underset{\lambda \to \lambda_c}{\sim} -(z'/b)(\lambda - \lambda_c). \tag{A.11}$$

The new branch of steady states bifurcates with a critical index of one in this case since c is linear in $(\lambda - \lambda_c)$. The new branch $C^* + c$ transverses the reference branch C^* at λ_c. Furthermore, by performing linear stability analysis from the new state as calculated from (A.11), one may easily show that the small deviations δc from the new state obey

$$d\delta c/dt = -z'(\lambda - \lambda_c)\delta c \tag{A.12}$$

and hence while the reference branch C^* is stable for $\lambda < \lambda_c$, the new branch is stable for $\lambda > \lambda_c$. Thus the two branches exchange stability.

Consider the simple example for which $\mathscr{R} = \lambda C - C^2$. The system has a steady state at $C^* = 0$. Here $L = \lambda$ and $\mathscr{N} = -C^2$. Thus $z' = 1, b = -1$, and the above theory predicts a new branch will bifurcate at $\lambda_c = 0$ in the form $c = \lambda$ which is obviously correct for this simple case.

The simple theory presented here for illustrative purposes may be extended to multicomponent systems and to spatio-temporally varying behavior and may be applied to the theory of homogeneous chemical oscillations [see Hopf (1942) and the Appendix of DelleDonne and Ortoleva (1977)], chemical waves (Kopell and Howard, 1973; Ortoleva and Ross, 1974; Hanusse *et al.*, 1978) and static structures in bounded (Achmuty and Nicolis, 1975; Keener, 1976) and unbounded media and for surface localized phenomena [see Ortoleva (1974, unpublished) and Section V,C,1].

B. LIMIT-CYCLE PERTURBATION THEORY

We now discuss a perturbation theory for systems with a homogeneous limit-cycle oscillation. Let the equation

$$d\chi_c/dt = \mathscr{R}[\chi_c] \tag{B.1}$$

have a stable limit-cycle solution χ_c with frequency ω_c. Then what is the effect of adding a correction term γG to the equation? Thus we seek C as a solution to

$$dC/dt = \mathscr{R}[C] + \gamma G[C] \tag{B.2}$$

and in particular seek periodic solutions in the limit as $\gamma \to 0$. First it is convenient to work with 2π-periodic functions. We let

$$\tau \equiv \omega(\gamma)t \tag{B.3}$$

where $\omega(\gamma)$ is the γ-dependent frequency. This γ dependence of ω is important since the coefficients of the series development of the 2π-periodic solutions $\psi(\tau)$ to (B.2),

$$\psi = \chi_c(\tau) + \sum_{n=1}^{\infty} \psi_n \gamma^n, \tag{B.4}$$

may be determined only if a solubility condition involving the expansion coefficients ω_n,

$$\omega = \omega_c + \sum_{n=1}^{\infty} \omega_n \gamma^n, \tag{B.5}$$

is satisfied. To lowest order in γ we obtain

$$\omega_c(d\psi_0/d\tau) = \mathcal{R}[\psi_0] \tag{B.6}$$

which is known to have a limit-cycle solution χ_c.

In next order we find

$$[\omega_c(d/d\tau) - \Omega(\tau)]\psi_1 = -\omega_1(d\psi_0/d\tau) \tag{B.7}$$

$$\Omega(\tau) \equiv (\partial\mathcal{R}/\partial C)_{C=\chi_c(\tau)}. \tag{B.8}$$

However we note that $[\omega_c(d/d\phi) - \Omega]d\psi_0/d\tau = 0$ and hence ψ_1 can only be determined as a 2π-periodic function if the right-hand side of (B.7) is appropriately orthogonal to $(d\chi_c/\partial\tau)$. Hence $\omega_1 = 0$ and thus the frequency shift due to $|\gamma| > 0$ occurs in order γ^2. The coefficient ω_2 is then determined in the next order by assuming that the right-hand side of

$$[\omega_c(d/d\tau) - \Omega]\psi_2 = -\omega_2(d\psi_0/d\tau) + (\partial^2\mathcal{R}/\partial C^2)_{C=\chi_c(\tau)}\psi_1\psi_1 \tag{B.9}$$

is orthogonal to the adjoint eigenvector f corresponding to $d\chi_0/d\tau$ (see below). We obtain

$$\omega_2 = \frac{1}{2\pi} \int_{-\pi}^{\pi} [f(\tau), (\partial^2\mathcal{R}/\partial C^2)_{c(\tau)}\chi_1(\tau)\chi_2(\tau)]\,d\tau \tag{B.10}$$

where (A, B) implies a sum over components and f is the solution of

$$\omega_c(df/d\tau) = \Omega^T(\tau)f(\tau) \tag{B.11}$$

where Ω^T is the transpose of Ω. The function f is 2π periodic and is normalized such that

$$\frac{1}{2\pi} \int_{-\pi}^{\pi} (f, d\chi_c/d\tau)\,d\tau = 1. \tag{B.12}$$

This type of expansion has been applied to develop spatio-temporal solutions to reaction diffusion equations with limit-cycle kinetics. With it both planar and more complex geometric chemical waves have been studied as long wavelength extensions of the limit-cycle solution [see (Kopell and Howard, 1973; Ortoleva and Ross, 1974; Hanusse *et al.*, 1978), and Section IV,C].

ACKNOWLEDGMENT

I wish to thank M. DelleDonne for several important comments that I have incorporated in the manuscript. I would also like to express my appreciation to Professor John Ross for the many pleasant years that we worked together at the Massachusetts Institute of Technology in the study of far from equilibrium phenomena. This work has been supported in part by a grant from the National Science Foundation.

References

Achmuty, J. F. G., and Nicolis, G. (1975). *Bull. Math. Biol.* **37**, 322.
Aronson, D. G., and Weinberger, H. F. (1975). "Nonlinear Diffusion in Population Genetics, Combustion and Nerve Propagation." Springer-Verlag, Berlin and New York.
Bimpong-Bota, K., Ortoleva, P., and Ross, J. (1974). *J. Chem. Phys.* **60**, 3124.
Bimpong-Bota, K., Nitzan, A., Ortoleva, P., and Ross, J. (1977). *J. Chem. Phys.* **66**, 3650.
Brakaw, C. J., and Benedict, B. (1968). *J. Gen. Physiol.* **52**, 283.
Caesari, L. (1971). "Asymptotic Behavior and Stability Problems in Ordinary Differential Equations." Springer-Verlag, Berlin and New York.
Chadam, J., Schmidt, S., and Ortoleva, P. (1978). "On the Morphological Stability of Growing Bodies," (in preparation).
Chance, B., Pye, E. K., Ghosh, A. K., and Hess, B. (eds.) (1973). "Biological and Biochemical Oscillations." Academic Press, New York.
Chandrasekar, S. (1964). "Hydrodynamic and Hydromagnetic Instability." Oxford Univ. Press, London and New York.
Creel, C. L., and Ross, J. (1976). *J. Chem. Phys.* **65**, 3779.
Davidson, R. C. (1973). "Methods in Nonlinear Plasma Theory." Academic Press, New York.
Defay, R., and Prigogine, I. (1956). "Surface Tension and Chemiadsorption." Wiley (Interscience), New York.
DeGroot, S. R., and Mazur, P. (1962). "Nonequilibrium Thermodynamics." North-Holland Publ., Amsterdam.
deLevie, R. (1970). *J. Electroanal. Chem.* **25**, 257.
DelleDonne, M., and Ortoleva, P. (1977). *J. Chem. Phys.* **67**, 1861.
DelleDonne, M., and Ortoleva, P. (1978). *Z. Naturforsch.* (in press).
Dreitlein, J., and Smoes, M.-L. (1974). *J. Theor. Biol.* **46**, 559.
Feinn, D., and Ortoleva, P. (1977). *J. Chem. Phys.* **67**, 2119.
Feinn, D., Ortoleva, P., Scalf, W., Schmidt, S., and Wolff, M. (1978). *J. Chem. Phys.* (submitted).
Fitts, D. D. (1962). "Nonequilibrium Thermodynamics." McGraw-Hill, New York.
Flicker, M. R., and Ross, J. (1974). *J. Chem. Phys.* **60**, 3458.
Gardiner, C. W., McNeil, K. J., Walls, D. F., and Matheson, I. S. (1976). *J. Stat. Phys.* **14**, 307.
Gilbert, R., Ortoleva, P., and Ross, J. (1973). *J. Chem. Phys.* **58**, 3625.
Glansdorff, P., and Prigogine, I. (1971). "Thermodynamic Theory of Structure, Stability and Fluctuations." Wiley (Interscience), New York.

Glass, L. (1973). *Science* **180**, 1061.

Haase, S., Chadam, J., and Ortoleva, P. (1978). "A Theory of Oscillatory Zoning in Plagoclase Feldspars." (in preparation).

Hanusse, P., Ortoleva, P., and Ross, J. (1978). *Adv. Chem. Phys.* (in press).

Hedges, E. S., and Myers, J. E. (1926). "The Problem of Physico-Chemical Perioicity." Longmans Green, New York.

Heiney, E., and Ortoleva, P. (1977). "Liesegang Band Regularization and Front Propagation Under Imposed Electric Fields." (in preparation).

Herschkowitz-Kaufman, M. (1978). "On the Bifurcation Diagram of a Model System." *Ann. N.Y. Acad. Sci.* (in press).

Hess, B., and Boiteaux, A. (1971). *Annu. Rev. Biochem.* **40**, 237.

Hopf, E. (1942). *Ber. Math. Phys. Sachs. Acad. Wiss. Leipzig* **94**, 1; see also Stakgold (1971) and Sattinger (1973).

Jaffe, L. F., Robinson, K. R., and Nuccitelli, R. (1974). *Ann. N.Y. Acad. Sci.* **238**, 372.

Keener, J. P. (1976). *Stud. Appl. Math.* **55**, 187.

Keizer, J. (1975). *J. Chem. Phys.* **63**, 5037.

Kopell, N., and Howard, L. (1973). *Stud. Appl. Math.* **52**, 291.

Langer, J. S. (1976). "Studies in the Theory of Interfacial Stability, II." (preprint).

Langer, J. S., and Turski, L. A. (1976). "Studies in the Theory of Interfacial Stability, I." (preprint).

Lefever, R., Herschkowitz-Kaufman, M., and Turner, J. (1977). "Dissipative Structures in a Soluble Nonlinear Reaction Diffusion System." (preprint).

Liesegang, R. E. (1896). *Phot. Archiv.* **21**, 221.

McQuarrie, D. (1967). *J. Appl. Prob.* **4**, 413.

Malek-Mansour, M., and Nicolis, G. (1975). *J. Stat. Phys.* **13**, 197.

Montrol, E. W. (1972). *In* "Statistical Mechanics. New Concepts, New Problems and New Applications" (S. A. Rice, K. F. Freed, and J. C. Light, eds.). Chicago Univ. Press, Chicago, Illinois.

Mullins, W. W., and Sekerka, R. F. (1967). *In* "Crystal Growth" (H. S. Peiser, ed.), p. 691. Pergamon, Oxford.

Murray, J. D. (1977). "Biological Oscillations: Spatial Structure and Nonlinear Wave Propagation in Reacting Systems." (preprint).

Nicolis, G., and Portnow, J. (1973). *Chem. Rev.* **73**, 369.

Nitzan, A., and Ross, J. (1973). *J. Chem. Phys.* **59**, 291.

Nitzan, A., Ortoleva, P., and Ross, J. (1974a). *J. Chem. Phys.* **60**, 3134.

Nitzan, A., Ortoleva, P., and Ross, J. (1974b). *Faraday Symp. Chem. Soc.* **9**, 241.

Nitzan, A., Ortoleva, P., Deutch, J., and Ross, J. (1974). *J. Chem. Phys.* **61**, 1056.

Noyes, R. M., and Field, R. J. (1974). *Annu. Rev. Phys. Chem.* **25**, 95.

Noyes, R. M., and Field, R. J. (1975). *Adv. Chem. Phys.* **29**.

Oppenheim, I., Shuler, K. E., and Weiss, G. H. (1977). "Stochastic Theory of Nonlinear Rate Processes with Multiple Stationary States." (preprint).

Ortoleva, P. (1972). "Phase Correlations in Limit Cycle Systems." (unpublished notes).

Ortoleva, P. (1974). "Nonlinear Perturbation Methods in the Theory of Physico-Chemical Surface Waves and Patterns." (unpublished notes).

Ortoleva, P. (1976). *J. Chem. Phys.* **64**, 1395.

Ortoleva, P. (1978a). "On the Theory of Self-Electrophoretic States in Cellular Systems." *J. Theoret. Biol.* (in press).

Ortoleva, P. (1978b). "Renormalized Transport and Effective Kinetics in Catalysist Suspensions." (in preparation).

Ortoleva, P. (1978c). "On a Scattering Theory for Chemical Waves." (in preparation).

Ortoleva, P., and Ross, J. (1972). *J. Chem. Phys.* **56**, 4397.
Ortoleva, P., and Ross, J. (1973a). *J. Chem. Phys.* **58**, 5673.
Ortoleva, P., and Ross, J. (1973b). *Dev. Biol.* **34**, F19.
Ortoleva, P., and Ross, J. (1973c). *Biophys. Chem.* **1**, 87.
Ortoleva, P., and Ross, J. (1974). *J. Chem. Phys.* **60**, 5090.
Ortoleva, P., and Yip, S. (1976). *J. Chem. Phys.* **65**, 2045.
Ostwald, W. (1925). *Kolloid-Z.* **36**, 380.
Overbeek, J. Th. G. (1971). MIT Press, Cambridge, Massachusetts.
Ponmarenkov, Iu. B. (1968). *Prikl. Mat. Mekh.* **32**, 244.
Prager, S. (1956). *J. Chem. Phys.* **25**, 279.
Prigogine, I., and Lefever, R. (1968). *J. Chem. Phys.* **48**, 1695.
Pringle, J. W. S. (1968). *Adv. Insect Physiol.* **5**, 763.
Ressler, O. (1976). *Z. Naturforsch.* **31a**, 1168.
Rinzeland, J., and Keller, J. B. (1973). *Biophys. J.* **13**, 1313.
Ross, J. (1976). *Ber. Bunsenges. Phys. Chem.* **80**, 1113.
Ruth, V., and Hirth, J. P. (1964). *J. Chem. Phys.* **41**, 3139.
Sattinger, D. H. (1973). "Topics in Stability and Bifurcation Theory." Springer-Verlag, Berlin
    and New York.
Schmidt, S., and Ortoleva, P. (1977). *J. Chem. Phys.* **67**, 3771.
Schmidt, S., and Ortoleva, P. (1978). "Chemical Waves in Ionic Systems: Propagating Planch
    Potentials and Imposed Field Effects." (in preparation).
Segal, L. A. (1965). *J. Fluid Mech.* **21**, pt. 2.
Shymko, R. M., and Glass, L. (1974). *J. Chem. Phys.* **60**, 835.
Stakgold, I. (1971). *SIAM Rev.* **13**, 289.
Toong, T. Y. (1972). *Combust. Flame* **18**, 207.
Turing, A. M. (1952). *Philos. Trans. R. Soc. London, Ser. B* **237**, 37.
Van Kampen, N. G. (1976). *Adv. Chem. Phys.* **34**, 245.
Walkind, D. J., and Segel, L. A. (1970). *Philos. Trans. R. Soc. London* **268**, 351.
Wigglesworth, V. B. (1972). "The Principles of Insect Physiology." Chapman & Hall, London.
Yamada, T., and Kuramoto, Y. (1976). *Prog. Theor. Phys.* **55**, 2035.

Subject Index

287

A
B
C 8
D 9
E 0
F 1
G 2
H 3
I 4
J 5